INTRODUCTION TO SPACE:

The Science of Spaceflight

SECOND EDITION

by
Thomas D. Damon
Pikes Peak Community College

with Foreword by
Edward G. Gibson

KRIEGER PUBLISHING COMPANY

MALABAR, FLORIDA
1995

Original Edition 1989
Second Edition 1995

Printed and Published by
KRIEGER PUBLISHING COMPANY
KRIEGER DRIVE
MALABAR, FLORIDA 32950

FROM A DECLARATION OF PRINCIPLES JOINTLY ADOPTED BY A
COMMITTEE OF THE AMERICAN BAR ASSOCIATION AND COMMIT-
TEE OF PUBLISHERS:

This Publication is designed to provide accurate and authoritative information
in regard to the subject matter covered. It is sold with the understanding that
the publisher is not engaged in rendering legal, accounting, or other professional
service. If legal advice or other expert assistance is required, the services of
a competent professional person should be sought.

Library of Congress Cataloging-In-Publication Data

Damon, Thomas.
    Introduction to space : the science of spaceflight / by Thomas
Damon ; with foreword by Edward G. Gibson. — 2nd ed.
        p.   cm. — (Orbit , a foundation series)
    Includes bibliographical references and index.
    ISBN 0-89464-056-9 (alk. paper). — ISBN 0-89464-053-4 (pbk. :
alk. paper)
    1. Space flight.   I. Title.
TL791.D36   1995
629.4—dc20   94-34253
                                                           CIP

10 9 8 7 6 5 4 3 2

To Jimmie and Jacob . . . may they one day walk on Mars.

Series editors
Edwin F. Strother, Ph.D.
Donald M. Waltz

Contents

Preface to the Second Edition

Five years have passed since this book was first published. A lot has happened in space research and exploration in those five years. The Space Shuttle is back in business. The Hubble Space Telescope was launched, found to be flawed, and repaired on orbit. Venus and the Moon have been completely mapped, one by radar and the other by multispectral sensors. Voyager left the Solar System. New spacecraft are bound for the Sun, Jupiter, and Saturn, and a few have failed. Space defense resources came into play in the Persian Gulf War, employed as they had never been before in actual combat. The Global Positioning System is completed. Direct broadcast satellites are about to become a reality. Biomedical research in space is leading to a better understanding of how the human body functions in orbit and on Earth. And more. . . .

Remarkable changes have taken place in world politics and economics as well, changes thought to be impossible in 1988. The Soviet Union disintegrated and the 45-year cold war came to an end. Russian space scientists and engineers are joining with their counterparts in the United States, seeking to work together on cooperative projects. The possibility that all spacefaring nations will collaborate in building and operating an international space station is not as far-fetched as it seemed only five years ago. Can international colonization of the Moon and human exploration of Mars be far behind? Successful cooperation in the space station could be a model for successful cooperation in other endeavors, both on Earth and in space. It could be the cornerstone for joining all peoples of Earth into permanent peaceful coexistence.

This second edition has been updated with the latest information in all aspects of space research and exploration. A special effort has been made to include space activities of all nations, past, present, and future. Space defense is no longer treated as a separate subject; the information in the previous Chapter 7 has been distributed to other appropriate chapters throughout the book. On the other hand, the discussion of astronomy from space has been removed from Chapter 6 to a separate chapter, Chapter 7 in this edition. Otherwise, the organization of the book is essentially the same as it was in the first edition.

Some of the **MATHBOXES** have been revised and five new ones have been added. There are numerous new or revised illustrations, more of them in color. For example, with printouts from new computer software, the orbit diagrams in Chapter 3 are clearer and more readable. Several new discussion questions have been added and the additional reading lists are updated and expanded.

I would like to acknowledge receipt of a grant from the Pikes Peak Community College International Education Committee to enable me to expand on international space activities in my course and, consequently, in this book. Special thanks to John Bally of the University of Colorado at Boulder for the remarkable Space Telescope image of the Orion Nebula. Special thanks also to my dear wife, Anita, for her encouragement, patience, and understanding during the past six months.

Preface to the First Edition

Most of what the general public "knows" about space has been learned from watching *Star Trek, Star Wars,* and the evening news. Unfortunately, these sources of "information" often lead to misconceptions about the laws of nature with regard to what is scientifically and technically possible. For example, in a classroom discussion concerning the hazards of rocket debris and dead satellites that orbit the Earth, one student suggested that the Space Shuttle could simply fly around picking up the pieces of space junk, drawing them into the cargo bay with a tractor beam. The student seriously believed that the Star Trek tractor beam really exists.

Several colleges and universities around the country offer courses in space science and technology, heavy in math and oriented toward upper division or graduate students in science and engineering. Because of the high interest in space activities in Colorado Springs, we at Pikes Peak Community College felt the need for a general science, general education course primarily for liberal arts students who would like to know, among other things, how spacecraft get into orbit, why astronauts float, what satellites are doing up there, and if permanent colonies can be built in space.

We searched for an appropriate textbook, but none was available. Considerable effort has gone into space education curriculum development at the grade school and junior high level, but ours may be the only course of its kind for nonspecialist college students and adults. At first we assembled a collection of NASA pamphlets, government publications, and contractor reports, along with a few freshly written sections on basic science. As the course became more popular, the need for a more formal textbook became obvious. This book is the result.

I have tried to avoid writing a trivia reference book. This is, I hope, a book of scientific and technical substance, not just a collection of interesting facts. On the other hand, I have avoided any "heavy" math, but have included a number of MATHBOXES for those students and classes that can handle simple algebra. MATHBOXES are set apart from the rest of text and may be omitted without disrupting the continuity of the book.

The subjects included here are those that have been of greatest interest to my students. A brief history is summarized in Chapter 1. The basic science of propulsion, orbital mechanics, and the space environment is covered in Chapters 2, 3, and 4. These topics are fundamental to understanding the rest of the book. The next three chapters discuss unmanned spacecraft with particular emphasis on remote sensing. Chapters 8, 9, and 10 deal with man in space, particularly the Space Shuttle, our only manned space program. The final three chapters project into the future with solid facts and some speculation about humankind's ventures into the final frontier.

The thirteen chapters can be covered at the average rate of one per week during a 15 week semester, with time left for tests and other subjects. Some topics, particularly Chapters 2 and 3, may require more than a week; some others can be covered more quickly.

I must acknowledge the help of many people in this project. Particular thanks go to Dr. Mario Iona, Professor Emeritus of Physics at Denver University, and to Dr. William Bennington, PPCC biologist, for their review and critique of certain chapters. The numerous individuals and groups that provided background information, artwork, and photographs are mentioned by organization, if not personally, in the captions of the illustrations. This book could never have been completed without their assistance. Special thanks go to Mary Roberts, editor, whose words of encouragement always came when they were most needed, and to Robert Krieger for his willingness to venture into the unknown.

Foreword

This time we've really gone and done it!

Regardless of the means—feet, horse, boat, covered wagon, car, or plane—we humans have always felt compelled to push outward, to take that next step, to surge into the space around us, to explode into every fresh frontier that offers growth and challenge. But this time, with the technology of the rocket, we've pierced gravity's cocoon, stepped outward and come face to face with our ultimate frontier, an insatiable frontier—the remainder of our universe.

True, so far we've made just a small, tentative step, maybe even just a shuffle. Yet, make no mistake about it, our first step into space is as important as those first steps we've made onto land, across mountains, over oceans, and into air. For we've just made that first step that must precede the strides and dashes to planets, the marathons between stars and that matter-of-fact galactic travel that someday will fill eons yet to come. No longer just observers of the heavens, we can now leave Earth and in only minutes become participants.

Intimidating? No doubt about it.

For after looking back at Earth from the Moon, we've responded by pulling back to regroup, to sort out the motives, methods, and timing of our next steps. Before beginning Tom Damon's clear, concise, and exciting presentation of the science of spaceflight, let's raise some key issues.

In every venture into a new frontier, we humans always benefit by proceeding in the same orderly manner: explore, develop, and use. Is space to be the same?

Yes. With manned and unmanned spacecraft we've explored the edge of our new frontier—low Earth orbit—and have now developed the most logical means to use it—space stations. The United States is planning to end its visitor status in space with its Space Station, as the Soviet Union has already done, and to reap the benefits of a permanent presence.

In the short term, Space Station provides a Space Research Center for science and technology, as well as for the development of new commercial products and industries. Occupancy in low Earth orbit provides researchers and entrepreneurs with an environment unequaled in any earthbound facility: near zero gravity, near perfect vacuum, and a vantage point from which to study Earth as well as the undistorted image of the universe.

In the long term, Space Station will be an assembly base, a stepping stone, a central gateway in an evolving infrastructure that enables our future expeditions to the Moon, Mars, and farther out. Behind our expeditions, colonies will spread permanent human presence throughout the Solar System. We can feel confident that with time, modest upgrades to our technology, and our permanent presence on space stations, this next outward surge will occur.

But let's be "prudent and economical" about this new frontier. Let's just develop it remotely from the ground or dart back and forth to it now and then and avoid the expense of a space hotel. We don't really have to stay on-board Space Station day after day, do we?

Yes! In earthbound research and industrial facilities, our human perception, logic, creativity, and dexterity are, without question, unique and indispensable. The value of this hands-on direction and support is undiminished by the relocation of our facilities into space. In truth, it is our continuous presence that permits us to most profitably operate, repair, maintain, and upgrade every facility and gives us the time and experience to evolve every concept, design, instrument, and operation to its fullest potential.

Sould we do anything more now than give space token attention?

Yes! The time for space is now. At any time in the past, only a few nations have had the strength and opportunity to shape human history. Not so today. Many nations are strong and have access to space. Those nations who do shape our future and, indeed, those who remain free of military or economic domination will be those who understand that response to challenge enhances strength and that, from this time onward, strength in space is closely tied to strength on Earth. Our decision can be delayed, our future cannot.

But didn't we say that once we've left gravity's cocoon behind we'd be ready for "our entire universe"? That's a bit pretentious. What about the distances and the technology required to cover them? Sure, new frontiers have always shocked us with their remoteness, until technology advancements reduce this to inconvenience. But beyond our Solar System loom invincible expanses that take years to travel, even at light speed. Can interstellar distance also be overcome?

Yes! But here our confidence requires more than a direct extrapolation of today's technology and time; it also requires a faith inspired by the history of human achievement and an awareness of what shaped it—humanity's expanding intellect and incessant drive.

Lastly, let's go back to the basic premise: ". . . compelled to push outward, to take that next step, to surge into the space around us. . . ." All froth and no substance? Do we need to allow anything other than economics, politics, or other concrete intellectual considerations to enter our deliberations on space?

Once again—yes! In the exploration and development of previous frontiers, we humans have always been obsessed with lists of scientific objectives and practical benefits. Yet when we look back, we realize that our rationalizations, often even our imaginations, fell far short of predicting the frontier's true importance. We really accepted the challenge for more basic reasons.

So it is now with our presence in space. We may underestimate its importance to us and our descendants as well as our basic motivation. Data, information, pictures, or other nourishments of the mind, by themselves, ring hollow. They can only thicken our intellectual shell, only toughen our educated hide. Inside us, as always, the essential spark resides, alive and well with all the primordial drives and desires, itches and urges. In time, we must break the physical limits of our turf and go there, to see and feel the new territory up close, to take physical measure of it with our own person before it can become a true part of our world. To deny this drive is to deny our humanity.

Tom Damon provides for you, the reader, a lucid understanding of the science and technology of entering space, what we are finding, and some of our plans for the future. Enjoy it! It is an extraordinary treatment of an exciting subject, an outstanding introduction to mankind's greatest, open-ended adventure.

EDWARD G. GIBSON
ASTRONAUT, SKYLAB 4

Chapter 1

History of Spaceflight

When the history of this time is written, what will be considered the most important of the daily multitude of momentous events to characterize this era? When those future historians chronicle our most significant contribution, I believe they will say that this was the time when humans took their first tentative steps out of the cradle, when they left Earth to seek greener pastures, to start anew, to try again. Space is the new and final frontier.

Americans have already impressed footprints in the dust of the Moon, put robots on the surface of Mars, sent spacecraft to fly past the other planets and moons in our Sun's family, and sent people commuting to work in orbit. The former Soviet Union has also sent robot explorers to the far reaches of the Solar System and has built a permanently manned space station where cosmonauts have set endurance records of more than 1 year living in space. Only 66 years, less than an average lifetime, took us from the Wright brothers' first 120 foot airplane flight in 1903 to a walk on the Moon in 1969.

The exploration and colonization of other worlds will be the most exciting undertaking of all time. It is inevitable, as inevitable as the exploration and colonization of America.

That is what spaceflight and this book are all about.

Rocketry: The Means of Getting There

The history of spaceflight begins with the history of rocketry; without rockets there would be no spaceflight.

The history of rocketry begins in China. The Chinese invented gunpowder which exploded violently when ignited in an enclosed space. However, if gunpowder were ignited in the open, it just burned rapidly. By putting the gunpowder into a chamber with one end open so the gases formed by the combustion could escape, the whole thing took off "like a rocket" instead of exploding. Indeed, it was a rocket. As early as 1212, Chinese armies fired rockets carrying flaming materials into Mongol encampments to start fires.

About the year 1500, according to legend, a Chinaman named Wan Hu made the first attempt to build a rocket-powered vehicle. He attached 47 rockets to some sort of cart and at a given signal 47 coolies lit the 47 rockets simultaneously. In the explosion that followed the entire vehicle disappeared in a cloud of smoke and Wan Hu was never seen in this world again.

Artillery rockets carrying bombs became highly developed in Europe by 1800. They had been used on a large scale against British troops in India, and in response, Sir William Congreve developed the modern stick-rocket in 1805. These new weapons of war were used by most world armies of that time. The British used them against Napoleon in Europe and against the newly organized United States of America. In 1814 during the 25 hour long British bombardment of Baltimore, Francis Scott Key, watching the action from a ship in Baltimore harbor, wrote a poem later to become our national anthem. "And the rocket's red glare . . . gave proof through the night that our flag was still there."

Science fiction writers have long written stories about people travelling to far-off planets. Propulsion systems included flocks of birds, balloons, and just closing your eyes and wishing. A cannon was the choice of Jules Verne who gave us *From the Earth to the Moon and Around the Moon* in the mid-1800s.

Russian schoolteacher, Constantin Tsiolkovski, was a space dreamer. He first published some of his ideas in 1883 and in 1903 published a treatise "Exploration of the Universe With Rocket Propelled Vehicles." By the time of his death in 1935 he had considered every aspect of spaceflight. Although he was considered mad by many during his lifetime, a monument to his memory now stands in Moscow.

Robert Goddard was the first to build successful liquid-fueled rockets. Using gasoline and liquid oxygen for propellants and a blowtorch for an igniter, he launched the first successful flight from a farm in Auburn, Massachusetts, on March 16, 1926. It lasted less than 3 seconds at a maximum speed of 60 miles per hour, covered a distance of 184 feet, and reached an altitude of 41 feet. The rocket weighed less than 6 pounds and carried about 5 pounds of propellant.

Goddard was a quiet person and, when he suggested that

rockets would be the way to leave Earth and fly to the Moon and Mars, he met with such ridicule that he never sought public attention again. *The New York Times* said Goddard did not have the knowledge of a high school student. The *Times* printed a retraction 49 years later when Apollo 11 landed on the Moon. In 1930, with financial backing from Charles Lindbergh and from the Guggenheim Foundation, Goddard moved to New Mexico where he continued his experiments until 1941.

At Goddard's request, his papers describing interplanetary flight were held secret by the Smithsonian Institution until 1970, twenty-five years after his death. Four papers covered nearly all the basic science and engineering principles that were needed for flight to the planets. He held 214 patents for his many inventions. In 1960 the U.S. government agreed that the large military rocket engines used for intercontinental ballistic missiles infringed on Goddard's patents and paid his widow a million dollars in damages. She gave half of it to the Guggenheim Foundation.

In Europe, the story was different. Unlike Goddard's bad public experience in the United States, Germans were highly interested in rocketry and the prospect of space travel. Rockets were being built by German amateur rocket societies, especially Verein fur Raumschiffahrt (Society for Space Travel). Hermann Oberth in 1923 published a book, *The Rocket into Interplanetary Space*. In Russia, Tsiolkovski's works were eagerly read and rocket societies were actively experimenting in the 1920s and 1930s. In England, the British Interplanetary Society did serious studies on all aspects of spaceflight. One of its early presidents was Arthur C. Clarke who later wrote *The Sentinel*, a short story which was made into the movie *2001: A Space Odyssey* by Stanley Kubrick.

The technology needed to actually fly into space and return safely was the outgrowth of several technologies that developed in mid-twentieth century. The German military, limited by the World War I Treaty of Versailles, became interested in the rockets being built by amateur rocket societies. Military research with rocket-powered artillery led to the first ballistic missile in 1943 and culminated in the V-2 "vengeance weapon" that carried warheads from Peenemünde, along the Baltic coast of Germany, to England in the closing days of World War II. It was a fearful weapon because there was no sound, no forewarning before the warhead hit and exploded. The V-2 rocket boosted the 1 ton warhead to 3,500 miles per hour and burned out. The warhead continued on a ballistic trajectory to a range of 200 miles and fell on its target. By contrast, there was some warning of an attack by airplanes and by the earlier guided missiles whose engines continued running until they reached their targets. Fortunately for the Allies, the V-2s came too late in the war to have any effect on the outcome.

While the Russian army was advancing across Germany at the end of the war, the German rocket scientists and engineers from Peenemünde sought out the American Army and surrendered. Some were captured by the Russians and taken to the Soviet Union. The Americans confiscated some of the V-2 rockets and took them along with Wernher von Braun and co-workers to White Sands, New Mexico, where they continued their research and development work. When asked about the design of their V-2, the Germans said to ask Robert Goddard; it had been copied from a rocket Goddard flew in 1939.

By 1948 the German engineers in the United States had flown a two-stage rocket to an altitude of 244 miles. Von Braun dreamed of space travel. He knew that the work he and his colleagues were doing for the Army could lead to rockets which could fly into orbit, but any time the subject came up it was vetoed. The Army was not in the space business. Nonetheless, without the military rocket development programs, the advances in rocketry necessary for spaceflight would not have come about as they did.

While the U.S. Army was interested in artillery, the U.S. Air Force was working toward the development of very long range rocket-powered missiles. Both the United States and the Soviet Union saw rockets as a means to deliver nuclear weapons against an enemy. Successful development of intercontinental ballistic missiles (ICBMs) was dependent on three things: the perfection of reliable high powered rocket engines, guidance systems, and reentry vehicles. The problem of reentry into Earth's atmosphere was particularly important. Without protection from the searing heat of atmospheric friction, the nuclear weapons would melt and vaporize before reaching their targets.

At about the same time, jet aircraft were being perfected and experiments with rocket-powered aircraft were beginning (Figure 1.1). These craft were able to fly to the upper fringes of the atmosphere, where a human could not survive without special protective equipment, the forerunners of the modern space suit and life support equipment.

The center of aircraft research was at Edwards Air Force Base in the desert northeast of Los Angeles. There, in the Bell X-1 rocket plane, Chuck Yeager became the first man to fly faster than sound in spite of broken ribs suffered in a fall from a horse a few days before (which he didn't tell the doctor). The X-1, shaped like a bullet, was powered by four liquid rockets and was carried to high altitude strapped under a B-29.

Further experiments with rocket airplanes led to the X-15 which flew 199 flights up to six times the speed of sound. Neil Armstrong, later to become the first man to walk on the Moon, was an X-15 test pilot.

Sputnik: Dawn of the Space Age

On October 4, 1957, the United States was shocked when the Soviet Union launched the first satellite into orbit. Sput-

Figure 1.1 A research rocket plane is dropped from a B-29 as a jet fighter chase plane follows.
Courtesy of NASA; artist William F. Phillips.

nik I was a small sphere carrying a radio transmitter which "beeped" a signal for all the world to hear of the supremacy of Russian technology. A month later, on November 3, Sputnik II was orbited carrying the space-faring dog, Laika.

In November 1957, President Eisenhower went on television and showed the first man-made object that had survived the heat of reentry and was recovered from orbit. The solution to the reentry problem was one of the main keys to successful manned spaceflight as well as successful ballistic missiles. Eisenhower emphasized that we were preparing to launch a satellite but that our major effort focussed on the development of ballistic missiles, not on space. In 1958 the president established a civilian space program and created NASA, the National Aeronautics and Space Administration, to oversee it.

American incentive was spurred and, after several spectacular failures of the Vanguard project, Explorer 1 was successfully launched on January 31, 1958. It was a 30 pound steel cylinder containing two detectors for micrometeoroids and one for high energy particle radiation. Their information was transmitted to the ground until May 23. Although small and simple, Explorer 1 made a major discovery, the Van Allen radiation belts around Earth.

Man in Space: A Russian

The Soviets also beat the Americans in putting a man into space. Yuri Gagarin orbited Earth in a Russian spacecraft on April 12, 1961. Soviet preeminence in these space spectaculars can be attributed to their very large rockets which had been built to carry their heavy hydrogen bombs. In the United States a breakthrough in nuclear weapons technology by Dr. Edward Teller made possible the construction of relatively lightweight warheads. Thus, American ICBM booster rockets were not large enough to carry heavy loads into orbit.

Mercury: Man in a Can

The first U.S. manned spaceflight program was called Mercury. Flights of chimpanzees preceded manned flights; Ham was the first passenger in the Mercury capsule on January 31, 1961. The first astronauts were all military test pilots, some of them from the rocket aircraft research at Edwards AFB. Alan Shepard became the first American in space when he flew a 15 minute 22 second suborbital flight on May 5, 1961 (Figure 1.2). He reached an altitude of 116.5 miles and a maximum speed of 5,180 miles per hour. He

Figure 1.2 Mercury capsule, Freedom 7, sits atop the Redstone booster rocket to carry Alan Shepard on the first American flight into space. The towerlike structure at the top is equipped with small rockets which would pull the capsule away from the booster in case of an emergency. The rocket tower was jettisoned if everything was operating properly. *Courtesy of NASA.*

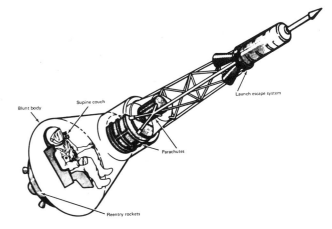

Figure 1.3 Diagram of Mercury capsule. *Courtesy of NASA.*

made; the final one flew 22 orbits in more than 34 hours on May 15 and 16, 1963.

Gemini: Twins in Orbit

Named for the constellation of stars which includes the twins Castor and Pollux, the Gemini spacecraft, Figure 1.4, was a two man capsule with a volume about the size of the front part of a compact car. It measured 19 feet long and 10 feet in diameter and had the same basic design as the Mercury capsule. Equipment and expendables that were not needed for the return to Earth were put outside and left behind, providing more room in the cabin. As with Mercury, the capsule was designed for one-time use and landed at sea. Although development of a capability for landing the spacecraft on land was begun in the United States, it was never completed. The Soviets, on the other hand, recovered their manned spacecraft on land right from the beginning.

Titan, a modified Air Force ballistic missile, carried Gemini spacecraft to orbit. Ten flights were made in 1965 and 1966. In those cramped quarters, one of the flights spent just a few hours short of two weeks in orbit, a world record at that time. Gemini astronauts learned how to maneuver, change orbit, and rendezvous and dock with other spacecraft. Gemini VI launched on December 15, 1965, and accomplished the first space rendezvous with Gemini VII which had been launched on December 4, seen in Figure 1.5. They flew together for over 5 hours at distances of 1 foot to 295 feet.

The first space walk was on June 5, 1965, when Edward White opened the hatch and floated into space, attached to Gemini IV by a 23 foot tether line.

Apollo: To the Moon

Apollo was the project that took men to the Moon and back. To meet President Kennedy's challenge, dozens of technological breakthroughs had to be made. The spacecraft car-

landed in the Atlantic Ocean, 302 miles from his starting point at Cape Canaveral, Florida.

Twenty days later President Kennedy stated the national goal of placing a man on the Moon and returning him safely to Earth before the end of the decade. Nearly a year went by before the first American orbited Earth. John Glenn made a three orbit flight on February 20, 1962.

The Mercury capsule, shown in Figure 1.3, was crowded with only one person inside. The astronaut barely had room to turn his head and move his arms. It measured only 6 feet 10 inches long and 6 feet 2-1/2 inches in diameter at its base. An escape tower was attached to the top to pull the capsule off the booster rocket in case of emergency. It was jettisoned if not needed. Two different rockets carried the Mercury capsule: Redstone, a modified Army missile, and Atlas, an Air Force ballistic missile.

A retro-rocket motor slowed the vehicle to take it out of orbit and return it to Earth. Attached to the blunt end of the capsule was a heat shield which absorbed the 3,000 °F heat generated by atmospheric friction during reentry. The capsule landed in the ocean, its descent slowed by parachutes, and was recovered by ships waiting at the site. Each capsule was used only once. Six successful manned flights were

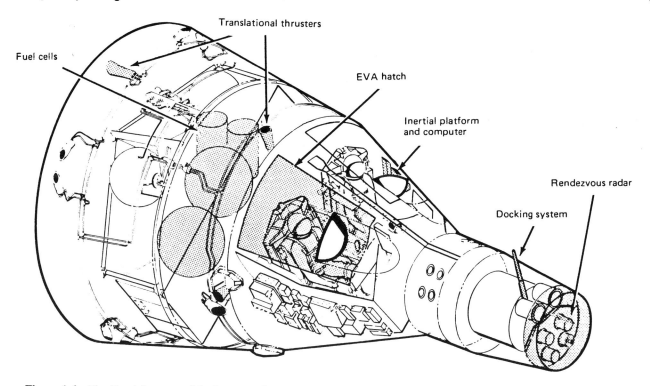

Fuel cells

Translational thrusters

EVA hatch

Inertial platform
and computer

Rendezvous radar

Docking system

Figure 1.4 The Gemini spacecraft had a crew of two. Fuel cells generated electricity to power the craft. Astronauts could exit the vehicle on a "space walk" through the EVA hatch. Thrusters allowed the spacecraft to maneuver, rendezvous, and dock with other spacecraft. *Courtesy of NASA.*

ried three men in a volume about the size of the inside of a minivan. It contained a shirtsleeve environment; that is, the astronauts could take off their space suits and move around a bit. They also had hot water aboard to prepare meals.

The huge Saturn V rocket which launched the Apollo spacecraft on its trajectory to the Moon is shown in Figure 1.6 with the previous manned space vehicles drawn to the same scale. Saturn V stood 363 feet high and weighed 6.5 million pounds. It burned liquid oxygen and kerosene, 15 tons per second at liftoff. The second and third stages used liquid hydrogen and liquid oxygen propellants.

The Apollo spacecraft itself is shown in Figure 1.7 compared to Gemini. The cone-shaped command module in the center carried the astronauts. It was 10 feet 7 inches high and 12 feet 10 inches in diameter. The cylindrical service module at the top carried electrical equipment, oxygen tanks, and the rocket engine for leaving lunar orbit to return to Earth. The spider-like 23 foot tall lunar module (LM) at the bottom was the vehicle that descended from lunar orbit carrying two of the three astronauts to the surface of the Moon (Figure 1.8). Because it didn't have to operate in the atmosphere (the Moon has none) it was not necessary to streamline it; it was actually the first true manned space vehicle. When the activity on the surface was finished, the lunar module rejoined the command and service modules, still in lunar orbit, for the return trip home. Only the command module with its three man crew landed back on Earth.

Figure 1.5 The Gemini VII spacecraft as seen through the window of Gemini VI during the first space rendezvous. *Courtesy of NASA.*

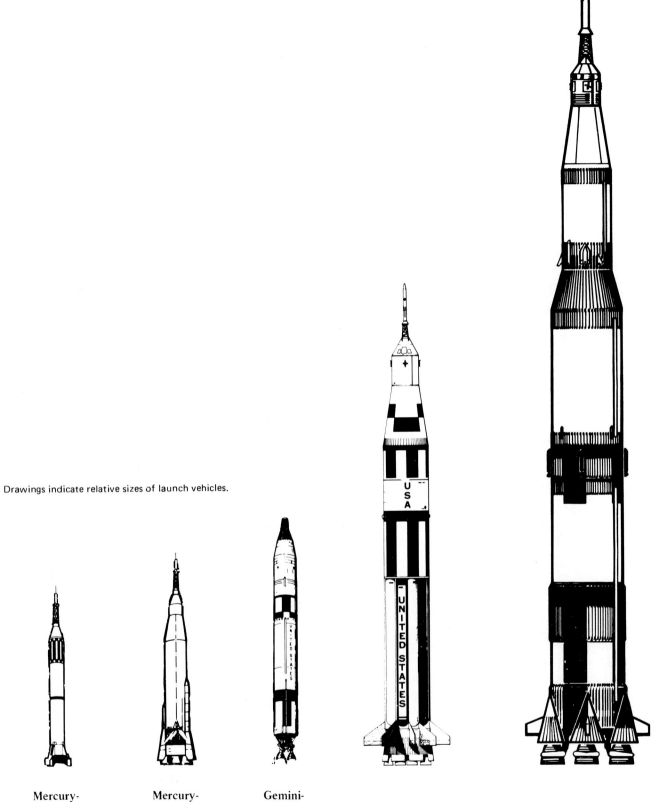

Drawings indicate relative sizes of launch vehicles.

Mercury-
Redstone

Mercury-
Atlas

Gemini-
Titan

Saturn IB

Saturn V

Figure 1.6.   Comparison of manned launch vehicles. *Courtesy of NASA.*

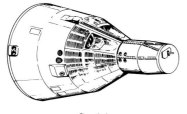

Gemini

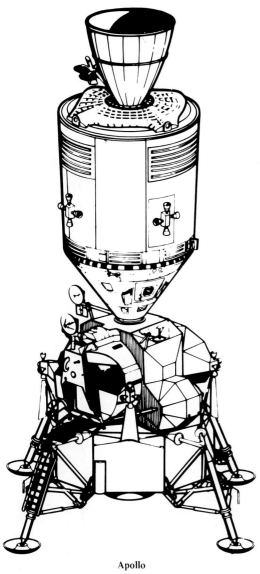

Apollo

Drawings indicate relative sizes of spacecraft.

Figure 1.7 Apollo and Gemini spacecraft drawn to the same scale. *Courtesy of NASA.*

As with Gemini and Mercury, Apollo landed in the ocean (Figure 1.9). Nothing, not even the command module, was reused.

For safety, lunar landings had to be made on smooth, level terrain, often uninteresting geologically. An electric pow-

ered car, the lunar rover shown in Figures 1.8, 1.10, and 1.11, was carried to the Moon on the last three flights so the astronauts had transportation while on the surface. The rover allowed them to drive up to 6 miles to hills, cliffs, and craters to take photos and bring back samples. Each of the four wheels had its own electric motor so that if one or two failed, the rover could still operate. Power was supplied by silver-zinc batteries. The astronauts covered a total of almost 60 miles on the three flights. The rovers were left behind along with just about everything else that was taken to the surface.

Disaster struck during the early days of the Apollo program. Three astronauts lost their lives in a launch pad fire on January 27, 1967, during preliminary checkout of the Apollo. During an actual flight, the command module contained an atmosphere of pure oxygen at a pressure of 5 pounds per square inch, about one-third normal atmospheric pressure. At that pressure things burn as they would in a normal oxygen-nitrogen atmosphere. But during the checkout the command module was pressurized to near normal sea level pressure with pure oxygen. Under these conditions materials burn explosively. A 2-1/2 month investigation resulted in a design for a fireproof spacecraft. Other changes were made, delaying the program about a year and a half, but the goal set by President Kennedy was met. Two lunar landings were made before the decade of the 60s was up.

Apollo 11 landed on the Moon July 20, 1969. Neil Armstrong's words "one small step for man, one giant leap for mankind" are part of history. A plaque was left behind, engraved with the words "Here men from the planet Earth first set foot upon the moon July 1969 A.D. We came in peace for all mankind."

Apollo 13 nearly ended in disaster when on April 13, 1970, the third day out, an explosion in one of the oxygen bottles blew a hole in the side of the service module and started a leak in a second oxygen bottle. Without oxygen, the fuel cells could not generate electricity and the command module was disabled. The three-man crew squeezed into the two-man LM which became their lifeboat. There was no way to turn back; they had to continue on around the Moon, without landing, and back to Earth. The LM was designed to support two men for 2 days; but it would take about 6 days to get back to Earth. The temperature in the LM dropped to 38°. Most of the food was dehydrated and required hot water to rehydrate; but no hot water was available. By careful minimal consumption of food, water, oxygen, and electricity, and by using the LM rockets for course adjustment, they managed to limp home. The three men lost a total of 31.5 pounds, but did not lose their lives. In that respect, the mission was a success; in spite of a serious problem the astronauts, with advice from the ground, used their limited equipment and supplies with great ingenuity and creativity to make it back.

Apollo was not only an exciting and spectacular adven-

Figure 1.8 An Apollo landing site. The lunar module is at the center and a lunar rover at the right. Because there is no air on the Moon, the flag staff was fitted with a horizontal bar to keep the flag "flying." *Courtesy of NASA.*

Figure 1.9 Three parachutes gently lower an Apollo spacecraft into the sea. *Courtesy of NASA.*

ture, it was also a scientific, technological, and political success. From 1969 to 1972 three manned flights were made around the Moon (Apollo 8, 10, and 13) and six landings were made (Apollo 11, 12, 14, 15, 16, 17). The locations are shown in Figure 1.12. Ten landings were originally planned but three were deleted because of NASA budget cuts. Twelve American astronauts left footprints in the lunar soil during 166 man-hours of surface exploration. About 850 pounds of soil and rocks were brought to Earth and nearly 100 experiments were carried out in orbit and on the surface.

As a result of the Apollo lunar exploration scientists have a pretty good understanding of the history of the Moon from its formation at the same time Earth formed, through a period of severe impacts by asteroids and meteoroids, to the present quiet. They also know the composition of lunar rock and soil so when a colony is established on the Moon we know what kinds of raw materials are available for construction and mining. This is discussed in Chapter 12.

Equally important were knowledge and technology gained from the research and development needed to make Apollo succeed: building huge reliable rockets, developing life support systems, refining orbital mechanics, miniaturizing electronic devices, and advancing computer technology. This kept America preeminent in space for the following decade.

Figure 1.10 An astronaut and his rover on the Moon. *Courtesy of NASA.*

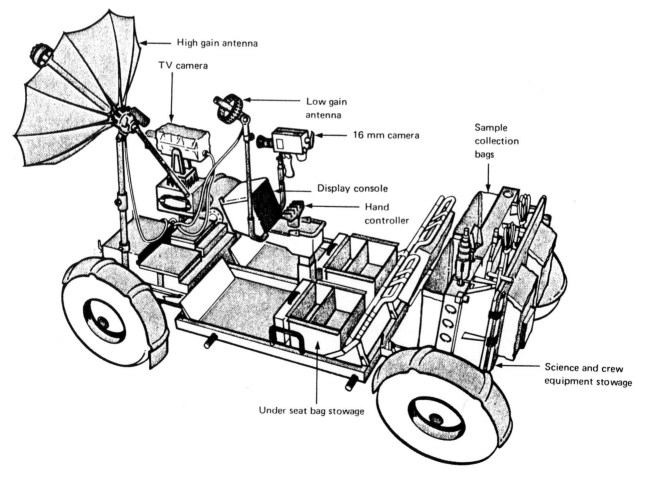

High gain antenna

TV camera

Low gain antenna

16 mm camera

Sample collection bags

Display console

Hand controller

Science and crew equipment stowage

Under seat bag stowage

Figure 1.11 Diagram of the lunar rover. *Courtesy of NASA.*

Figure 1.12 The six Apollo landing sites. The Soviet Union put two unmanned spacecraft on the Moon, Luna 16 and 20, which picked up samples of lunar soil and returned to Earth. *Courtesy of NASA.*

The former Soviet Union planned manned flights to the Moon also, but because of a number of booster failures, never succeeded in putting a man on the Moon. The first stage booster consisted of 30 rocket engines all burning siumltaneously. Apparently the problem was in maintaining reliability and stability while keeping the thrust of the 30 engines balanced. By comparison, the Saturn V first stage had a cluster of five engines.

The Soviets did have a very comprehensive and successful unmanned lunar exploration consisted of 24 flights including the first soft landing on the Moon; the first flight around the Moon with photos of the back side which never faces the Earth; two Lunokhod roving vehicles which landed on the Moon, drove about for more than a year examining the soil and sending back 100,000 photographs of the lunar surface; and several flights which returned samples of the lunar soil to the Earth.

Soyuz-Salyut: Soviet Space Stations

Meanwhile, the Soviets concentrated on manned flight in low Earth orbit. The Soyuz spacecraft was the workhorse of their manned space activities. The Soyuz spacecraft consists of three segments which can be seen in Figure 1.13. On one end a module carries equipment and instruments and two solar panels extend from its opposite sides. On the other end is a spherical-shaped living and laboratory module. The dome-shaped segment in the center is the reentry vehicle which carries the crew back to Earth. The crew consists of

Figure 1.13 Artist's concept of the hookup of Apollo command and service modules (left) with Soviet Soyuz spacecraft on the right. *Courtesy of NASA.*

up to three cosmonauts. While Mercury, Gemini, and Apollo spacecraft landed in the ocean, Soyuz returns to dry land. Small retrorockets fire just before touchdown to bring the craft to a gentle landing. At least 60 Soyuz flights were made since 1965.

The seven Salyut space stations were launched beginning in 1971. At first, they were placed in low orbits which decayed in just a few months. Later models remained in orbit for years. Salyut's main section was about 40 feet long by 14 feet in diameter and weighed just over 20 tons. It enclosed a laboratory and living section, an equipment-propulsion-control section, and an airlock equipped for the Soyuz vehicles to dock. Soyuz acted as a shuttle craft to bring crews to and from Salyut.

Skylab: U.S. Space Station

Built from the empty third stage of a Saturn V rocket, the 100 ton Skylab was about the size of a small three bedroom house. (See Figure 1.14) It was launched on May 14, 1973, into a 270 mile orbit. In 1973 and 1974 Skylab was manned by three crews of three men. Crews and supplies were ferried to the space station using Apollo vehicles. The crew would rendezvous and dock with Skylab and enter the space station; the Apollo command module and service module would remain attached until it was time to go home. The last Skylab mission still holds the record for the longest American spaceflight: 84 days. The crew included Edward Gibson, William Pogue, and Gerald Carr.

Much research in solar physics, space physics, earth science, and human biology was done. It gave the United States its best experience in long duration space flight, especially the effects of weightlessness on the human body. Medical research was given top priority. Many interesting problems arose: How do you weigh a weightless person? Answer: the period of oscillation of a seat attached to springs is propor-

Figure 1.14 The Skylab space station, photographed by the crew as they departed for Earth in their Apollo capsule. Skylab was damaged during launch and had to be repaired on orbit by the astronauts. Solar cells cover the X-shaped structure and the panels to the lower right; the left solar cell array is missing. A sunshield was improvised over the top of the workshop where the micrometeoroid shield was torn off. *Courtesy of NASA.*

tional to the persons weight. How do you eat, sleep, and take a shower? And the most asked question, "How do you go to the bathroom in space?" the title of a fascinating book by William Pogue. These and other questions will be discussed in later chapters.

Apollo-Soyuz

The Apollo program ended in July 1975 with history's first international spaceflight when an Apollo spacecraft linked up with a Russian two-man Soyuz spacecraft. The Apollo capsule was equipped with a docking device which allowed it to connect to the Soyuz. An artist's concept of the hookup is shown in Figure 1.13. For nearly 2 days the astronauts and cosmonauts moved between the two vehicles, did joint experiments, and displayed good fellowship. Although 28 scientific experiments were conducted, this event was more of a political achievement than a scientific or technological breakthrough.

Unmanned Spacecraft

Meanwhile, satellites were invading the space surrounding the Earth. By way of definition, a *spacecraft* is a self-contained vehicle designed for spaceflight. A spacecraft is called a *satellite* when it is in a closed orbit around some celestial body such as the Earth or a planet.

Figure 1.15   An artist's concept of a spacecraft passing Saturn with the Sun and other planets in the background. *Courtesy of NASA.*

The first spacecraft were research satellites which studied the space environment. Instrumented satellites have measured the content of near-Earth space, electrons, protons, atmospheric gases, the solar wind, radiation from the sun, aurora, the magnetic field, and micrometeorites. Dozens of communications satellites were launched to relay telephone and television signals from one side of the Earth to the other. Other orbiting satellites looked down at the Earth, measured its size and shape, took its picture in visible light and invisible infrared, observed and reported the weather, and watched for rocket launches and signs of military activity. A series of satellites prepared the way for a global navigation system.

Spacecraft have been sent to all the planets of our Solar System except Pluto. See Figure 1.15. In 1962, Mariner 2 was the first to fly close to another planet, Venus. Mariner 4 flew by Mars in 1965 and two Viking spacecraft landed on the red planet in 1976. The Voyager spacecraft sent back spectacular pictures of the giant gas planets, Jupiter, Saturn, Uranus, and Neptune, from 1979 to 1989. Mariner 10 flew by Venus in 1974 and by Mercury three times in 1974 and 1975. Magellan, in orbit around Venus, made radar maps of the surface through the dense clouds that blanket that planet. Others are still on their way to their targets. Galileo is on its way to Jupiter for a closer examination of the largest planet and its moons, Cassini will be a return mission to the Saturnian system, and Ulysses is a mission to study the polar regions of the Sun. A flyby of Pluto is planned for early in the next century.

Astronomical instruments have also been sent into orbit enabling astronomers to view the sky unhindered by the Earth's atmosphere.

We will examine the specifics of unmanned satellites and spacecraft in the following chapters.

Space Shuttle

The Space Shuttle, Figure 1.16, is a reusable spacecraft that takes off like a rocket, flies in orbit like a spaceship, and returns to Earth like an airplane. The emphasis is on reuse in order to reduce the cost of transportation into orbit. The Shuttle is the mainstay of the U.S. manned space program through the turn of the century. We will discuss it in detail in following chapters. The Soviets built what appears to be a duplicate of the Space Shuttle orbiter which was flown on suborbital flights only

DISCUSSION QUESTIONS

1. What do you suppose were some of the technical problems with the Apollo-Soyuz rendezvous in space?

2. Why couldn't an automobile gasoline engine have been used on the lunar rover?

Figure 1.16 Space Shuttle lifts off. This and the Russion Soyuz are the two vehicles capable of carrying people into space. *Courtesy of NASA.*

3. What would have been some of the considerations in choosing the landing sites on the Moon?

4. Why did the returning manned spacecraft land in the oceans?

5. Do you think that exploration of space should continue, or should the government concentrate its resources on other more earthly problems?

ADDITIONAL READING

Bainbridge, William S. *The Spaceflight Revolution*. Krieger Publishing Co., 1983.

Baker, D. *The History of Manned Space Flight*. Crown Publishers, 1982.

Benford, Timothy B., and Brian Wilkes. *The Space Program Quiz and Fact Book*. Harper and Row, 1985.

Caprara, Giovanni. *The Complete Encyclopedia of Space Satellites*. Portland House, Crown Publishers, 1986. Every military and civil satellite of the world from 1957 to 1986.

Cortright, Edgar M., editor. *Apollo Expeditions to the Moon*, NASA SP-350, Government Printing Office, 1975.

Dunne, James A., and Eric Burgess. *The Voyage of Mariner 10*. NASA SP 424, Government Printing Office, 1978. Complete well-illustrated story of the Mariner 10 mission to Venus and Mercury.

Fimmel, Richard O., et al. *Pioneer Venus*. NASA SP 461, Government Printing Office, 1983. Complete well-illustrated story of the "assault" on Venus in December 1978 by ten U.S. and Soviet spacecraft.

Hart, Douglas. *The Encyclopedia of Soviet Spacecraft*. Exeter Books, 1987. Describes all Soviet space programs from Sputnik I to 1987. Well illustrated.

Kerrod, Robin. *The Illustrated History of Man in Space*. Mallard Press, 1989. Picture book.

Logsdon, John M., and Alain Dupas. "Was the Race to the Moon Real?" *Scientific American*, June 1994.

Miller, Ron. *The Dream Machines: An illustated history of the spaceship in art, science and literature*. Krieger Publishing Co., 1993.

NASA. The First 25 Years, 1958–1983. Government Printing Office, 1983.

Pogue, William R. *How Do You Go to the Bathroom in Space?* Tom Doherty Associates, 1985. What its like to live in space, written by a Skylab astronaut.

Taylor, G. Jeffrey. "The Scientific Legacy of Apollo." *Scientific American*, July 1994.

Wilson, Andrew, ed. *Interavia Space Directory, 94–95*. Jane's Information Group, Inc., 1994.

Yenne, Bill. *The Pictorial History of NASA*. Gallery Books, 1989.

NOTES

Chapter 2

Propulsion

How does a rocketship get off the ground? You have probably heard the terms *thrust*, *force*, and *acceleration*. These words have precise physical meaning but are often misunderstood and misused. To understand them, you need to understand a little physics. In this matter, we will follow Albert Einstein's advice, "Everything should be made as simple as possible, but no simpler." That is, we will explain the laws of motion as clearly as possible without oversimplification.

Mass and Weight

First, we must distinguish between mass and weight. The *mass* of an object is related to the quantity of matter that goes to make up the object, the number of protons, electrons, and neutrons, plus the energy that binds them together. An objects mass is the same wherever it is in the universe. You have seen pictures of astronauts floating around the cabin of their spaceship. It is said that they are "weightless." But they are not "massless." They still have the same mass, the same number of particles of matter, that they had when they were on Earth.

The *weight* of an object, on the other hand, depends on gravity. Weight is the force that the object exerts downward as a result of the gravitational attraction between its mass and the Earth's mass. The farther you are from the center of the Earth, the less is the force of gravity. At the top of Pikes Peak you weigh less than you do in Houston, although the difference is very small. If you weigh 150 pounds in Houston, you would weigh 149.8 pounds on Pikes Peak. At 100 miles altitude, the decrease in weight is about 5 percent. MATHBOX 2.1 shows how these calculations are made.

If gravity were zero, then weight would be zero. But gravitation is a natural and universal force of attraction between *any* two masses; it cannot be turned off. It does decrease with distance, but even in deep space, far from Earth, there is a force present due to the gravity of the planets, the Sun, and the stars.

Scales are devices which measure the force due to gravity. When you stand on a scale it tells your weight, that is, how much force you are exerting toward the center of the

Earth. If it reads 150 pounds, then you are pushing on the Earth (through the scale) with a force of 150 pounds. You can feel that force on your posterior as you sit in your chair. If you and your chair were on the Moon your weight force would be only about one-sixth of what it is on the Earth, that is 25 pounds instead of 150 pounds. A scale would show your reduced weight. The gravitational force between you and the Moon is smaller because the Moon is so much smaller and less massive than the Earth. So weight, unlike mass, depends on where in the universe you are.

Incidentally, the type of scale you use can confuse the issue. Figure 2.1 shows that the ball weighs 4 ounces. On this type of scale, weights are put into the right pan until the scale balances. The two downward forces must then be the same, that is, the ball must weigh the same as the weights. Question: Where was the picture taken? It could

Figure 2.1 A pan balance compares masses.

MATHBOX 2.1

Change of Weight with Altitude

The weight of an object decreases as it moves farther from Earth. Because the force of gravity, which determines weight, decreases with the square of the distance from the center of Earth, the formula is

$$W = \frac{R^2}{(R + r)^2}\, w$$

where R is the radius of Earth (3,960 miles) r is the altitude of the object above the surface of Earth, w is the weight of the object at the surface, and W is the weight of the object at altitude r.

Example: If an astronaut has a weight of 120 pounds on Earth, how much does she weigh at an altitude of 100 miles?

$$W = \frac{(3960)^2}{(3960 + 100)^2}\, 120 = 114 \text{ pounds.}$$

---

have been taken on Earth, but when we take our experiment to the Moon, the picture would be the same. Both objects weigh only one-sixth of their Earth weight, but both still have the same mass. The scale actually compares the masses of the two objects. If they have the same mass, the scale is in balance.

Acceleration

Acceleration describes a change in motion. When accelerated, a motionless object will begin to move; an object already in motion will change its speed or its direction of motion.

Whenever you accelerate you get a sensation that your weight changes. For example, when you begin to descend in an elevator. As the elevator accelerates downward, that is, as it starts moving from a stop, you feel momentarily lighter. Indeed, if you were standing on a scale, it would indicate that your weight is less. Similarly, as the elevator accelerates upward, you feel momentarily heavier, and a scale would register an increase in weight. Notice that your weight seems different to you only while the elevator is changing speed, that is, accelerating, either upward or downward. As long as the elevator's speed is constant, you feel nothing unusual. Of course, it isn't a change in gravity that makes you feel that way; the force of gravity inside an elevator is no different than the force of gravity outside the elevator. It is the brief acceleration that gives the momentary sensation of a change in weight.

You feel a similar effect, horizontally instead of vertically, in a car when you press the accelerator to increase your speed. You are pushed back against the seat while your speed is changing, that is, while you are accelerating. When you once again reach a steady speed, when your acceleration is zero, you no longer feel the push against the back of the seat.

Forces

A *force* is simply defined as a push or a pull. When a force is applied to an object, the object will accelerate unless the force is cancelled out by other forces. If Jim gives you a shove on your right arm, you move to the left. If Mike shoves on your left arm, you move to the right. If both push at the same time with equal but opposite forces, you do not move. (You may wish they would quit, but in the interest of science you sit there and contemplate the laws of physics.) The sum of all forces acting on an object is called the *net force*. Two equal forces acting in opposite directions cancel each other out. They add to a net force of zero.

Sitting on your chair, you exert a force downward, your weight. Why do you not accelerate downward if you are exerting a force downward? There must be some balancing force such that the net force is zero. Fortunately, the interconnections between the molecules in a solid substance are a somewhat flexible lattice network. The lattice that makes up the chair "stretches," similar to the way a spring stretches, and it exerts a force upward on your bottom-side equal to the downward force the Earth's gravity exerts on you. The net force is zero and you remain still. Similarly, when you stand on the floor, the floor exerts an upward force to balance the downward force of gravity. If you push on a wall, the wall pushes back, and so on.

But suppose you sat down on a shoe box. The molecular

Figure 2.2   The molecules of wood in a yardstick apply an upward force to balance the downward weight force of the object. If the weight exceeds the force that holds the lattice structure together, then it comes apart.

lattice of the shoe box probably could not exert an upward force equal to your weight without coming apart. The downward weight force exceeds the upward force from the box; the net force is not zero so you accelerate down to the floor. If the net force is not zero, then the forces are said to be unbalanced.

Figure 2.2 illustrates the principle. The molecular lattice of the yardstick stretches, applying an upward force to the weight, trying to hold together, until the downward weight exceeds what the molecules can hold and the yardstick breaks.

Isaac Newton

All of this was summarized by Isaac Newton (1642–1727) in a book titled *Philosophiae Naturalis Principia Mathematica*, usually shortened to just *Principia*, published on July 5, 1687.

Newton's first law of motion says that an object at rest will remain at rest unless acted on by an unbalanced force. If the net force is zero, the object does not move. If the net force is not zero, the object will accelerate in the direction of the net force. A second part to this law states that an object in motion will continue in straight line motion at constant speed unless it is acted on by an unbalanced force.

Newton's first law is sometimes called the law of inertia because it points out that a force must be applied to change the motion of a body, that is, to accelerate it.

Newton's second law of motion tells us how to calculate the acceleration of an object: simply divide the net unbalanced force by the mass of the object. His *third law* explains the principle of rocket engines which we will encounter in a few paragraphs.

(And may the net force be with you!)

Freefall

Now suppose that you are back in the elevator. The floor of the elevator is pushing up against the bottoms of your feet with a force equal to your weight. The net force is zero and everything is OK. But what happens if the cable breaks? The floor of the elevator falls away under you and no longer exerts an upward force on your feet. Gravity is accelerating you and the elevator downward at the same rate; you are both in free fall. If you could manage to put a scale under your feet, it would register zero because it too is in free fall accelerating downward at the same rate. The gravitational force of the Earth still pulls on you, but because you, the scale, and the elevator are all in free fall together, you are weightless.

Bouncing on a trampoline, jumping from a diving board, and sky diving all give similar sensations. Because you have no support, no upward force to balance your downward weight force, you are in free fall.

This is precisely the situation with the astronauts in their space ship. The spacecraft, astronauts and all, are in free fall toward the Earth. Therefore they have the feeling of weightlessness. More about this in the next chapter when we discuss orbits.

Liftoff!

Consider the Space Shuttle in Figure 2.3 sitting on the launch pad attached to its external tank and booster rockets. Fully loaded and ready to go it weighs about 4.4 million pounds. That is, it is exerting a 4.4 million pound force downward on the launch pad. The launch pad, a structure of concrete and steel, is able to support the spacecraft; it can exert an upward force of 4.4 million pounds. The forces are in balance, the net force is zero, and the spacecraft just sits there. Now, how do you make it accelerate, preferably upward?

Think for a moment about a more manageable problem. The book in Figure 2.4a is lying on the table and you want to put it up on a shelf. The book weighs 1 pound; gravity, acting on the mass of the book, is exerting a 1 pound force downward on the table. The table is exerting a one pound force upward on the book so the net force on the book is zero and it just sits there. Now pick it up in your hand (Figure 2.4b) and hold it stationary. Like the table, you are exerting a 1 pound upward force to balance the downward weight force. It does not move because the forces are balanced and the net force is zero.

If you tighten your muscles and exert an upward force greater than the 1 pound downward weight force, the net force on the book is no longer zero. The net unbalanced upward force causes the book to accelerate upward (Figure 2.4c).

So it is with the Shuttle on the launch pad. All we need to start it accelerating upward is to apply an upward force greater than its weight. Where do you find a force greater than 4.4 million pounds? Rocket engines! As the rocket engines fire and the Shuttle lifts off, the launch pad no longer provides support. The engines must provide the necessary 4.4 million pounds of upward force, and then some. To accelerate upward, there must be an unbalanced force, that is, a net upward force, greater than 4.4 million pounds.

When discussing the capabilities of rocket engines, we use the term *thrust* rather than *force*. A rocket engine rated at 100,000 pounds thrust, therefore, can lift 100,000 pounds off the ground. This does not mean, however, that the engine could put 100,000 pounds into orbit. In addition to lifting the spacecraft, the rocket engine must lift itself and its fuel.

Figure 2.3 Space Shuttle poised on its launch pad, preparing to go. *Courtesy of NASA*

As fuel is burned and exhausted out the nozzle, the vehicle becomes lighter and requires less force to keep it accelerating.

The interaction of these and other variables becomes a complex mathematical problem. However, we can make a simple first estimate of the acceleration of the Space Shuttle as it lifts off the launch pad. *Newton's second law of motion* states that to calculate the acceleration of an object we divide the net force by the mass of the object. MATHBOX 2.2 shows the calculations.

The conclusion from this calculation is that the Shuttle accelerates at about 10 miles per hour per second; that is, after 1 second it is going 10 miles per hour, after 2 seconds its velocity is 20 miles per hour, after 5 seconds it will be going 50 miles per hour, and so on. At the end of 5 minutes, 300 seconds, it will be moving at a speed of 3,000 miles per hour. To go into orbit, the spacecraft must be travelling at more than 17,000 miles per hour.

In violation of Einstein's advice, we made this calculation too simple. During the flight, none of the numbers that we used remain constant. They all change as the Shuttle heads for orbit. Its mass decreases as fuel is burned and thrown overboard. Parts of the vehicle detach and drop away, further decreasing the mass. The thrust of the rocket engines is made to vary during the flight, as we will see. As the vehicle pitches over to horizontal its thrust no longer is directed directly against gravity. Also, we did not include atmospheric frictional drag. Air drag increases as the vehicle increases speed but then decreases as it reaches higher altitudes where the air is less dense. In addition, the force of gravity decreases as the spacecraft moves away from the Earth. These factors are significant and must be taken into account in a "real-world" calculation.

Using computers, engineers make a step by step, second by second calculation, taking into account the changing mass, forces, and gravity. The result of their calculation is a predicted flight profile showing where the Shuttle should be, how fast it should be going, and in what direction it should be heading. During a flight, then, the actual position and speed are compared to the predicted values to see if everything is running OK.

Reaction Engines

A rocket is a *reaction engine*. It is basically a very simple device, consisting of a chamber in which propellants (solid, liquid, or gas) are accelerated to high speed and expelled through the nozzle at the open end. In reaction to the motion of the propellant, the engine is accelerated in the opposite direction.

A toy balloon is a simple example of this principle. When you blow up the balloon, the air inside is contained under pressure. When you release it, the air is expelled from the nozzle and the balloon speeds off in the opposite direction. In Figure 2.5A the balloon is tied shut. The air inside pushes outward while the atmospheric pressure on the outside pushes in. The force in each direction is balanced by a force in the opposite direction; the net force is zero and the balloon does not accelerate.

However, in Figure 2.5B, the nozzle of the balloon is open so air escapes. The force at C is no longer balanced by an equal and opposite force at D; therefore the balloon accelerates in the direction of the arrow at C.

Figure 2.4 Forces in balance (a and b) and unbalanced (c).

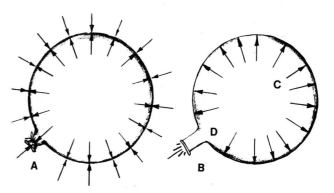

Figure 2.5 Forces in balance and out of balance in a balloon. *Courtesy of Estes Industries.*

MATHBOX 2.2

Newtons Second Law of Motion Applied

The Space Shuttle sitting fully loaded on the launch pad has a weight of 4.4 million pounds. It has three main engines and two solid rocket boosters. Rated at 375,000 pounds thrust each, the three main engines apply a total of 1,125,000 pounds force. The two boosters provide 3,300,000 pounds of thrust each. Total thrust from the five engines at takeoff, then, is 7,725,000 pounds!

Newton's second law states that the acceleration of an object is equal to the net force acting on it divided by its mass. Mathematically,

$$a = \frac{F}{m}.$$

To solve this we must find the mass of the Space Shuttle on the launch pad. The mass of any object is found by dividing its weight by g, the acceleration of gravity, which is 32 feet per second per second at the surface of the Earth. For our present problem,

$$\text{mass} = \frac{\text{weight}}{g} = \frac{4,400,000 \, \text{lb}}{32 \, \text{ft/sec}^2} = 140,000 \text{ slugs.}$$

The slug is the unit of mass in the English system of measurement. Now, apply Newton's second law. In the case of the Shuttle, the downward force is its weight, 4.4 million pounds and the upward force is the total thrust of the rockets, about 7.7 million pounds. The net force is, then, 3.3 million pounds in an upward direction. See Figure 2.2.1.

$$a = \frac{F}{m} = \frac{3,300,000 \, \text{lb}}{140,000 \, \text{slugs}} = 24 \, \text{ft/sec}^2 = 16 \, \text{mi/hr/sec.}$$

This means that for each second that passes, the Shuttle accelerates, increasing its speed by 16 miles per hour.

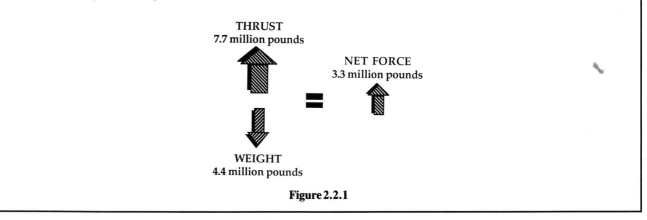

THRUST
7.7 million pounds

NET FORCE
3.3 million pounds

WEIGHT
4.4 million pounds

Figure 2.2.1

The motion of balloons and rockets also was explained by Newton. His *third law of motion* states that for every action there is an equal and opposite reaction. The acceleration of the propellants out the engine in one direction causes an acceleration of the engine in the opposite direction.

See Figure 2.6. Because there is an opening in the rear, the forces of the gases in the combustion chamber forward and rearward are unbalanced. Forces on the sides are balanced by forces on the opposite sides. But in the fore-aft direction there is a net force forward. The thrust of the rocket is, then, the unbalanced force on the forward wall of the combustion chamber which accelerates the vehicle forward. The thrust of any reaction engine is determined primarily by the amount of propellant expelled out the nozzle and how fast the propellent is moving. To increase thrust, simply throw more propellant out the nozzle at higher speed.

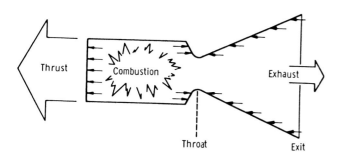

Figure 2.6   Rocket engine, a reaction engine. *Courtesy of NASA.*

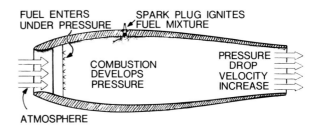

Figure 2.7   Ramjet engine. *Courtesy of Estes Industries.*

A ramjet engine is also a reaction engine. Air is taken into the front; fuel is burned and exhausted out the rear (Figure 2.7). The action of the jet of exhaust gas against the air being "rammed" in at the front produces a reaction which drives the engine forward (along with whatever is attached to it). A ramjet engine does not work at low speeds; it cannot start and take off from the ground. The craft must be moving to compress the air coming into the engine.

A rocket engine is a reaction engine also, but it does not intake air. It carries its own oxygen supply in addition to the fuel. Therefore, it can operate in space where there is no air.

Chemical Rockets

Combustion is a chemical process in which a fuel oxidizes, that is, combines with oxygen. For example, wood burns in air as in a campfire. An automobile's internal combustion engine burns (oxidizes) gasoline with air brought into the cylinders through the carburetor. A jet engine burns jet fuel mixed with air. In these examples, the oxygen supply comes from the surrounding air. But a chemical rocket engine carries both fuel and oxidizer. It does not depend on air to support the combustion.

In the combustion chamber of a rocket engine, Figure 2.6, the confined gases from the burning fuel build up tremendous *pressure*, exerting force against the walls of the chamber. Pressure is the force divided by the area over which the force is applied. The pressure is highest in the combustion chamber. On the way out, the exhaust gases pass through the throat, a constriction or reduced area which increases their speed. Once through the throat, the gases expand rapidly in the nozzle, accelerating further and reducing their internal pressure. Of course, pressure must be maintained in the combustion chamber in order to keep exhaust gases flowing out the nozzle. The burning of fuel heats the gas. Pressure is maintained by the high temperature and the continuous flow of propellants into the chamber.

Ideally, when the exhaust reaches the end of the nozzle, its pressure should have dropped to equal the pressure of the outside air. At that point, the velocity of the gases would be maximum and the engine is most efficient. If the outside air pressure were greater, it would inhibit the flow of gases out-

ward. If the exhaust pressure were greater, some of the energy of the exhaust would be wasted, producing no useful thrust.

It must be pointed out emphatically that a rocket does not obtain its forward motion by pushing against the outside air. If that were the case, it would not operate in the near vacuum of space. In fact, the outside air only hinders the motion of the rocket, reducing its efficiency in two ways. First, it causes a frictional drag which reduces the forward acceleration. Equally important, the gases expanding out the nozzle are retarded in their expansion by the surrounding atmosphere. Thus, pressure at the nozzle exit cannot drop as low as it could in a vacuum. Consequently, the pressure difference is not as great, the net force is not as great, and the thrust is lower. A rocket operating in a vacuum is significaly more efficient than one operating in the atmosphere. That is one of the major reasons for launching straight up, to get out of the dense lower atmosphere as quickly as possible.

To go into orbit a spacecraft must be accelerated to a speed of over 17,000 miles per hour, requiring a lot of fuel producing high thrust for a long time. In fact, the weight of fuel and oxidizer in the Shuttle at liftoff is nearly 3.8 million pounds, 86 percent of the total weight! It is all thrown overboard to produce the thrust. Only 5 percent is the orbiter which goes into orbit. The remaining 9 percent is engines and tanks which drop off along the way.

Specific impulse is a number which indicates the effectiveness of a rocket fuel. It may be thought of as the time that 1 pound of fuel will burn while it is producing 1 pound of thrust. The specific impulse of some typical fuel-oxidizer combinations is shown in Table 2.1. A typical solid fuel, gunpowder for example, has a specific impulse of 350 seconds; that means that 1 pound of gunpowder will produce 1 pound of thrust for 350 seconds. A higher number indicates a more effective thrust-producing fuel. Various fuel-oxi-

TABLE 2.1   Specific Impulse of Various Propellants

Propellants	Specific Impulse
Typical solid fuels, e.g., gunpowder	350 seconds
Kerosene-LOX	363 seconds
Liquid hydrogen-LOX	462 seconds
Liquid hydrogen-fluorine	483 seconds
Hydrogen heated by nuclear reactor	800 seconds
Hydrogen gas heated by electric arc	1,300–2,000 seconds

dizer combinations will be discussed further in the following sections.

Solid Propellant Rockets

The first rockets, built by the Chinese, were propelled by a mixture of solids similar to black gunpowder: potassium nitrate, sulfur, and charcoal. These three ingredients burn together without the need for air. Oxygen for combustion is supplied by the oxidizer, potassium nitrate. (" . . . ate" at the end of a chemical name generally indicates that oxygen atoms are contained in the chemical molecule.) When combined in certain proportions and ignited in a closed space, these three ingredients will explode. The problem is to control the explosion, to slow it down so the energy is released over a longer period of time rather than a quick bang. Varying the percentage of charcoal provides a control over the speed of the combustion; increasing the charcoal content slows the burning.

Products of the combustion in the confines of the rocket engine include carbon dioxide, nitrogen, sulfur dioxide, and oxides of nitrogen. These gases, heated by the combustion process, are expelled at high speed out the rocket nozzle to provide the forward thrust. Control of thrust can be achieved by varying the fuel-oxidizer ratio. As mentioned before, increasing the proportion of charcoal in black powder slows the burning.

Look for a moment at the flight of a model rocket shown in Figure 2.8. The components are shown in Figure 2.9. Some model rocket engines are constructed with low charcoal content at the start of the burn for maximum thrust (see Figure 2.10). Then the charcoal is increased to yield a slow burn of constant thrust. After a delay period while the rocket coasts to maximum altitude, an explosive charge at the top of the engine ignites to blow off the nose cone and eject the parachute for descent. This can be seen in the graph of thrust versus time in Figure 2.10.

Modern solid propellants include potassium perchlorate (oxidizer) with asphalt (fuel), ammonium perchlorate (oxidizer) with aluminum powder (fuel), and nitrocellulose with nitroglycerine (both oxiders and fuel). The ingredients are mixed with a binder to hold them in a desired shape, a stabilizer to keep them from decomposing, and sometimes a catalyst to speed up the reaction. The mixture is molded into a *grain* of the desired shape and size to fit into the outer casing or shell which provides structural integrity.

A grain of solid propellant burns on its exposed surface and produces hot gases. Besides varying the fuel-oxidizer mixture, thrust can also be controlled by the way the grain is shaped. Thrust depends on the rate of burning, that is, how fast the hot gas is being produced and exhausted through the nozzle. That, in turn, depends on the exposed surface area. Refer to Figure 2.11. A cylindrical-shaped grain which completely fills the casing burns like a cigarette from one

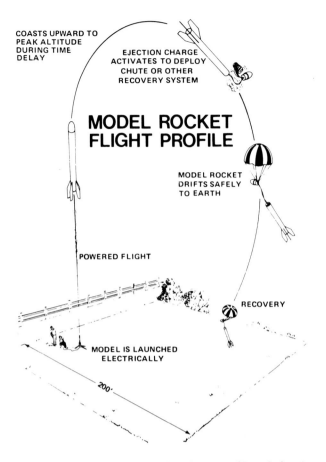

Figure 2.8 Flight of a model rocket. *Courtesy of Estes Industries.*

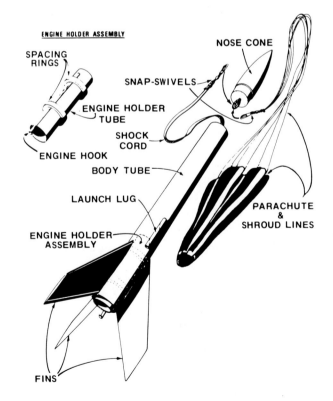

Figure 2.9 Parts of a model rocket. *Courtesy of Estes Industries.*

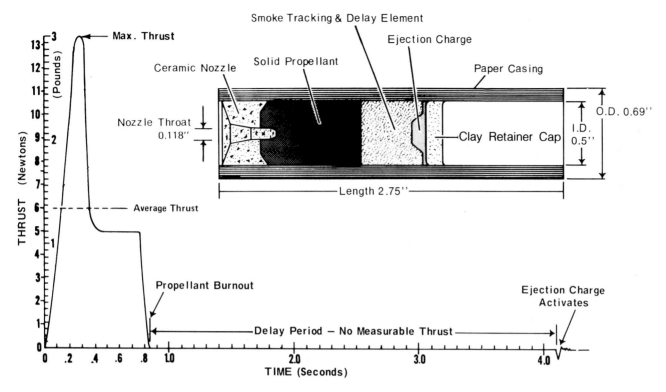

Figure 2.10 Cross section diagram of model rocket engine and its time-thrust curve. Thrust is given in both pounds and newtons (metric units of force). Note that the thrust rises quickly to maximum in about 0.3 second, and that the thrust is constant at about 1.1 pounds (5 newtons) from 0.4 to 0.8 second. The nose cone is blown off and the parachute is ejected about 4 seconds after ignition. *Courtesy of Estes Industries.*

end to the other. Because the burning surface area remains constant throughout the burn, the thrust is constant, a *neutral burn*. A problem that must be considered when designing a rocket with a cylindrical grain is that it becomes top-heavy as the bottom burns away.

Some grains are molded with holes of various shapes and sizes running down the center, called *perforations*, in which the burning takes place. See Figure 2.12. As the burn progresses, the surface area increases and so does the thrust, a *progressive burn*. Yet another type of grain is made smaller in diameter than the outer casing and held centered in the casing by metal spiders. See Figure 2.13. In this type, the outer surface of the grain is the burning surface. Because this surface area decreases as the burn progresses, the thrust decreases with time, a *regressive burn*.

Figure 2.14 shows several grain cross sections. The end-burning cylinder has a smaller burning area than those of the same size but with perforations; therefore, its thrust is lower and constant. A star-shaped perforation has a large burning area producing a high thrust for a shorter time; perhaps surprisingly, it has a neutral burn. The multiperforated grain is progressive-regressive with a large increasing exposed area at first, then a decreasing area as the tubes merge and the burn comes to an end. Try to figure out what kind of burn is produced by the cruciform grain; consider step by step how the burning area is changing.

An *inhibitor* is coated over the surfaces of the grain that are not supposed to burn, the ends of those with perforations for example. A simple but effective inhibitor is latex paint.

A major limitation of a solid rocket is that once it is ignited it cannot be turned off and restarted. Thrust can be terminated, however, by blowing off a top cap and allowing

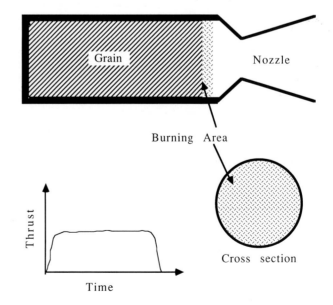

Figure 2.11 A neutral burn, cylindrical grain.

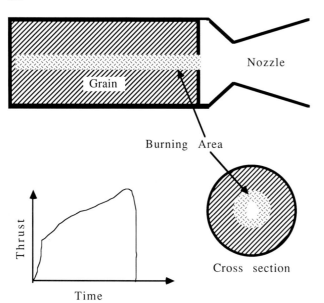

Figure 2.12 A progressive burn, bored cylinder grain.

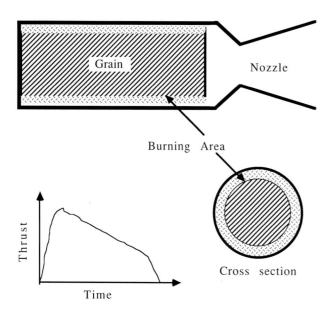

Figure 2.13 A regressive burn, supported cylinder grain.

the gases to escape from both ends, i.e., an unbalanced force no longer exists.

Solid Propellant Boosters

Solid propellant rockets are often used as booster rockets. They provide a desired amount of thrust for a specific amount of time and are discarded when they burn out. There are a number of advantages to solid rockets. They are simple and easy to construct, relatively low in cost, highly reliable, easy to store, and can be made ready to fire in a short time. For these reasons, they are used for most military rockets. The Space Shuttle uses two solid rocket boosters (SRBs) which can be seen in Figure 2.3, one on either side of the

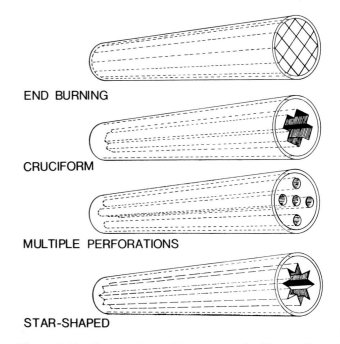

Figure 2.14 Cross sections of four types of solid propellant grains. *Courtesy of Estes Industries.*

spacecraft. Standing nearly 150 feet tall and over 12 feet in diameter, they are the largest ever built and the first used for manned spaceflight. Table 2.2 shows their propellants. They will be described in more detail in Chapter 8.

Very large solid propellant rockets must be constructed in several segments rather than in one piece. A one-piece grain the size of the Shuttle SRB would likely crack while hardening and curing. Furthermore, it would be nearly impossible to move such a large rocket in one piece from the manufacturer in Utah to the launch site in Florida without damage. On arriving at Kennedy Space Center, the segments making up the SRB are stacked and connected together.

TABLE 2.2 U.S. Manned Spaceflight Rocket Engines

Vehicle	Fuel	Oxidizer	Thrust
Mercury-Redstone	Alcohol	LOX	78,000 lb
Mercury-Atlas	RP-1	LOX	365,000 lb
Gemini-Titan			
First stage	Aerozine 50	Nitrogen tetroxide	430,000 lb
Second stage	Aerozine 50	Nitrogen tetroxide	100,000 lb
Saturn 1B			
First stage	RP-1	LOX	1,600,000 lb
Second stage	Liquid hydrogen	LOX	200,000 lb
Saturn V			
First stage	RP-1	LOX	7,760,000 lb
Second stage	Liquid hydrogen	LOX	1,150,000 lb
Third stage	Liquid hydrogen	LOX	230,000 lb
Space Shuttle			
Main engines	Liquid hydrogen	LOX	375,000 lb
Solid rocket boosters	Aluminum powder	Ammonium perchlorate	2,650,000 lb

The joints between segments must be tightly sealed against the enormous pressure of the combustion taking place inside the rocket to prevent hot gases from leaking through. A leak would melt the rocket casing, reduce the internal pressure and thrust of the motor, and would produce unwanted thrust in a direction opposite the leaking gas.

The top of each segment has a U-shaped groove, called a *clevis*, into which the bottom of the next segment, called a *tang*, fits. Figure 2.15 is a simplified diagram of the joint. Because metal against metal does not seal well, two flexible, rubber-like seals, *O-rings*, are set into slots cut into the clevis. In the original design, zinc chromate putty filled the space between the two segments. When the booster ignites, the high pressure gas inside is supposed to push against the putty to compress the air ahead of it and force the O-rings into the gap between the tang and clevis making a leakproof seal. This seal must take place in less than 1 second or the hot gas will blow by the O-ring.

That is exactly what caused the *Challenger* accident in January 1986. After 24 successful Shuttle flights, one booster failed catastrophically. It was a cold morning in Florida and the O-rings were cold and stiff. One did not seal properly and the hot gas leak caused the accident. The puff

of smoke in Figure 2.16 occurred at the point where the O-ring did not seal. First the putty came through the gap, then came the hot gases from the burning propellant which caused the destruction of the external tank and the orbiter. After the accident, the joint was redesigned to incorporate the third O-ring, zinc chromate putty was replaced by bonded insulation, and electric heaters were incorporated into each joint to keep the O-rings warm and flexible.

Liquid Propellant Rockets

Solid propellant rockets have been around a long time; the Chinese recorded their use before A.D. 1200. But it wasn't until March 16, 1926, in Auburn, Massachusetts, that the first successful liquid fuel rocket was flown. Engineered by Robert Goddard, the pioneer flight lasted 2.5 seconds, reached an altitude of 41 feet, a range of 184 feet, and a speed of 60 miles per hour. The rocket weighed less than 6 pounds and carried about 5 pounds of propellant.

The development of liquid propellant rockets was not easy. Two formidable problems had to be overcome: the pumping of extremely cold (*cryogenic*) liquid propellants

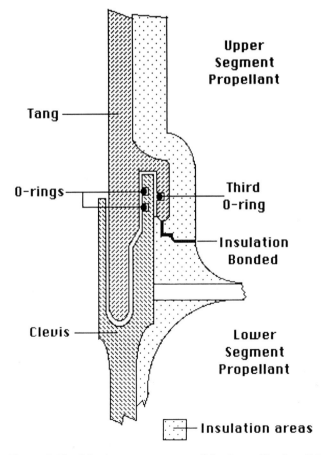

Figure 2.15   Joint between segments of the Space Shuttle solid rocket booster. *After NASA diagram.*

Figure 2.16   The puff of smoke when the *Challenger* boosters ignited shows where the seal failed and hot gases leaked through the joint. *Presidential Commission Report, NASA photo.*

into the combustion chamber and the high temperature of the burning fuel. The propellants used first by Goddard were gasoline for the fuel and liquid oxygen (LOX) as the oxidizer. An assistant ignited the rocket with a blowtorch.

Handling the gasoline was not particularly difficult. There had been considerable experience with automobile engines to draw on. It could be stored in steel tanks, easily pumped, and rubber gaskets could be used for seals. Liquid oxygen was another matter. It boils at −297 °F and must be kept colder than that or it boils away and returns to a gas. Metal valves and pipes contract and crack at that low temperature. Rubber freezes and shatters. Atmospheric water vapor freezes on and in everything. Lubricants solidify, oils congeal, and pumps freeze up. On the other hand, when LOX and gasoline burn, temperatures in the combustion chamber reach 5,560 °F. Steel melts at that temperature. The hot gases flowing out the combustion chamber at high speed can melt through the throat of the nozzle where the heat transfer from the exhaust gas is greatest. All these engineering problems and many more took time, talent, and money to solve.

Liquid propellant rockets have been developed to a high degree of performance and reliability. Table 2.2 shows the propellant combinations used in American rockets for manned spaceflight. RP-1 is a kerosene-like fuel. Note that liquid oxygen (LOX) is most often used as the oxidizer. *Oxidation* refers to a type of chemical reaction, not necessarily involving oxygen. In that context, there are oxidizers that are chemically more energetic than oxygen, e.g., fluorine. However, fluorine is extremely corrosive and toxic. When burned with hydrogen, the exhaust includes hydrofluoric acid which reacts with most common substances, including glass. Thus, LOX is the most widely used oxidizer because it is safer and easier to handle and produces less hazardous exhaust products. Fluorine has been used only in experimental rockets.

Nitrogen tetroxide was used in Titan missiles because, unlike LOX, it is easily stored. Aerozine 50 and nitrogen tetroxide are *hypergolic*, that is, they react chemically on contact, like Alka Selzer in a glass of water. No igniter is needed. These features are most desirable for a military rocket which remains in "storage" for its entire life (we hope) yet must be ready to go instantly if necessary. Hypergolic propellants are also used in rocket engines and thrusters used for rotating and maneuvering while in orbit because they can be turned on and off repeatedly with high reliability.

A much simplified diagram of a liquid propellant rocket is shown in Figure 2.17. The fuel and oxidizer are carried in separate tanks and pumped into the engine's combustion chamber by turbine pumps which are powered by a gas generator. Some of the fuel and oxidizer is burned in the gas generator to produce the pressure needed to run the pumps.

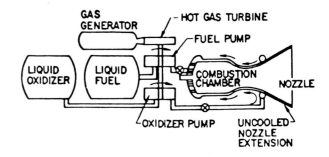

Figure 2.17 Simplified diagram of liquid propellant rocket engine. *Couresy of NASA.*

Fuel and oxidizer enter the combustion chamber through *injectors* where they are ignited and burn. The primary purpose of the injectors is to break up the propellants into fine streams or droplets and spray them into the combustion chamber. They perform a function similar to an automobile carburetor which sprays fuel from the gas tank into an incoming stream of air. Injectors come in a variety of designs, some of them quite complex. Early injectors were nothing more than metal plates punched with fine holes at the forward end of the combustion chamber. Later models looked something like shower heads or garden hose nozzles. Figure 2.18 shows some types and arrangements of injectors. Fuel injectors have replaced carburetors in many automobile engines.

It is important that the burning take place in the center of the chamber so the side walls and injectors do not overheat and melt. It is also important that the propellants be thoroughly mixed in the proper proportions. The fineness of the spray depends on the size of the injector holes and on the difference in pressure between the pipes and the combustion chamber. Smaller holes and greater pressure difference produce a finer spray.

The fuel is ignited by any of a number of devices, commonly a spark plug similar in principle to an automobile spark plug. Once burning begins, the temperature rises rapidly, fuel ignites as it reaches the burning volume, and the igniter is no longer needed to keep the combustion going. As mentioned above, hypergolic propellants do not need igniters; they react on contact.

The high pressure produced by the burning fuel in the combustion chamber opposes the propellants spraying into the chamber through the injectors. Therefore, the propellant pumps must produce an even higher pressure than the combustion produces. Designing and building reliable high velocity, high pressure pumps to pump liquid oxygen at near −300 °F was not an easy task.

To keep the combustion chamber from melting, one common solution is to run the cold LOX feed lines through pipes or a cooling jacket around the combustion chamber. This arrangement can be seen in Figure 2.17. An automobile en-

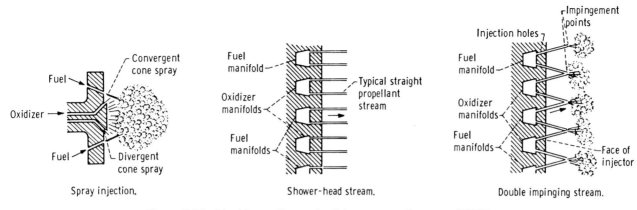

Figure 2.18 Liquid propellant rocket injector types. *Courtesy of NASA.*

gine has a similar arrangement where a water-antifreeze mixture circulates in a jacket around the cylinders to absorb the heat, then flows through the radiator where it loses the heat to the cool air passing through. In a rocket engine the cold propellant circulates around the combustion chamber and nozzle. This accomplishes two things: it cools the engine and preheats the propellant at the same time. It is important that the fuel and oxidizer be at nearly the same temperature. For example, if cold LOX and RP-1 are brought into contact, the RP-1 would freeze solid.

Thrust of a liquid rocket engine can be controlled by varying the rate at which the propellants are pumped into the combustion chamber, thereby varying the rate at which exhaust gases are produced. A throttle works on the same principle as the accelerator pedal on your car. Also, some liquid rocket engines have been designed so they can be be turned off and back on again as often as needed, using hypergolic fuels for reliable ignition.

The Space Shuttle has three liquid rocket engines at the rear of the orbiter, visible in Figure 2.19. Propellants are fed to them from the large external tank between the solid rocket boosters. Pumps on the Space Shuttle carry 64,000 gallons of LOX and liquid hydrogen (at −423 °F) per minute to the engines through 17 inch pipes, an engineering problem of the highest order! They are probably the most complex rocket engines ever built.

When hydrogen burns in oxygen the result is water (H_2O). The cloud of exhaust coming from the three main engines in Figure 2.19 is a real cloud, nothing put pure water. The Shuttle has several other small liquid propellant rocket engines and thrusters which will be discussed in Chapter 8.

Other Reaction Engines

Actually, any device which throws mass overboard in one direction in order to accelerate in the opposite direction qualifies as a reaction engine. Besides the solid and liquid propellant chemical rocket engines we have been discuss-

ing, several other "exotic" engines have been designed, some prototypes have been built, and some have even flown.

The use of nuclear reactors to heat matter and expel it from an exhaust nozzle was under serious study from 1955 to 1972 in Project Rover. Figure 2.20 compares a nuclear

Figure 2.19 Space Shuttle on its way to orbit. Notice the exhaust from the three main engines is barely visible. They burn liquid hydrogen and liquid oxygen and the exhaust is pure water. By comparison, the solid rocket booster exhaust is a mixture of many different chemical compounds from the combustion of the aluminum fuel, ammonium perchlorate oxidizer, and other materials used in making the grain. *Courtesy of NASA.*

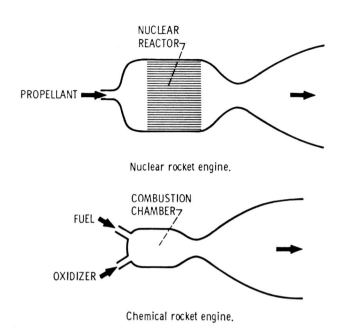

Nuclear rocket engine.

Chemical rocket engine.

Figure 2.20 Comparison of a nuclear rocket and a chemical rocket. They operate on the same principle, but the heat source in a nuclear rocket is a nuclear reactor while the heat in the chemical rocket comes from the combustion of the fuel and oxidizer. *Courtesy of NASA.*

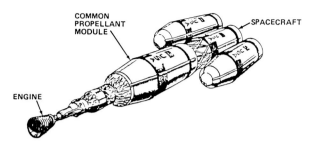

Figure 2.21 Sketch of vehicle using NERVA engine. *Courtesy of NASA.*

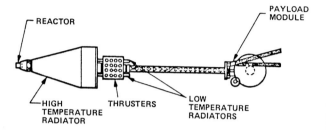

Figure 2.22 Simple concept sketch of a nuclear-electric propulsion vehicle. *Courtesy of NASA.*

high speed, however, making them suitable for interplanetary journeys.

and a chemical rocket. In the NERVA engine (Nuclear Engine for Rocket Vehicle Applications) liquid hydrogen was pumped through the reactor where it absorbed heat to reach about 4,000 °F before being ejected through the nozzle at high speed. The hydrogen acted as a moderator to slow neutrons from the nuclear reactions. The pump was also driven by hydrogen gas heated in the reactor. Figure 2.21 shows a concept of an interplanetary vehicle powered by a NERVA engine. Development of the flight test engine was just beginning when the project was cancelled in favor of Space Shuttle development.

There is some interest in reviving nuclear propulsion systems for space vehicles. Now, however, the concept may be different. The nuclear reactor could generate electricity which ionizes and accelerates some fuel, such as vaporized cesium, to high speed and expels it out the rocket nozzle. Ionized particles are electrically charged and can be accelerated in an electric field to speeds unattainable in chemical combustion. Figure 2.22 is a simple sketch of such a vehicle. Note that in Figures 2.21 and 2.22 the payload is far removed from the nuclear reactor to avoid a radiation hazard. There would seem to be two options: a heavily shielded reactor on a physically smaller spacecraft or a lightly shielded reactor on a long spacecraft.

Both of these engines produce low thrust for long periods of time. Refer back to Table 2.1 and note their high specific impulse. Because acceleration is low they would not be used to take off from Earth. They eventually can achieve very

DISCUSSION QUESTIONS

1. If an astronaut weighs 120 pounds at an altitude of 100 miles above the Earth, then why is she weightless in orbit?

2. Explain what is meant by miles per hour per second or feet per second per second as units of acceleration.

3. Why does a toy balloon fly an erratic path but a rocket ship flies straight?

4. How does the turbojet engine on an airliner work? Is it a reaction engine?

5. What precautions would have to be taken when using a nuclear-powered rocket?

ADDITIONAL READING

Estes Industries. *The Classic Collection, Model Rocketry.* Estes Industries. Nonmathematical technical notes and reports.

Muolo, Michael J., et al. *Space Handbook, Volume 1, A Warfighter's Guide to Space* and *Volume 2, An Analyst's Guide.* Air University Press, December 1993. Mostly descriptive with some algebra.

NASA. *Exploring in Aerospace Rocketry.* Government Printing Office, 1971. Algebra level mathematics.

Newgard, John J., and Myron Levoy. "Nuclear Rockets." *Scientific American,* May 1959. Early work, before nuclear rockets were shelved.

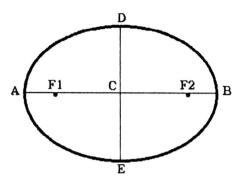

Figure 3.6 The ellipse. A to B is the major axis; D to E is the minor axis. The center is at C and the two focal points (foci) are at F1 and F2.

An *ellipse* is an oval shape as shown in Figure 3.6. You can draw an ellipse by using two thumb tacks and a piece of string. See Figure 3.7. Stick the thumbtacks into a board; they will become the two *focal points* or *foci*. Tie the string into a loop and hang it loosely around the thumbtacks; then insert your pencil into the loop and draw it up tightly. As you move the pencil around the thumbtacks while keeping the string taut, it will trace out an ellipse. Use a shorter piece of string or set the thumbtacks far apart for an elongated ellipse. Use a longer piece of string or set the thumbtacks close together for a near-circle. Set them on top of one another, i.e., use only one thumbtack at one focal point, and you will get a circle.

The *eccentricity* of an ellipse is a number between zero and one which tells just how elongated the oval is. It is calculated by dividing the distance between the foci by the length of the major axis as shown in MATHBOX 3.1. If the ellipse is a very elongated oval, then the eccentricity is close to one. At the other extreme, the eccentricity of a circle is zero.

Kepler's first law of planetary motion states that planets move in elliptical orbits with the Sun at one focus. The eccentricity of the Earth's orbit is very small, 0.017, indicating that the orbit is nearly circular. In fact, if you draw it to scale on a piece of paper, it looks like a perfect circle. Pluto's orbit has the greatest eccentricity of any of the planets, 0.248.

Satellite Orbits

Satellites in orbit around Earth also follow Kepler's first law, moving in elliptical paths with the center of Earth at one of the foci. The satellite's speed, direction, and distance from Earth at the instant the engines burn out determine the size and shape of the orbit. The only force acting is gravity, and left on its own, the satellite will follow its elliptical orbit and return to the point where the engines shut down.

The point of closest approach to Earth is called *perigee* and the point on the orbit farthest from Earth is *apogee*. (Geos is the Greek word for Earth.) In Figure 3.8, Earth is at F1, perigee is at A, and apogee is at B. Notice that perigee and apogee are on opposite ends of the major axis. When referring to an orbit, the major axis is called the *line of apsides*. For an orbit around the Sun, the nearest and farthest points to the Sun are called *perihelion* and *aphelion*. (The Greek word for Sun is helios.) The orbit is fixed in space with the spacecraft moving on it and Earth rotating inside it.

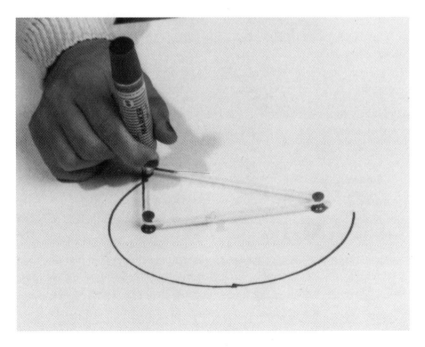

Figure 3.7 Drawing an ellipse.

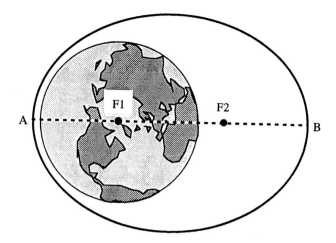

Figure 3.8   Elliptical orbit around Earth.

The speed of a spacecraft varies from point to point in an elliptical orbit depending on how far it is from Earth. The farther out it is, the slower it travels. Thus, a spacecraft's speed is greatest at perigee and slowest at apogee. As a consequence, its speed is increasing as it moves from apogee to perigee and it is slowing down as it moves from perigee back to apogee. This makes intuitive sense. A ball thrown upward, moving away from Earth, slows down; falling back, moving toward Earth, it speeds up. Of course, if the spacecraft's orbit is a circle, then its distance from Earth does not change, perigee and apogee are undefined, and its speed is constant. MATHBOX 3.2 shows how to calculate the speed of a spacecraft in circular orbit.

Kepler recognized these facts in his studies of planetary motion around the Sun. His second law states that equal areas are swept out in equal times. At first glance, that statement seems unrelated to our previous discussion about speed,

---

MATHBOX 3.1

Eccentricity

Let us represent the distance between the foci of an ellipse as f and the length of the major axis from A to B as d. The definition of eccentricity e is

$$e = \frac{f}{d}$$

In the ellipse of Figure 3.1.1 (next page), f is 2 inches and d is 5 inches. Then its eccentricity is

$$e = \frac{2}{5} = 0.4$$

In Figure 3.1.2 on the next page, $f = 4$ inches and $d = 5$ inches. Its eccentricity is

$$e = \frac{4}{5} = 0.8$$

Finally, the third ellipse in Figure 3.1.3 is really a circle with a diameter of 3 inches. The two foci are coincident so $f = 0$, and d is the diameter. The eccentricity of the circle is

$$e = \frac{0}{3} = 0$$

Note that if the two foci are close together, the distance between them is very small, and the eccentricity is a very small number. In the extreme case where there is only one focus, the ellipse becomes a circle.

On the other hand, if the foci are far apart, the ellipse is elongated and the eccentricity is a larger number. In an extremely elongated ellipse, the distance between the foci is nearly the same as the length of the major axis and the eccentricity is nearly 1.

So the eccentricity is a number with values between zero and one. It indicates how elongated the ellipse is.

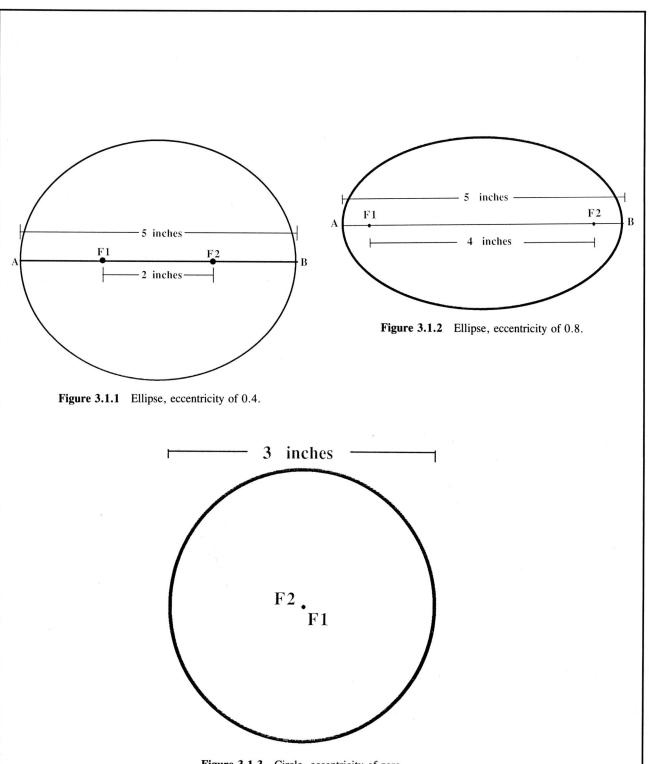

Figure 3.1.1 Ellipse, eccentricity of 0.4.

Figure 3.1.2 Ellipse, eccentricity of 0.8.

Figure 3.1.3 Circle, eccentricity of zero.

but look at Figure 3.9. In a certain time, say 10 minutes, the spacecraft moves from point a to point b near perigee, and the line from the spacecraft to the focus at the center of Earth sweeps out the area A1. Since it moves slower at apogee, the distance covered in 10 minutes will be only from point c to point d, sweeping out area A2. Kepler's second law states that A1 must equal A2. The fact that the speed of an orbiting spacecraft depends on its distance from Earth is a most important consideration in orbital maneuvering and rendezvous.

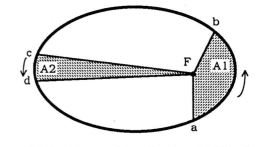

Figure 3.9 Kepler's second law of motion of orbiting bodies.

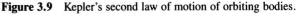

MATHBOX 3.2

Speed in Circular Orbit

According to Newton, an object will move in a straight line unless some force acts on it. Therefore, a force must be acting on a spacecraft to keep it moving in a circular orbit. That force, directed toward the center of the orbit, is gravity.

Any centrally directed force is called a *centripetal force*. Its magnitude is given by

$$F_c = \frac{mv^2}{r}$$

where m is the mass of the object, v is its speed, and r is its distance from the center.

In the case of an orbiting spacecraft, that force is supplied by gravity, Newton's law of gravitation tells us the magnitude of the force of gravity:

$$F_g = \frac{GmM}{r^2}$$

where M is the mass of the central body (Earth, Sun, etc.) and G is the universal constant of gravitation, a number which is the same everywhere in the universe. The m and r are the same as in the previous equation.

But the gravitational force is the centripetal force. therefore

$$\frac{mv^2}{r} = \frac{GmM}{r^2} \ .$$

If we solve this for v we get

$$v = \sqrt{\frac{GM}{r}} \ .$$

Now, G times M is a constant for any particular central body. For Earth it is 1.24×10^{12} if r is given in miles and the speed is in miles per hour. Let us make the calculation for an orbit with an altitude of 200 miles. Because r in the formula is the distance to the center of the orbit, we must add the radius of the Earth, about 4,000 miles. Then

$$v = \sqrt{\frac{1.24 \times 10^{12}}{4200}} = 17,180 \text{ miles per hour.}$$

Try this calculation for orbit at other altitudes. Remember, this equation is valid only for circular orbits.

G times M for the Sun is 4.13×10^{17}. How fast is the Earth moving, assuming the orbit to be circular?

The *inclination* of an orbit is the angle the orbit makes with the plane of the equator (Figure 3.10). An orbit directly over Earth's equator has an inclination of zero degrees; an orbit that passes over both poles has an inclination of 90 degrees. The angle is measured counterclockwise from the equator at the place where the spacecraft crosses the equator heading north, so it is possible to have an orbit with inclination greater than 90 degrees. The *period* of an orbit is the time it takes the spacecraft to make one complete orbit. MATHBOX 3.3 tells how to calculate the period of a circular orbit.

The *ground trace* of a spacecraft is the track it makes over the surface of the Earth. The inclination is the farthest north and south latitude that the spacecraft reaches on its ground trace. Ground traces, as we shall see, often make unexpected and unusual patterns when drawn on a flat map of Earth.

Useful Orbits

Described here are several particularly useful orbits.

1. *Low Earth orbit, about 175 miles above the surface, nearly circular, period about 90 minutes.* This is a typical orbit for the Space Shuttle, space stations, astronomy satellites, scientific research satellites, etc. When drawn to scale as in Figure 3.11 it appears remarkably low, just skimming

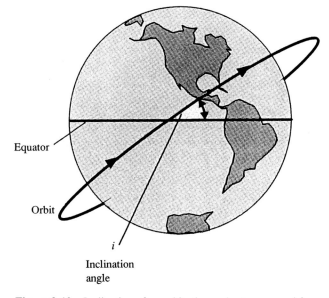

Figure 3.10 Inclination of an orbit, the angle, *i*, measured from the equator counterclockwise as the satellite heads north.

the surface. Figure 3.12 shows the ground trace for two revolutions around Earth. The ground trace looks like a sine wave that does not repeat itself. Starting at point 1, the satellite circles Earth twice to point 2. In space, the orbit closes on itself; points 1 and 2 coincide. However, when we view the path on a map it does not appear to do so because

---

MATHBOX 3.3

Period of a Circular Orbit

Kepler also had a third law which related the orbital period of a planet, its "year," to its distance from the Sun. In words, it says the square of the period is proportional to the cube of the distance from the Sun. It wasn't until Newton, however, that the constant of proportionality was found. The formula is

$$P = \sqrt{\frac{4\pi^2 r^3}{GM}}$$

where r is the radius of the orbit in miles and P is the period in hours.

Kepler's third law also applies to satellites in circular orbit around the Earth. Remember that the radius of the orbit must be measured to the center of the Earth.

Now, GM is a constant (see MATHBOX 3.2), and p is also a constant (3.14159). So for Earth satellites in circular orbit the formula becomes

$$P = 5.64 \times 10^{-6} \sqrt{r^3}.$$

For a satellite in a 200 mile orbit above the Earth, r is 4,200 miles. The period turns out to be 1.53 hours.

You can use the same formula to find the period of a planet given its distance from the Sun, or vice versa, but you must use the value of GM for the Sun. That number was given in MATHBOX 3.2. Try finding the period of the Earth.

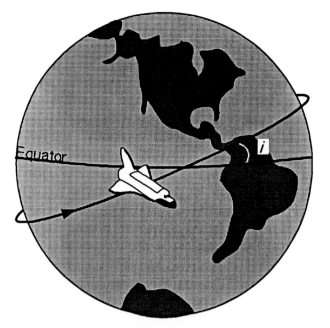

Figure 3.11 Circular low Earth orbit drawn to scale. Altitude is 190 miles, period is 91 minutes, and inclination is 28.5 degrees.

22.5 degrees farther west than the comparable point on the previous orbit. Notice that the inclination of the orbit is 28.5 degrees and the spacecraft reaches to 28.5 degrees latitude north and south of the equator. Thus, with the combined rotation of the Earth and motion of the spacecraft in its orbit, it eventually passes over all areas of the Earth between 28.5 degrees north and 28.5 degrees south. Because of its low inclination, it does not overfly areas to the far north or south.

2. *Polar orbit, inclination near 90 degrees, circular.* In contrast to the low inclination orbit, a polar orbit passes over the entire surface of the Earth in just a few days. The ground trace for a 90 degree orbit at an altitude of 500 miles and period of about 100 minutes is shown in Figure 3.13. Notice that while the satellite comes down over the north pole, heading nearly due south, the ground trace is a line tilted toward the west because of the rotation of the Earth. The same is true as it comes up from the south pole headed north. Figure 3.14 is the ground trace for an orbit with a 98 degree inclination. At 98 degrees, the westward motion of the satellite just keeps up with the Sun's westward motion so it passes over the Earth at the same local time on each pass. For that reason, the orbit is called "Sun synchronous." We will see the value of this orbit in Chapter 6. A polar orbit is used for observing the weather and mapping Earth resources.

3. *Geosynchronous orbit, 22,400 miles above the surface, circular, zero degree inclination, over the equator.* This orbit is drawn to scale in Figure 3.15. The satellite completes one orbit in 23 hours and 56 minutes, the same time as it takes

the Earth is rotating under the satellite orbit. In the 90 minutes that the satellite takes to complete the orbit, the Earth rotates 22.5 degrees of longitude, about 1,500 miles at the equator. Therefore, during those 90 minutes the flat Earth map shifts to the east by that amount and the satellite does not pass over the point where it began, but 22.5 degrees of longitude farther west. Each point on the ground trace is

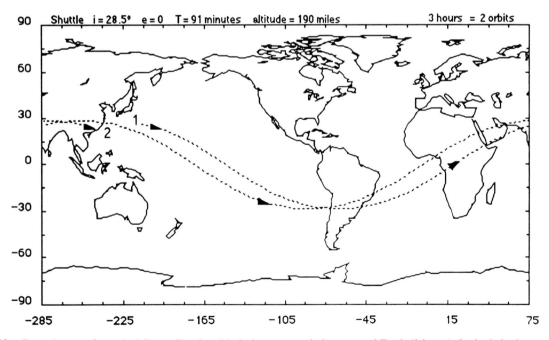

Figure 3.12 Ground trace of a typical Space Shuttle orbit during two revolutions around Earth (3 hours). Latitude is shown on the left; degrees south of the equator are negative. Longitude is measured eastward from Greenwich, England; negative numbers are longitudes west of Greenwich. Inclination, eccentricity, period, and altitude are given at the top of the map. Arrowheads on the ground trace are at 1 hour intervals.

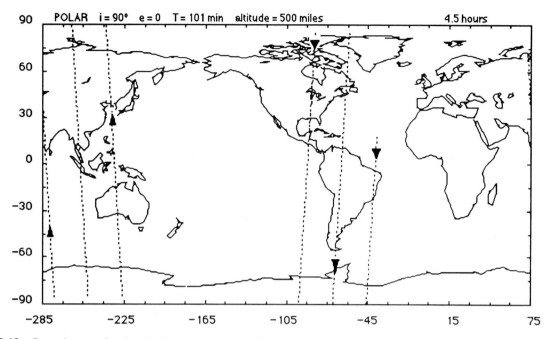

Figure 3.13 Ground trace of a circular 90 degree (polar) orbit. Notice that the spacecraft drifts westward on each orbit. The trace shows 2 1/2 orbits. Arrowheads are at 1 hour intervals.

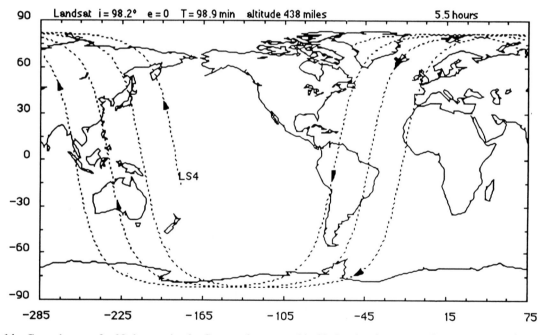

Figure 3.14 Ground trace of a 98 degree circular Sun synchronous orbit. Notice that the spacecraft moves westward on each orbit, and that it reaches to 82 degrees north and south latitude. The trace shows 3 1/3 orbits. Arrowheads are at 1 hour intervals.

Earth to rotate once on its axis. Therefore, from a vantage point on Earth, the satellite appears to remain fixed in the sky. Consequently, the orbit is sometimes called a *geostationary orbit*. The ground trace, Figure 3.16, is simply one spot on the equator. This orbit is particularly useful for communications satellites because the antennas can be pointed at that fixed spot. From that altitude the satellite transmitter can cover nearly one-third of the Earth; it cannot reach far

north countries, however. The geostationary orbit is also valuable for warning of missile attacks because the satellite sensors can continuously watch large areas of Earth.

There are some interesting variations on the geosynchronous orbit. If the orbit has an inclination other than zero, the satellite's ground trace is a line extending north and south of the equator to the latitude equal to the inclination

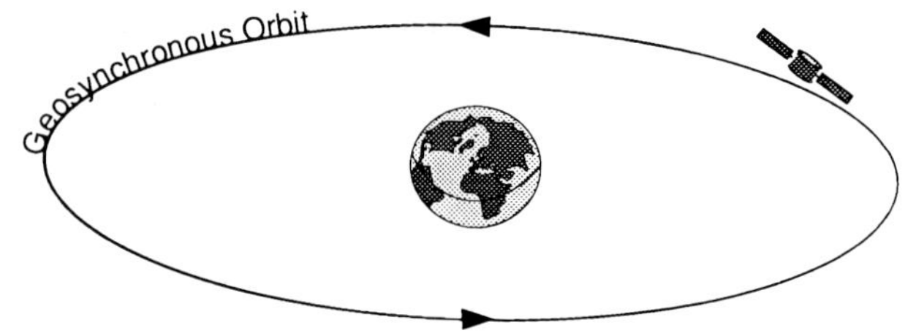

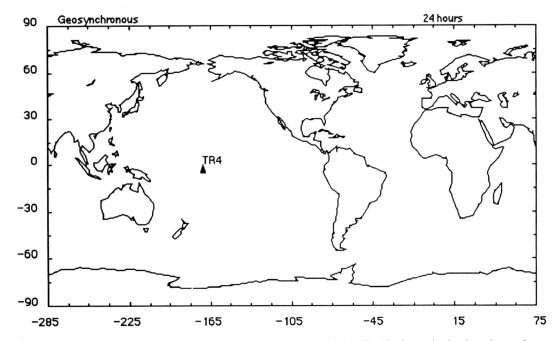

Figure 3.15 Geosynchronous orbit drawn to scale. The circular orbit is at an altitude of 22,300 miles over the equator. Period is 23 hours 56 minutes.

of the orbit. If the orbit is not circular, then, because the satellite speeds up at perigee and slows down approaching apogee, its ground trace is an east-west line along the equator. If the orbit is both noncircular and inclined, the ground trace may be an oval or a figure eight. See Figure 3.17.

4. *High eccentricity, with perigee at about 300 miles and apogee at about 25,000 miles above Earth.* The orbit in Figures 3.18 and 3.19 has an inclination of 64 degrees, an eccentricity of 0.69, and a period of 11 hours and 58 minutes. The arrowheads on the ground trace are at 1 hour intervals. Notice that they are far apart in the southern hemisphere, indicating that the satellite is travelling at high speed, while they are so close together in the northern hemisphere that they form a continuous line. Because apogee is near geosynchronous altitude, the satellite moves very slowly, hovering over one point on Earth for a long time, then moves very rapidly past perigee to return to apogee and hover again. This is the orbit of a Russian Molniya communications satellite. It remains nearly stationary for more than 8

hours at apogee over northern Siberia; then it speeds up, drops down past perigee, and returns to apogee in less than 4 hours. Meanwhile, Earth will have rotated half a turn so apogee is over Canada. Twelve hours later it is back over Siberia. This orbit is particularly useful for communications in the far north where geosynchronous satellites over the equator cannot be seen.

Orbital Perturbations

So far, our discussion of orbits has been somewhat theoretical. In the real world, orbits are never perfect ellipses or circles. The gravitational attraction of Earth is the dominant factor in shaping the orbit. But the Earth is not a perfect sphere; it bulges at the equator and is flattened at the poles. This causes the line of apsides to drift westward, a perturbation called *regression*.

The gravity of the Moon and Sun may affect high altitude satellites by changing their inclination or their altitude of perigee. The effect is small because the Sun and Moon are so far away, but may become significant as it accumulates over a long period of time.

Atmospheric drag, friction with the air, also changes an orbit. The effect is most pronounced at perigee when the spacecraft is closest to Earth. Each time the spacecraft passes perigee, the drag reduces its energy slightly so it cannot make it out as far as its previous apogee. On each orbit, then, apogee is lowered slightly while perigee remains at about the same altitude. The eccentricity decreases as the orbit becomes more circular. After a period of time apogee also lowers into the denser air so that the spacecraft expe-

Figure 3.16 The ground trace of a geosynchronous satellite over a spot in the Pacific Ocean is simply a dot on the equator.

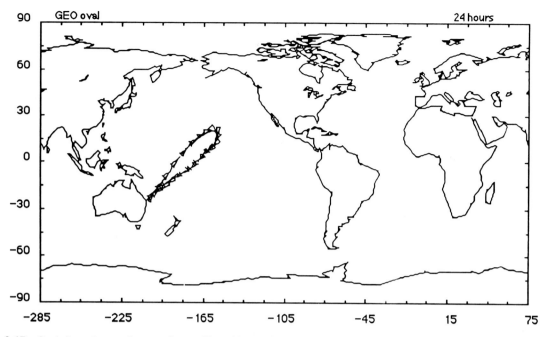

Figure 3.17 Oval-shaped ground trace of a satellite with a period of 1 day, an inclination of 20 degrees, and eccentricity of 0.2.

riences continuous frictional drag. Finally it cannot remain in orbit and it falls back to Earth, usually burning up in the atmosphere on its way down. The lowest of the satellite orbits decay rather rapidly over a period of a few weeks or months. Atmospheric drag varies seasonally, and with disturbances on the Sun and in the space environment, a subject to be covered in the next chapter.

Orbital Maneuvering

Once in space, the orbit of a satellite can be modified or changed by applying thrust in the proper direction and at the proper time to change the speed and energy of the spacecraft. In general, the orbit will increase in size if energy is increased by firing the engines normally. If energy is decreased by turning around and firing the engines in the direction of motion, called *retrofiring*, then the orbit decreases in size.

Changing Eccentricity

Look at Figure 3.20. If the rockets are fired at perigee the satellite's speed and energy increase. It will go into a larger orbit, farther out beyond its original apogee. The orbit has become more eccentric. Refer now to Figure 3.21. Suppose the satellite is at apogee when the engines are ignited. Once again, it transfers to a larger orbit because it has increased energy, but note that this time the orbit becomes less elliptical. If just the proper amount of energy is added, the orbit becomes a circle. This technique, called *apogee kick*, is used to circularize the orbits.

Note, also, that firing the engines anywhere in a circular orbit will increase the eccentricity of the orbit. Any change from circular increases the eccentricity.

Hohmann Transfer

Next, suppose a satellite is to be raised to a higher altitude by transferring it from one circular orbit, position 1 in Figure 3.22, to another higher circular orbit, position 4. The move is accomplished by igniting the engines twice. First, when the satellite is at point 2, the rockets are fired to increase its speed and put it into an elliptical transfer orbit with apogee at 3, the desired final altitude. When it reaches point 3 the engines are ignited once more to circularize the orbit at that altitude. This maneuver is called a *Hohmann transfer*, named after the German who, in 1925, first came up with the idea.

An alternate method of moving from one circular orbit to another is shown in Figure 3.23. Engines are fired at position 2, but with a greater increase in energy than for a Hohmann transfer. The elliptical transfer orbit is therefore larger and intersects the desired circular orbit (4) at position 3. When the satellite reaches 3 the engines are ignited again, but this time greater energy is required because the direction as well as the speed must be changed. The satellite has to turn a corner. This method is faster than the Hohmann transfer, but requires more fuel. The Hohmann method uses the least fuel but requires more time to complete.

Moving from a larger, higher altitude orbit to a lower, smaller orbit may be accomplished by a Hohmann transfer, also. Retrofiring the rockets moves the satellite into an elliptical transfer orbit. Then a second retrofiring at the peri-

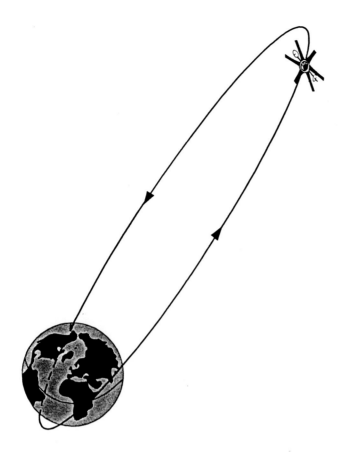

gee of the transfer ellipse circularizes the orbit at the desired altitude. Again, there are faster ways of doing it, but the Hohmann technique requires the least fuel.

Rendezvous

How can one spacecraft catch up and rendezvous with another? Sounds like a simple problem. Luke Skywalker and Hans Solo had no problem in the Star Wars movies. Just point at the Empire tie fighter or cut inside his turn and "step on the gas." That is the way jet fighter pilots do it.

In the real world of space it is not so easy. A jet fighter is supported by its wings moving through the air and the air in turn produces a frictional drag on the aircraft. By increasing the drag more on one side than the other the plane turns in the direction of the greater drag. This is accomplished with ailerons, elevators, and rudders. But in space there is no air to "push against."

The first successful orbital rendezvous came in December 1965, when Wally Shirra and Tom Stafford in Gemini VI came within 6 feet of Gemini VII carrying Frank Borman and Jim Lovell and flew side by side for over 5 hours. Figure 1.5 is a photograph of their success.

Imagine you are in orbit in a spaceship and want to join up with a space station just ahead of you. You are both in the same orbit and therefore going at the same speed. You aim toward the space station, fire your engines to catch up, and discover that you fall farther behind! Why?

Look again at Figure 3.23. Suppose your spaceship is in the inner orbit and you are at perigee. Fire your engines and

Figure 3.18 Molniya orbit drawn to scale. Perigee is at about 300 miles in the southern hemisphere. Apogee is at about 25,000 miles in the northern hemisphere.

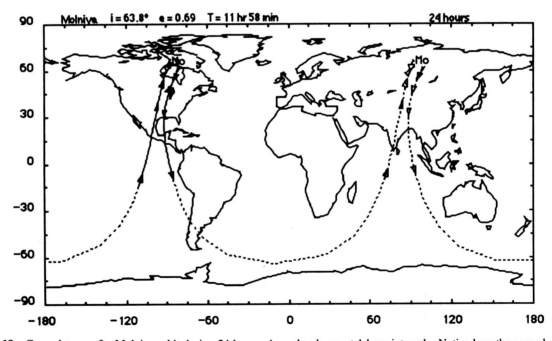

Figure 3.19 Ground trace of a Molniya orbit during 24 hours. Arrowheads are at 1 hour intervals. Notice how they are close together in the northern hemisphere, showing a slow speed near apogee. In the southern hemisphere the arrowheads are far apart indicating a higher speed near perigee. Thus, the satellite hovers for 9 to 10 hours alternately over Siberia and over Canada and spends only about 2 hours in the southern hemisphere.

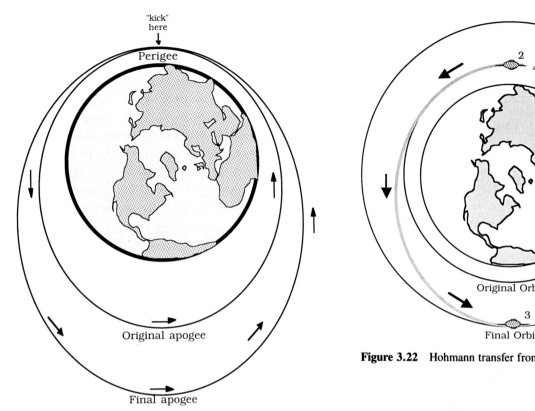

Figure 3.20   Firing rockets at perigee to increase eccentricity.

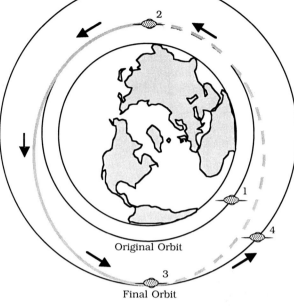

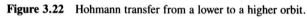

Figure 3.22   Hohmann transfer from a lower to a higher orbit.

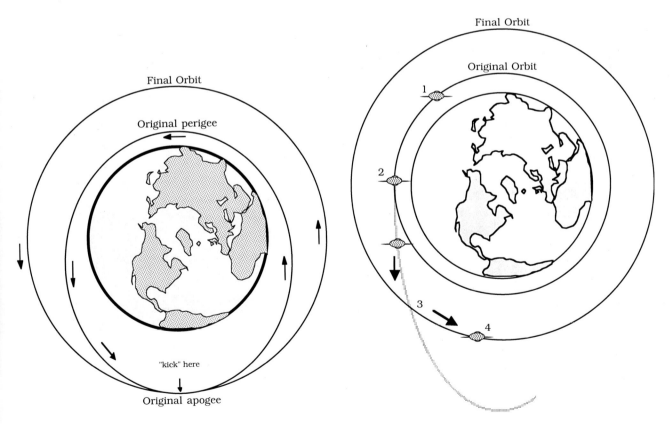

Figure 3.21   Apogee "kick" decreases eccentricity.

Figure 3.23   Fast transfer from a lower orbit to a higher one.

you increase your speed and your energy. Therefore, you will not remain in that orbit, but will move to the larger, higher orbit. To further complicate the situation, because you have transferred to a higher orbit, you move farther from Earth and consequently you slow down, even though you just fired your engines to speed up.

Rendezvous with another spacecraft, then, is rather tricky. Increasing your energy to catch up will move you to a higher orbit and you will instead fall behind. Imagine you are at point Y1 in Figure 3.24 and want to rendezvous with the space station now located at point S1. Ignite your engines to move to the elliptical transfer orbit. The proper increase in energy will put you into an orbit which touches the space station's orbit at point 2, the apogee of your transfer orbit. When you reach 2 you again fire your engines, an apogee kick, to transfer to the space station's orbit. This is actually a Hohmann transfer.

The tricky part is to time your engine ignitions in such a way that you and the space station both reach the rendezvous point, 2, at the same time. Because the space station is in a higher orbit, it is moving at a slower speed. Therefore, you must wait until it is closer to the rendezvous point than you are because you are travelling at a higher speed. Calculating the exact time is quite involved and we will not go into it here.

Other Maneuvers

It is also possible to change the inclination of an orbit, change the location of perigee, rendezvous with spacecraft in different orbital planes, and perform other maneuvers.

Each requires the application of the proper amount of thrust in the proper direction and at the proper time.

Deorbit

Reentry and landing back on Earth are accomplished by turning the spacecraft around and firing the engines to apply thrust in the opposite direction. Retrofiring reduces the spacecraft's energy, transferring to a lower orbit which intersects Earth, increasing speed as it falls. In Figure 3.25, retrofiring at the place shown, Y, transfers the spacecraft to the more elliptical orbit which intersects Earth at the place of landing. Timing and length of burn must be precise so the new orbit intersects Earth at the desired landing spot. The atmospheric drag during reentry distorts the orbit from a true ellipse which must be taken into account also.

Trajectory to the Moon

The Moon is in orbit around the Earth just like a spacecraft, so going to the Moon is similar to a rendezvous maneuver. There is one important difference, however; the Moon is so large that its gravitational attraction will control the orbit of the spacecraft as it draws near.

First consider a Hohmann transfer. The rockets are ignited to inject the spacecraft from a low Earth orbit to an elliptical transfer orbit with apogee at the Moon. As the spacecraft heads for apogee it slows down, as usual. But as it approaches the Moon and comes under the lunar gravitational influence, it deviates from its transfer ellipse and falls toward the Moon with increasing speed. One of two things may happen: (1) it hits the Moon or (2) it misses but because of its speed, it curves around the Moon and escapes.

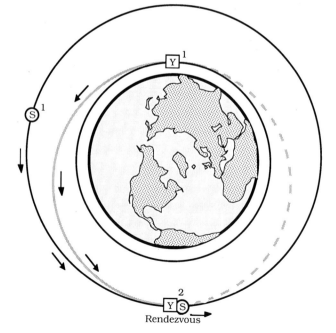

Figure 3.24 Rendezvous in space using a Hohmann transfer.

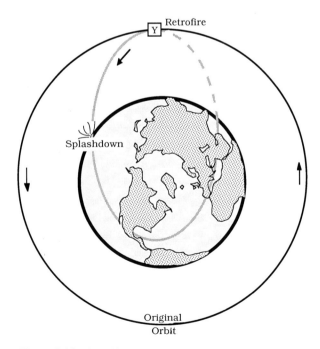

Figure 3.25 Deorbit maneuver.

Continuing on, it is then in a different elliptical orbit around Earth. Figure 3.26 shows this situation. If our goal was to land on the Moon, we did something wrong. We succeeded in our rendezvous, but our speed was too great. The solution is to slow down at the proper moment so the spacecraft is captured by the Moon and goes into a lunar orbit.

Figure 3.27 is the Apollo mission profile, not drawn to scale. First, the Saturn V rocket carried the command module, service module, lunar module, and a booster rocket into a low parking orbit. After everything was checked out and the vehicle was at the correct position in orbit, the booster rocket fired to inject it into the transfer orbit to the Moon (lower path in Figure 3.27). If there was any error in the transfer orbit, engines were fired again for a midcourse correction. The vehicle coasted for nearly 3 days and as it approached the Moon, the gravitational attraction caused it to leave the transfer orbit and curve around the Moon. Retrofiring the rockets slowed the vehicle and injected it into an elliptical lunar orbit. Another rocket burn circularized that orbit. The lunar module with two astronauts separated from the command and service module and its rockets retrofired to take it out of orbit to a landing. As it approached the surface, rockets were fired to slow its descent and set it gently onto the lunar soil. After surface activities were finished, the astronauts again fired the lunar module rockets to take it back to orbit and a rendezvous with the command and service modules where the third astronaut was waiting (upper part of Figure 3.27). The lunar module was then separated and left behind. Engines were fired once more to inject the command and service modules into a transfer orbit back to Earth. A midcourse correction was made if necessary. As Earth gravity took over, the spacecraft deviated

from the transfer orbit for a direct entry into the atmosphere. Only the command module with the three astronauts splashed down in the ocean.

These were the most complicated manned spaceflights ever made. Count how many times rocket engines had to be fired! Each firing had to be made at precisely the right moment for precisely the correct period of time in exactly the proper direction. Computing those times and directions applied principles of orbital mechanics that had never before been used in manned flight. It was a tremendous accomplishment.

To Another Planet

Flying to another planet is similar to completing a rendezvous with a space station. The difference is that in near Earth orbits the gravitational attraction of Earth controls the orbit, while in interplanetary space the Sun is in control and toward the end of the flight the gravitational attraction of the target planet takes over. So we can examine the trajectory to a planet by separately considering these three segments.

Imagine a flight to Mars, as depicted in Figure 3.28. The Sun is at the focus of the orbits of both Mars and Earth. From our point of view high above the north pole, Earth and Mars are both moving counterclockwise in their orbits. The spacecraft is first injected into a low Earth orbit. At just the proper time the engines are ignited and the spacecraft is transferred to the dashed trajectory at a speed greater than escape velocity from Earth. (MATHBOX 3.4 shows how to calculate escape speed from Earth or any other celestial body.) This new trajectory is not a closed ellipse around Earth, but is an open curve called a *hyperbola*. It is curved at first but approaches a straight line as it gets farther from Earth. About a million miles out from Earth the gravity of the Sun becomes dominant and the spacecraft is in an elliptical orbit around the Sun, just like a planet.

Now, we can look at the path of the vehicle from two points of view: with respect to the Earth or with respect to the Sun. With respect to Earth it is on an escape trajectory, never to return. But with respect to the Sun it is in an elliptical orbit. Actually, the spacecraft was orbiting the Sun even when it was on the launch pad, since the Earth and everything on it is in orbit around the Sun.

When we fire the vehicle into the hyperbolic escape trajectory, we send it off in the same direction that the Earth is moving. It is then travelling faster than the Earth, which transfers it to an orbit larger than Earth's orbit toward the outer planets. This elliptical transfer orbit is like a Hohmann transfer except at the two ends where the gravity of Earth and Mars control the motion.

If the spacecraft is to land on Mars, another maneuver is necessary. Just as in the flight to the Moon, as the spacecraft approaches Mars it falls in toward the planet and its speed

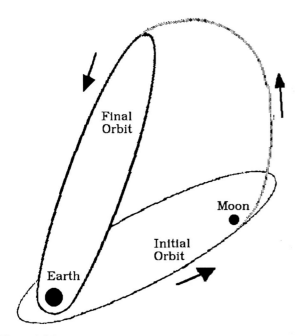

Figure 3.26 Trajectory past the Moon transfers the spacecraft to a new orbit around Earth.

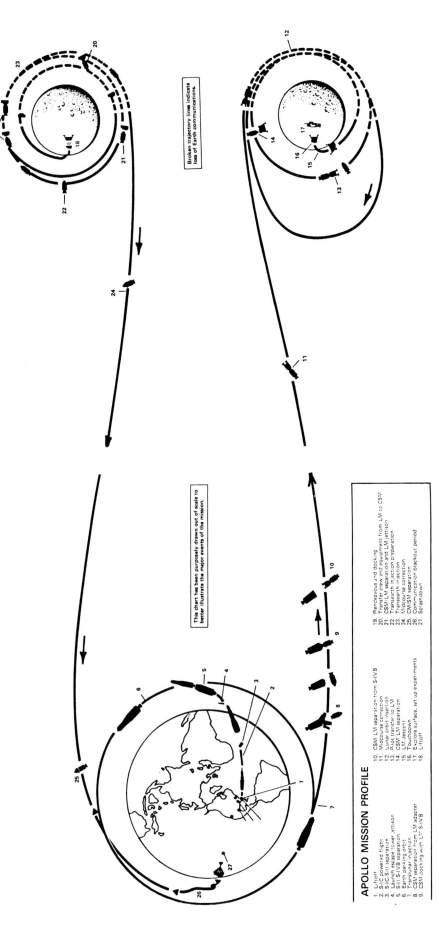

APOLLO MISSION PROFILE

1. Liftoff
2. S/C powered flight
3. S-IC S-II separation
4. S-II S-IVB separation
5. S-II/S-IVB separation
6. Earth parking orbit
7. Translunar injection
8. CSM separation from LM adapter
9. CSM docking with LM, S-IVB

10. CSM, LM separation from S-IVB
11. Midcourse correction
12. Lunar orbit injection
13. Pilot transfer to LM
14. CSM LM separation
15. LM descent
16. Touchdown
17. Explore surface, set up experiments
18. Liftoff*

19. Rendezvous and docking
20. Transfer crew and equipment from LM to CSM
21. CSM/LM separation and LM ettison
22. Transearth injection preparation
23. Transearth injection
24. Midcourse correction
25. CM/SM separation
26. Communication blackout period
27. Splashdown

Broken trajectory lines indicate loss of Earth communications.

This chart has been purposely drawn out of scale to better illustrate the major events of the mission.

Figure 2.27 Apollo mission to the Moon. *Courtesy of NASA.*

MATHBOX 3.4

Escape Speed

Throw a baseball up and its slows, stops, and comes back down. How fast do you have to throw it so it keeps going never to return? That speed is called escape speed. It is different for every planet or other central body and it can be calculated by the formula

$$v = \sqrt{\frac{2GM}{r}}$$

where r is the distance from the center of the central body and GM is the gravitation constant for the central body as explained in MATHBOX 3.2.

Let us find the escape speed from the surface of the Earth. GM for Earth is 1.24×10^{12} if r is given in miles and v is in miles per hour. At the surface we are about 4,000 miles from the center, therefore,

$$v = \sqrt{\frac{(2)(1.24 \times 10^{12})}{4000}} = 24,900\,\text{mi/hr} = 7\,\text{mi/s}.$$

We can calculate escape speed at any distance from any central body provided we know its mass so that we can calculate GM for that body. Again, refer to MATHBOX 3.2.

---

increases. Retrofiring the engines reduces its speed so it enters an orbit around Mars. Without this braking maneuver the vehicle would be deflected somewhat by Mars gravity and continue on a modified orbit around the Sun. Actually, if the vehicle missed the planet completely, it would just continue to orbit the Sun.

Going to Venus is similar, except that Venus's orbit lies inside Earth's orbit. Therefore, the vehicle must be accelerated to escape velocity on a hyperbolic trajectory from Earth, but in the direction opposite to the direction of

Earth's motion. Its speed with respect to the Sun is then less than Earth's speed and it moves into a smaller transfer orbit toward the inner planets. See Figure 3.29.

Many spacecraft have left Earth for encounters with other planets. We will discuss some of them in later chapters.

Figure 3.28   Transfer orbits to Mars.

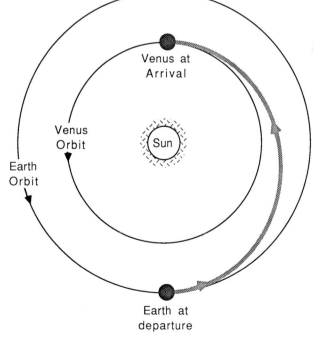

Figure 3.29   Hohmann transfer to Venus.

DISCUSSION QUESTIONS

1. According to Newton, an object accelerates (changes velocity) only as long as long as a force is applied. When a rocket engine shuts down there is no more thrust. Why then does a spacecraft change speed in an elliptical orbit? Why doesn't it change speed in a circular orbit?

2. If Earth rotated on its axis once in 12 hours instead of once in 24 hours, would a geosynchronous orbit be higher or lower?

3. What is the ground trace of a satellite orbiting at an altitude of 40,000 miles?

4. If a spacecraft is 10 miles directly above a space station, how would you get it down for rendezvous and docking?

5. What are the advantages and disadvantages of a Hohmann transfer to carry a human crew to Mars? a cargo ship to Mars?

ADDITIONAL READING

American Astronautical Society. *The Journal of the Astronautical Sciences*. Quarterly publication of the American Astronautical Society.

Bate, Roger R., et al. *Fundamentals of Astrodynamics*. Dover Publications, 1971. Calculus level mathematics.

Cortright, Edgar M., editor. *Apollo Expeditions to the Moon*. NASA SP-350, Government Printing Office, 1975. Detailed history of the Apollo program.

Melbourne, William G. "Navigation between the Planets." *Scientific American*, June 1976.

Mickelwait, Aubrey B., et al. "Interplanetary Navigation." *Scientific American*, March 1960.

Muolo, Michael J., et al. *Space Handbook, Volume 1, A Warfighter's Guide to Space* and *Volume 2, An Analyst's Guide*. Air University Press, December 1993. Mostly descriptive with some algebra.

Prussing, John E., and Bruce A. Conway. *Orbital Mechanics*. Oxford University Press, 1993. Vector calculus level mathematics.

Thomson, William T. *Introduction to Space Dynamics*. Dover Publications, 1986. Calculus level mathematics.

NOTES

Chapter 4

The Space Environment

When a spacecraft leaves Earth it enters a rather unfriendly place. At the surface of Earth we live in a warm comfortable environment with air to breathe, water to drink, land to walk on. Space offers none of these. We often hear and use the term "empty space." True enough, the content of space is pretty sparse when compared to our earthly environment. The atmosphere surrounding us contains about 2 pounds of air in each cubic yard at a pressure of 14.7 pounds per square inch. At 150 miles up a cubic yard contains only a millionth of a pound at a pressure of 0.07 pound per square inch, a better vacuum than can be produced by laboratory equipment on Earth.

But space is far from being truly empty. In the space environment around Earth we find not only particles, many of them electrically charged, but also electromagnetic energy in the form of waves: x-rays, ultraviolet, gamma rays, visible light, infrared, radio waves, and microwaves. Cosmic rays which are really particles, not rays, whiz past at high speed. In addition, meteoroids, which are small pieces of dust and debris, abound in space. Meteoroids are thought to be pieces of matter left over from the formation of the Solar System. Man-made debris from spaceflight activities, from flecks of paint to burned-out rockets and satellites, is scattered about in low Earth orbits.

The electromagnetic radiant energy and the few pieces of matter that are present in space could damage a spacecraft and could be lethal to an unprotected human. This chapter discusses these components of the space environment, their sources, and their potential hazard to spaceflight.

The Sun

The Sun is the source of all the particles and waves in the space environment except for cosmic rays, meteoroids, and man-made debris. Some cosmic rays come from the Sun, but most come from outside the Solar System; indeed, some of them may come from outside our Galaxy. To understand the contents of the environment in which our spacecraft and astronauts fly, we must first understand the structure and activity of the Sun. We also need to become familiar with the electromagnetic spectrum. Electromagnetic waves will

be of interest to us several times as we progress through this book.

The Sun is a star. We can see that it is round and through telescopes we can observe features on its surface. Other stars, even with the largest telescopes, appear as just points of light in the sky because they are so far away. The average distance from the Sun to Earth is about 93 million miles. Light from the Sun, travelling at 186,000 miles per second, takes 8.3 minutes to travel that distance. On the other hand, light from the next nearest star, Alpha Centauri, the brightest star in the constellation Centaurus, takes 4.3 years to reach us.

The Sun is a sphere about 865,000 miles in diameter, 110 times the diameter of Earth and over a million times the volume of Earth. It is a medium-sized star. Many stars are much larger; some giants are 400 times the diameter of our Sun. Like other stars, it is composed mostly of hydrogen with some helium and traces of other elements. Earth orbits the Sun with a period of 1 year; that is the definition of a year. The orbit is nearly circular with an eccentricity of 0.017, a perihelion of 91.3 million miles, and an aphelion of 94.4 million miles.

Atoms

Before going further, it is important to understand the structure of an atom. As long ago as 400 B.C. Greek philosophers believed that if an object were divided into smaller and smaller parts, it would eventually get down to a smallest piece that could not be divided any further. That smallest piece of matter was called an *atom*. Twenty-three hundred years later scientists discovered that atoms could be split further into three more elementary particles. *Electrons* are the smallest and they carry a negative electrical charge. *Protons* are about 1,800 times more massive than electrons and they carry a positive charge. *Neutrons* have about the same mass as protons but are electrically neutral.

All atoms in the universe are made up of these three elementary particles. See Figure 4.1. Each *element* such as oxygen, iron, gold, lead, hydrogen, and carbon has a unique combination of protons, neutrons, and electrons, different

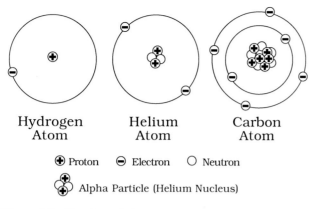

Figure 4.1 Structure of atoms.

from all other elements. The interior of an atom is something like a miniature solar system, with a nucleus composed of protons and neutrons located at the center and electrons in orbit around it. The simplest possible atom would be a one-proton nucleus and one electron in orbit; that describes an atom of hydrogen. The next more complex atom, helium, has two protons and two neutrons in the nucleus with two electrons in orbit. Gold has 79 protons and 118 neutrons with 79 electrons in orbit. The number of neutrons may vary. For example, carbon has six protons and either six or seven neutrons in the nucleus. It is the number of protons that determines which element it is.

Notice that in a normal atom the number of protons and the number of electrons are equal. Thus, with an equal number of positive and negative charges, the atom is electrically neutral. Under some circumstances one or more electrons may be removed from an atom. We say that it is *ionized*. If all the electrons are stripped away, nothing is left but the nucleus. A hydrogen nucleus is simply a proton. A helium nucleus—two protons and two neutrons—is given a special name, an *alpha particle*.

A Nuclear Furnace

Now back to the Sun. The Sun shines by its own light; the Moon, planets, and comets shine by reflected sunlight. The source of the Sun's energy is a nuclear "furnace" in its center, powered by a process called *nuclear fusion*. Deep in the Sun's interior, hydrogen is being converted to helium with a release of energy. To be more specific, four protons, hydrogen nuclei, fuse together to form an alpha particle. In this fusion process two of the original protons have become neutrons; that is, they lost their positive charge. If you add together the mass of the four protons, you find that they total more than the mass of an alpha particle: about 0.7 percent of the mass has disappeared in the fusion. According to Albert Einstein's famous equation, $E = mc^2$, the lost mass is converted to energy. See **MATHBOX 4.1** for an actual calculation. This mass-converted-to-energy is released in the form of a *gamma ray*, an electromagnetic wave.

Fusion requires temperatures in excess of 30 million degrees Fahrenheit and a pressure of a trillion tons per square inch. Such temperatures and pressures are common in the cores of stars. Once fusion begins, the heat produced keeps the process going until the hydrogen fuel runs out or the pressure drops. The necessary high pressure is maintained by the weight of the overlying mass of the star held together by gravity.

Nuclear fusion is the same process that takes place when a thermonuclear hydrogen bomb explodes. In a thermonuclear bomb, the fusion of a quantity of hydrogen into helium takes place in a fraction of a second and the energy is released as an enormous explosion causing a massive rearrangement of things in the immediate vicinity. It is a star on Earth for an instant. If it were possible to control the fusion, to slow it down so the energy would be released over a long period of time, it could be used beneficially, for example, to boil water to run steam electric generators.

In the Sun, fusion started about 4.5 billion years ago. Hydrogen is fusing to helium in the central core of the Sun at a rate of nearly 5 million tons a second. And with the bountiful supply of hydrogen available, it will continue to do so for another 5.5 billion years. We do not have to worry that it will blink out on us, at least not this weekend. This calculation is also shown in **MATHBOX 4.1**. Our Sun, then, is a middle-aged star, about halfway through its lifetime. Temperatures and pressures are great enough to support the fusion in only about the central 10 percent of the Sun. After the hydrogen in that region is consumed, the nuclear furnace will go out and the Sun will go through various phases until it dies. The life cycles of stars are a fascinating study in nuclear physics and are the subject of investigation by astronomers. Here we are only concerned with our Sun at its present stage of evolution.

Solar Structure

Figure 4.2 is a diagram of the surface and a cross section of the Sun. The *photosphere* is the visible surface of the Sun. Above the photosphere is the *chromosphere*, a region that can best be described as an atmosphere. Next is the *corona*, the upper atmosphere of the Sun, which extends out to the planets. It may truly be said that Earth lies in the solar atmosphere. The corona, for some unknown reason, is the hottest part of the solar atmosphere. Its temperature reaches several million degrees Fahrenheit compared to 11,000 °F in the photosphere and the lowest temperature in the Sun of 5,800 °F at about 660 miles above the photosphere in the chromosphere. Just why the corona is hotter than the lower regions of the Sun's atmosphere is one of the mysteries of solar physics.

The *core* is the nuclear furnace where the energy is produced. Temperature in the core is about 35 million degrees Fahrenheit and pressure is about a trillion tons per square

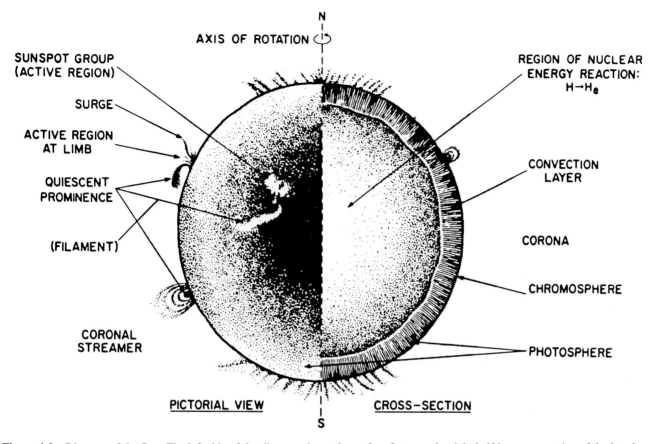

AXIS OF ROTATION

N

SUNSPOT GROUP
(ACTIVE REGION)

REGION OF NUCLEAR
ENERGY REACTION:
H→H$_e$

SURGE

ACTIVE REGION
AT LIMB

CONVECTION
LAYER

QUIESCENT
PROMINENCE

CORONA

(FILAMENT)

CHROMOSPHERE

CORONAL
STREAMER

PHOTOSPHERE

PICTORIAL VIEW   CROSS-SECTION

S

Figure 4.2 Diagram of the Sun. The left side of the diagram shows the surface features; the right half is a cross section of the interior. *Courtesy of the U.S. Air Force.*

inch. Gamma rays produced in the core radiate outward through the *radiation zone* as photons, bundles of electromagnetic energy. In free space all electromagnetic waves travel at the same speed, 186,000 miles per second. But in the dense interior of the Sun they travel much more slowly, so slowly that it takes more than 20,000 years for the gamma rays to reach the *convection zone*. Enroute, they lose energy to their surroundings. When they reach the convection zone, the electromagnetic energy is absorbed by the gases and converted to a form of heat energy. The hot gases rise upward while cooler gases from above sink into the deeper parts where they are heated and rise again.

These continuously circulating convection cells are very similar to the ones which produce cumulus clouds in Earth's atmosphere. Air in contact with the ground on a warm day is heated from below and rises, bubble-like, to higher altitudes, cooling as it goes. Meanwhile cooler air sinks toward the surface where it is warmed and rises. If sufficient water vapor is present then clouds form in the upward moving air. This continuous convective cell circulation can also be observed in a pot of water on the stove. Water heated by contact with the hot bottom of the pot can be seen to rise while cooler water sinks around the edges. If you look closely you can see the shimmering in the water. The heat is thus distributed by the circulating water.

Similarly, in the interior of the Sun, convective cells of circulating hydrogen carry the heat energy from the zone of radiation to the photosphere where it flows into space, mostly as light, infrared, and ultraviolet. It is indeed fortunate for us that the lethal gamma rays do not reach the surface of the Sun and radiate into space. If they did, life would be impossible on Earth or anywhere else in the Solar System.

The Sun rotates on its axis just as Earth and other celestial bodies do. But because it is a fluid body, it rotates differentially; that is, different parts of the Sun rotate at different speeds. The polar regions take about 32 days to make a complete rotation, while the equatorial regions rotate once in about 25 days. That leads to complicated flow patterns in the upper layers of the Sun and distortion of its magnetic field, leading to occasional explosions called flares. We will discuss flares in more detail later in this chapter.

Electromagnetic Waves

The Sun emits most of its energy in the form of light (41 percent of the total), infrared (52 percent), and ultraviolet (7 percent). Smaller quantities of x-rays, radio waves, and microwaves are also emitted. X-rays, gamma rays, ultraviolet, radio waves, television waves, and light are really all manifestations of the same thing, *electromagnetic waves*.

MATHBOX 4.1

Nuclear Fusion

When four protons combine into one alpha particle some mass is "lost," converted to energy.

The mass of atomic particles is measured in atomic mass units (amu). A proton has a mass of 1.0078, so four protons have a mass of 4.0312 amu. An alpha particle has a mass of 4.0026.

$$4.0312 - 4.0026 = 0.0286 \text{ amu converted to energy}$$

$$\frac{0.0286}{4.0312} = .0071 = 0.71\% \text{ of the mass is converted to energy}$$

One atomic mass unit is 1.7×10^{-26} kilograms. So 0.0286 amu is 4.8×10^{-29} kilograms. The energy produced by this mass is found from Einstein's famous equation

$$E = mc^2$$

where c is the speed of light, 3×10^8 meters per second.

$$E = (4.8 \times 10^{-29} \text{ kg}) (3 \times 10^8 \text{m/s})^2 = 4.3 \times 10^{-12} \text{ joules}$$

Now, that is not very much energy per fusion. It must be going on at a tremendous rate in the core for the Sun to be radiating energy away at the rate of 3.8×10^{26} joules per second.

$$\frac{3.8 \times 10^{26} \text{ j/s}}{4.3 \times 10^{-12} \text{ j/fusion}} = 8.9 \times 10^{37} \text{ fusions per second}$$

Let us go one step further. We found that 4.8×10^{-29} kg are converted to energy in each fusion. So the Sun must be losing

$$(4.8 \times 10^{-29} \text{ kg/fusion}) (8.9 \times 10^{37} \text{ fusions/s}) = 4.3 \times 10^9 \text{ kg/s}$$

In more familiar units, 4.3 billion kilograms per second is about 4.7 million tons per second of matter being converted to energy. At that rate, how long can the Sun last? The Sun has a mass of 2×10^{30} kg. About the inner 10%, 2×10^{29} kg, is involved in the fusion, and 0.71% of it will be converted to energy before the core goes out.

$$\frac{(2 \times 10^{29} \text{ kg}) (.0071)}{4.3 \times 10^9 \text{ kg/s}} = 3.3 \times 10^{17} \text{ seconds} = 10 \text{ billion years}$$

Since the Sun is about 4.5 billion years old, it still has enough fuel to run for another 5.5 billion years.

The only difference among them is their wavelength. Certainly you have seen water waves or ripples in a pond. The wavelength of a wave is the distance from one crest to the next crest, or from one trough to the next

Figure 4.3 lists various electromagnetic waves in order of wavelength from the longest to the shortest. Wavelengths are given in metric units. Remember that a meter and a yard are about the same, the meter being about 3 inches longer. A kilometer is about 0.6 mile. It takes about 25,000 micrometers to make an inch; a grain of sand is about 300 micrometers in diameter.

The length of a radio wave, measured from crest to crest

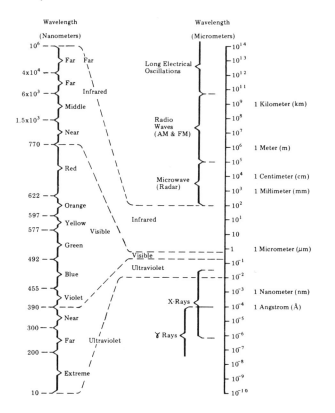

Figure 4.3 The electromagnetic spectrum. Wavelengths are given in metric units. In the right column they are millionths of a meter (micrometers); other perhaps more familiar units are indicated at some places, e.g., 10^6 micrometers is 1 meter. In the left column the wavelengths of infrared, visible light, and ultraviolet are given in billionths of a meter (nanometers). *Courtesy of NASA.*

or trough to trough, ranges from a tenth of a meter to many kilometers (a few inches to many miles). For example, channel 2 television waves are about 5 meters (15 feet) long. Microwave oven waves are about 1 centimeter (0.4 inch) long. Infrared waves range from a micrometer (millionth of a meter) to several hundred micrometers in length. (A micrometer is also called a micron.) Light waves, on the other hand, measure from 0.4 micrometer to 0.7 micrometer. Ultraviolet, x-rays, and gamma rays are even shorter. Gamma rays and x-rays come from the most violent, high energy events in the universe. The dividing line between the various types of waves is not as sharp as Figure 4.3 indicates. There is some overlap.

All electromagnetic waves are identical in character except for wavelength. As the name suggests, they have an electrical and a magnetic component. They are created when an electrically charged particle, such as an electron, oscillates or accelerates. For example, they are produced in atoms as the electrons change orbit around the nucleus, escape from the nucleus (ionization), or are captured by a nucleus (recombination).

Sound waves and water waves must have a medium in which to travel. Sound waves travel in air, through liquids, or through solids like walls and windows. But electromagnetic waves do not require a medium; therefore they can travel through empty space. All travel at the same speed, the speed of light.

Particles in Space: The Solar Wind

In addition to electromagnetic waves, charged particles also stream from the Sun, mostly electrons, protons, and alpha particles. These particles become so hot in the corona that some of them acquire energies greater than escape velocity from the Sun. A steady stream of these particles flows outward from the Sun in what has been named the *solar wind*. The solar wind, actually an extension of the solar corona, blows outward past the planets, filling the Solar System with charged particles. Usually the speed past Earth is about a million miles an hour with a density of about 1,600 particles per cubic inch, but the solar wind increases in both speed and density during periods of high solar activity. Exposure to the solar wind is like being exposed to low level radioactive material. Dosage does not usually reach a dangerous level.

Particles in Space: Cosmic Rays

Cosmic rays are particles, not rays, travelling in all directions at extremely high speed, some of them approaching the speed of light. They come from two different sources. *Solar cosmic rays* are protons which are accelerated to high speed during explosive eruptions on the Sun. These solar cosmic ray events are relatively infrequent, but are deadly to humans when they do occur. We will discuss them in more detail later in this chapter.

A steady stream of cosmic rays comes from outside the Solar System. These are referred to as *galactic cosmic rays*. When they were first discovered in the early part of this century they were thought to be electromagnetic radiation and were dubbed "rays." We now know they are the nuclei of atoms stripped of their electrons and therefore bearing a positive charge. They are 85–90 percent protons, 10–12 percent alpha particles, about 1 percent electrons, and 1 percent nuclei of heavy atoms such as oxygen, nitrogen, iron, and neon. Some perhaps had their origin in events during or shortly after the "big bang" that started the universe. Others come from exploding dying stars, called supernovas. A few may come from other galaxies.

Many galactic cosmic rays travel at such high speed that they pass right through the spacecraft, people and all, at a rate of perhaps a dozen per square inch per hour.

Sunspots

Dark spots are often visible in the Sun's photosphere (Figure 4.4). Although they appear tiny from our viewpoint, smaller *sunspots* may be about the size of Earth and largest

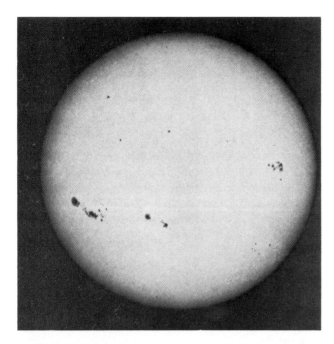

Figure 4.4 Sunspots. *Courtesy of NASA.*

ones are bigger than a hundred Earths. They vary in number from day to day and year to year. Sunspots were observed by Galileo in 1610 shortly after he first pointed a telescope at the heavens. Since then records have been kept of their number, although the records are quite erratic for the first two hundred years after Galileo discovered them. Actually observations of sunspots had been recorded centuries earlier in China on several occasions when the Sun was low on the horizon behind a veil of clouds and spots were large enough to be seen with the naked eye.

Sunspots may be observed with any telescope equipped with a device for reducing the light intensity by at least 99 percent. Looking through a telescope without some sort of light-reducing device will cause permanent eye damage. You have perhaps used a magnifying glass to focus the Sun onto a sheet of paper and watched the paper burst into flames. That is some indication of what happens to the retina of an eye focused on the Sun. Projection of the image of the Sun onto a white card held a couple of feet behind the telescope eyepiece also allows comfortable viewing.

Sunspots are "cool" regions in the photosphere. Photospheric temperature is nearly 11,000 °F; sunspot temperatures are about 7,600 °F. Although 7,600 °F is not very cool, the spots appear black against the hotter surroundings. Figure 4.5 is a closeup photo of a group of sunspots. If we could see a sunspot against the cold blackness of space, it would appear bright. The cause of the lower temperature is probably involved with the magnetic fields of the Sun. Whenever an electrically charged particle is in motion, it is accompanied by a magnetic field. The convection cells in

the Sun are mass motions of charged particles, mostly negative electrons and positive protons. Therefore, magnetic fields are present. When complex flow patterns occur in the convection zone, the magnetic fields become entwined and tangled. This prevents the usual rather smooth flow of heat from the interior to the photosphere. Consequently, with the heat flow interrupted, the region becomes cooler. The differential rotation of the Sun also contributes to the complexity of heat flow. This simplified explanation may not be completely precise. However, complex magnetic field structures are observed in sunspot regions and there is undoubtedly some relationship between them.

For the past two centuries daily observations of sunspots have been tabulated and studied. Spots occur in groups with as many as 100 spots or as few as one to a group. A standard method of counting sunspots originated in the 1800s. It was thought that the existence of a group might be significant. Therefore, the sunspot number for the day is equal to ten times the number of groups, plus the number of spots. An example is shown in Figure 4.6.

When sunspot numbers are averaged over 12 months and then plotted on a graph extending over many years, a pattern shows up. Figure 4.7 shows such a graph from Galileo's time to 1975. Most obvious is the cyclic pattern beginning about 1700 with a period of about 11 years. Few spots were seen in the 60 years from 1645 to 1705. Indeed they were so rare that many people doubted their existence. The most recent maximum occurred in July 1989 when the running average sunspot number reached 158.5, the third highest on record. The highest maximum ever recorded was 201.3 which occurred in March 1958. A minimum is expected in 1995 or 1996.

Sunspots by themselves do not do anything significant to the space environment, but as we will see, they often accompany other activity on the Sun which may adversely affect spaceflight. There have been many attempts to relate sunspots to events on Earth, particularly weather and climate. Statistical correlations have been found between the sunspot cycle and many other things: droughts on the high plains of the United States, the stock market, the length of women's skirts. But no cause-effect relationships have been found. That is, given that the stock market varies with the sunspot cycle, what is the cause? "Coincidence" is the only logical response.

The Chromosphere

Some telescopes are equipped with a special filter which allows only one wavelength of light to pass through, the red light of hydrogen-alpha. The main source of hydrogen-alpha light is the chromosphere, the lower atmosphere of the Sun. Figure 4.8 is a photograph taken through a hydrogen-alpha filter which reveals many interesting features. First, the bright areas that look like clouds are hot regions called

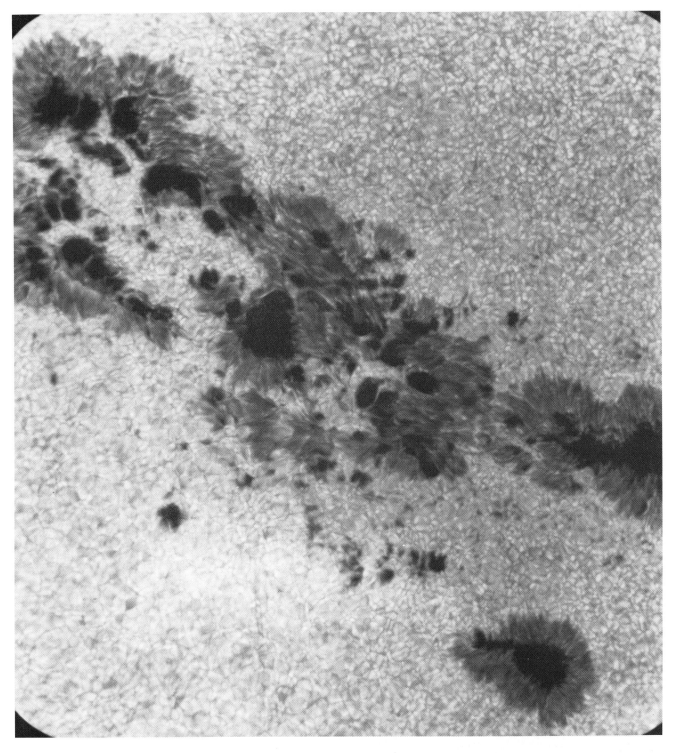

Figure 4.5   Closeup photo of a sunspot group.*Courtesy of National Solar Observatory, Association of Universities for Research in Astronomy, Inc.*

plages, the French word for beaches; they do have the appearance of a white sandy beach, although I wouldn't recommend them for a vacation spot.

The long black streaks are called *filaments*. They pro-

trude upward into the corona and are cooler than the photosphere. Therefore they appear black, the same as sunspots do. If a filament is on the *limb*, the edge of the solar disk, it appears bright against the blackness of space and is called a *prominence* because of the way it sticks out from the limb

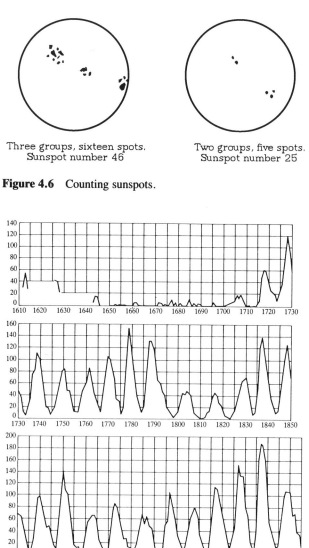

Three groups, sixteen spots.
Sunspot number 46

Two groups, five spots.
Sunspot number 25

Figure 4.6 Counting sunspots.

Figure 4.7 The 11 year sunspot cycle. *Courtesy of NASA.*

(Figure 4.9). Prominences appear in many different shapes with fanciful names: loops, arches, hedgerows. In a time-lapse movie they can be seen to grow, dissipate, change shape, oscillate, and sometimes explode into space. They are generally associated with *active regions*, near plages and sunspots.

Solar Flares

When the energy wrapped up in the magnetic fields of an active region can no longer contain itself, it is sometimes released suddenly in a gigantic explosion called a *flare*. Flares may last from a few minutes to a few hours. Even small ones may be larger than Earth and release as much energy as a billion hydrogen bombs. A large flare will release more energy than has been consumed by all humans in all history. The frequency of occurrence of flares follows the sunspot cycle. As the sunspots increase so does the num-

ber of flares. Up to 1,000 flares per month are observed at the time of sunspot maximum. However, large flares often occur after the sunspot cycle reaches a peak and begins to decline. The very large flare in Figure 4.10 occurred nearly 4 years after the 1968 peak in sunspot number.

The light from a flare can be seen in a telescope equipped with a hydrogen-alpha filter. Once or twice in an 11 year cycle a flare is so brilliant that it can be seen against the bright photosphere without a filter. Besides visible light, bursts of x-rays, ultraviolet, and radio noise often accompany the explosion on the Sun. In addition, particles, mostly electrons and protons, may be accelerated to high speed, exceed escape velocity from the Sun, and spew out into space. See Figure 4.11.

Occasionally during very intense flares, protons reach speeds up to one-fourth the speed of light and fit into the classification of solar cosmic rays. At that speed they reach Earth in half an hour, compared to the usual 3 to 4 days for particles in the solar wind. A solar cosmic ray event, also referred to as a *proton event*, may last from several hours to a day. Solar cosmic rays would be lethal to an unprotected astronaut. The U.S. government has set a standard for the maximum amount of radiation that workers may receive if they deal with radioactive material on a regular basis. During a proton event an astronaut could be exposed to more radiation in half an hour than is allowed a radiation worker in a year. Major solar flares in August 1972 and in October 1989 produced particle storms that would have incapacitated astronauts in the cargo bay of the Space Shuttle orbiter. For the most energetic of these events, even the cabin walls of the orbiter would not be sufficient shielding.

A high flying airplane in the polar regions of the Earth could experience a higher radiation dosage than the Space Shuttle in a low inclination orbit. Because the supersonic Concorde airplane flies above most of the atmosphere on the fringe of space, it carries detectors which warn of solar storm particles so that it can drop to a lower altitude if the particle radiation reaches a dangerous level. The detectors sounded an alarm on a Concorde flying over the north pole during a major storm on September 29, 1989.

We will discuss more about this particle radiation hazard in Chapter 9.

Sun-Earth Relationships

The Sun pours forth great quantities of particles and electromagnetic energy; Earth intercepts only a small percentage of it. Because of our distance from the Sun and the relatively small size of Earth in space, only about two-billionths of the total solar output falls on Earth. Yet nearly all energy consumed on Earth can be traced to the Sun. For example, plants harvest sunlight to build their "bodies" by photosynthesis, humans and animals eat plants, and humans,

Figure 4.8   The Sun in hydrogen-alpha light. The north pole is at the top, south at the bottom, east on the left side, and west on the right. Many filaments and plages are scattered over the surface. Four large spots and some smaller ones are visible. A small prominence can be seen on the left (east) side. Little activity takes place at the equator. *Courtesy of National Solar Observatory, Association of Universities for Research in Astronomy, Inc.*

in turn, eat the animals. The base of the entire food chain begins with photosynthesizing plants. Fossil fuels are really stored solar energy. Oil, coal, and natural gas come from ancient living organisms which died, were buried, and were transformed into those fossil fuels. Wind is caused by uneven heating of Earth's surface by the Sun. Windmills, therefore, derive their energy from the Sun. Rain is produced by solar energy evaporating water from lakes and soil. The water vapor condenses into clouds and raindrops which fill the reservoirs from which falling water drives electric generators. Photovoltaic cells, commonly called solar cells, convert sunlight directly into electricity with no moving parts. All of these examples of useful energy on Earth originate with the energy of the Sun.

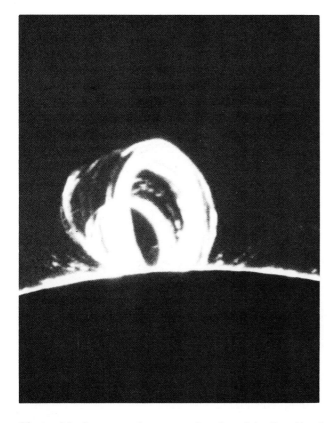

Figure 4.9 Loop prominences on the edge of the Sun. To take this photo, a disk was placed in the telescope to block out the light from the surface of the Sun. Otherwise, the nebulous features on the edge would not be visible. *Courtesy of National Solar Observatory, Association of Universities for Research in Astronomy, Inc.*

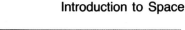
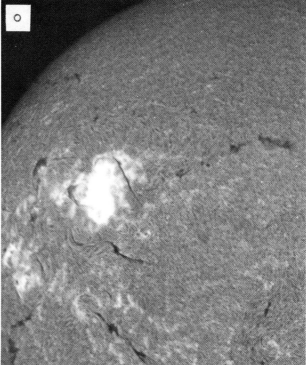

Figure 4.10 A large solar flare as seen through a hydrogen-alpha filter, August 7, 1972. For comparison, the circle at the upper left is approximately the size of the Earth. *Courtesy of National Solar Observatory, Association of Universities for Research in Astronomy, Inc.*

The only kinds of energy used on Earth that cannot be traced to the Sun are tidal energy, nuclear energy, and geothermal energy from the heat deep in Earth's crust. Even the tides are partially due to the Sun, though the Moon is the primary cause. The gravity of the Sun is enormous, but the Moon's closeness makes its gravitational force more significant in producing tidal action.

Life on Earth is well protected from lethal electromagnetic waves, charged particles, meteoroids, and cosmic rays. Earth's atmosphere and magnetic field act as a shield against these hazards.

X-rays and long wavelength ultraviolet are absorbed in the *ionosphere*, the ionized top part of the atmosphere above about 50 miles. The energy of these electromagnetic waves ionizes the oxygen and nitrogen atoms, removing an electron from them. Therefore, the ionosphere is called a *plasma*, a mixture of neutral and ionized atoms.

Ultraviolet is also absorbed by the ozone layer some 30 miles high. Ozone is made of three atoms of oxygen bonded together to form a single molecule. Normal oxygen in the air around us consists of only two atoms bonded together. Short wave ultraviolet can break the bond of an ordinary oxygen molecule, splitting it into two individual atoms. One

of those atoms then attaches itself to normal oxygen to form ozone. The ozone layer, then, is effective in preventing the ultraviolet rays from reaching the ground.

Thus, hazardous x-rays and ultraviolet do not reach the surface of Earth to damage living organisms. On the other hand, the atmosphere is transparent to visible light, the wavelengths that the Sun emits with greatest intensity and that our eyes respond to. A good arrangement!

The particles travelling from the Sun carry an electrical charge and electric charges in motion are always accompanied by a magnetic field. Earth has a permanent magnetic field, as if a giant bar magnet were buried deep in its interior. Figure 4.12 is a photograph of small bits of iron in close proximity to a bar magnet. Notice how they appear to follow curved lines which run from one end of the magnet to the other end. Compare this with Figure 4.13, a diagram of Earth's magnetic field. Remember that the solar wind blowing past Earth is composed of charged particles in motion and therefore bears a magnetic field. The interaction of the solar wind with Earth's magnetic field causes it to be distorted: compressed on the sunward side and stretched out downstream. It carves out a region of space called the *magnetosphere*. Solar wind particles follow the field lines and are deflected around the magnetosphere. Thus, the magneto-

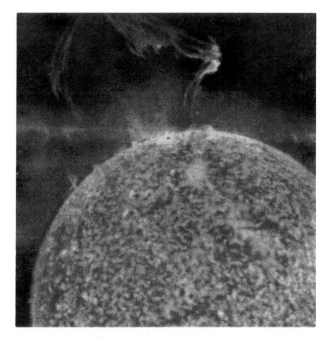

Figure 4.11 Photographed from the Skylab space station, a huge solar eruption throws high energy particles out into space. *Courtesy of NASA.*

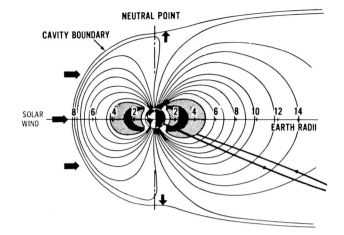

Figure 4.13 Earth's magnetic field. It is distorted by the solar wind, the flow of charged particles from the Sun; the cavity in space thus provided is called the magnetosphere. The distance scale on the diagram is marked in "Earth radii"; the radius of the Earth is about 4,000 miles. The orbit of a spacecraft which investigated the magnetosphere extends to the lower right of the Earth. *Courtesy of NASA.*

sphere acts as a shield, protecting life on Earth from deadly particle radiation.

Some particles do make it into Earth's atmosphere, however. There are weaknesses in the magnetosphere in the polar regions where particles can enter. In Figure 4.13 these weak places are shown by the two arrows and broken line marked "neutral point." Also, cosmic rays travel at such high speed that they can rip through the magnetic shield and collide with atoms of gas in the upper atmosphere, smashing them into bits. These bits are secondary radiation which does reach the ground. We are all constantly exposed to this low level radiation; that is part of living. It is more intense at higher altitudes than at sea level.

The Earth's magnetic field often becomes disrupted a few days following a solar flare, a phenomenon called a *geomagnetic storm*. The rapidly varying magnetic field during a geomagnetic storm produces unwanted currents in electric power lines. In March 1989 a severe geomagnetic storm produced such high current in the power lines across Canada that transformers melted.

Solar Activity and Atmospheric Density

An important aspect of solar activity on spaceflight is its effect on the density of the air. The upper atmosphere is heated by the collisions with solar wind particles and by absorbing electromagnetic energy, particularly x-rays and ultraviolet. When solar activity increases, the temperature of the upper atmosphere increases. You usually hear that heating causes things to expand and therefore you would expect the density of such substances to decrease. Not so in the upper atmosphere. When the temperature goes up, the atmosphere does indeed expand, but at orbital altitude this

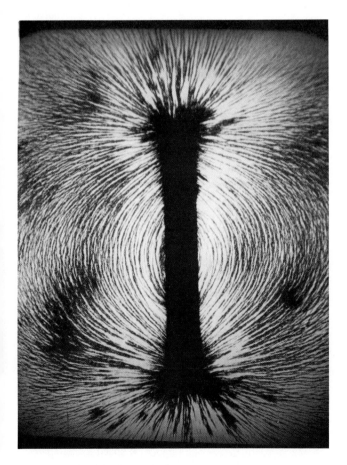

Figure 4.12 The magnetic field of a bar magnet made visible by sprinkling iron filings on a piece of glass over the magnet.

expansion brings up more dense air from below. Thus, at a given altitude the density increases.

The effect on a satellite orbit can be dramatic, particularly when there is a sudden increase in the solar wind. The result, as described in Chapter 3, is a more rapid decay of the orbit and an early burn in of the satellite. This is not all bad because it cleans up the low orbits as spent rocket tanks, dead payloads, and other space junk fall out of orbit and burn up in the atmosphere. Less than 10 percent of the large objects survive to hit the Earth, and those usually land in the oceans which cover 75 percent of the Earth's surface. At such times the orbits of satellites at low altitude may change so rapidly that those who monitor and control the satellites may lose track of them. Within 5 days after a storm it is possible to lose track of as many as 1,300 to 1,400 objects.

Trapped Particles: The Van Allen Belts

Two donut-shaped regions circling Earth inside the magnetosphere trap charged particles and hold them. These regions are called the Van Allen radiation belts. See Figure 4.14. Their existence was unknown until James Van Allen of the University of Iowa discovered them from data recorded by the first U.S. satellite, Explorer I.

Particles remain trapped in the Van Allen belts until a change in the solar wind causes them to dump out and "rain" into Earth's polar regions. When they strike the upper atmosphere they cause it to glow, producing the phenomenon commonly called the northern lights, more technically named the aurora borealis in the northern hemisphere and aurora australis (southern lights) in the southern hemisphere. Particles dumping out of the Van Allen belts also heat the upper atmosphere causing the density to increase as described in the last section.

The trapped particles are very energetic and high in number. They would be deadly to astronauts traversing them, but their locations in space are well mapped and they can be avoided. The inner belt begins at 250 to 750 miles, depending on latitude, and extends to about 6,200 miles. It is occupied mostly by protons which have a peak count over 60,000 per square inch per second at about 2,000 miles. The outer belt contains electrons and extends from the top of the inner belt, 6,200 miles up to 37,000 to 52,000 miles depending on solar activity. Peak electron count is at about 9,900 miles.

The Van Allen belts are far enough from Earth that they are no problem for vehicles in low Earth orbits except over the South Atlantic Ocean just east of Brazil. Because of an irregularity in Earth's magnetic field in that region, they dip to their lowest altitude. In a 275 mile orbit, a thousandfold increase in radiation is noted by spacecraft as they pass through; a hundredfold increase is seen at 140 miles.

Spacecraft at higher altitudes such as in geosynchronous

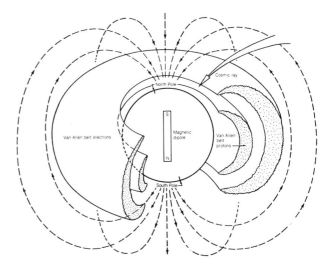

Figure 4.14   Van Allen radiation belts. Protons are trapped in the inner belt and electrons are found in the outer belt, while the dashed lines represent Earth's magnetic field. Cosmic rays can penetrate the magnetosphere especially in the polar regions. Note that Earth's north and south poles are tilted with respect to the magnetic field. *Courtesy of NASA.*

orbits, highly elliptical orbits, and on trajectories to the Moon or other planets must either avoid the belts completely or be prepared to cope with the trapped particles.

Meteoroids, Meteors, Meteorites

From earliest recorded history people have watched streaks of light flash across the night sky—shooting stars they call them or, more technically, *meteors*. At least a few appear every night, but there are certain dates every year when they occur in such profusion we call the event a meteor shower.

Occasionally we hear reports of a rock or chunk of iron falling from the heavens, sometimes a whole shower of them; on rare occasions someone is hit by one. In the early morning hours of March 4, 1960, near Edmonton in western Canada, a bright fireball crossed the sky and was followed a few minutes later by a thundering noise which rattled windows. The next day farmers found more than 180 pounds of black stones scattered over the snow in a 2 by 3 mile area. Such objects are called *meteorites*.

A mile-wide crater in northwestern Arizona was made by the impact of a large chunk of iron-nickel which fell from the sky about 30,000 years ago. Because of Arizona's desert climate Meteor Crater has not eroded and we see it today much as it looked when it was created.

These happenings are related. They are all caused by pieces of space debris that fall to Earth. The only difference is the size of the piece. While they are still in space we call these objects *meteoroids*. They come in all sizes, some so small you need a microscope to see them (*micrometeor-*

oids), some the size of small planets (*asteroids*). By far the vast majority of them have a mass less than a thousandth of a gram. (It takes nearly 30 grams to make 1 ounce.)

It is thought that meteoroids originated at the beginning of the Solar System, that they are pieces of debris left over when the Solar System formed. They are still being swept up from space by passing planets and moons. There is evidence of impacts on the Moon by all sizes of objects. Large craters where asteroid-sized objects impacted are obvious in photographs of the Moon, and the Apollo astronauts brought back Moon rocks that bear microscopic impact craters.

The Long Duration Exposure Facility (LDEF), shown in Figure 4.15, was a passive spacecraft that was dropped off by the Space Shuttle *Challenger* in a 247 nautical mile orbit on April 7, 1984. LDEF's twelve sides and ends were covered with panels of metals, plastics, various paints, reflective coatings, and epoxy composites. The purpose was to leave it orbiting for 18 to 24 months, then retrieve it and study the panels to determine the effects of the space environment on each of the materials. One end of LDEF was always pointed toward space and the other toward Earth. Because of the *Challenger* accident, LDEF wasn't brought back to Earth until January 12, 1990, after it had spent 5 1/2 years in orbit. Micrometeorite impacts were counted and studied in detail. Ten times as many microcraters were found on the end facing space than on the end facing Earth. Impacts on the end toward space were caused by micrometeorites. Figure 4.16 shows one such crater. By analyzing these pockmarks and extrapolating the data, researchers estimate that about 45 tons of the microscopic material arrive at Earth annually.

We see few meteorite craters on Earth because of our atmosphere and our weather. First, few meteoroids make it through the atmosphere to the ground. They burn up as meteors on the way down because of friction with the air. Those that do reach the ground, now called meteorites, must be quite large before their impact will leave a crater. Second, growing plants, running water, weathering, and erosion wear and tear the crater away so it is no longer recognizable after a few thousand years. Because our Moon has no atmosphere or weather, craters appear as they did when they formed, most of them 3 to 4 billion years ago.

Periodic meteor showers are probably associated with comets. Comets have been described as dirty snowballs, composed of water ice and dry ice with various kinds of dust and debris frozen in. Some comets are periodic, orbiting the Sun in elliptical orbits with periods of a few years to a few centuries. Each time they approach the Sun they heat up. The ices melt or sublimate and blow off some of the dust and debris into space. It is this material that forms the comet's tail. When the comet leaves the vicinity of the Sun, it once again freezes up until its next approach. The material

Figure 4.15 The LDEF spacecraft being held in place by the remote manipulator arm on the Shuttle orbiter. *Courtesy of NASA.*

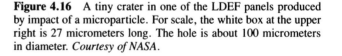

Figure 4.16 A tiny crater in one of the LDEF panels produced by impact of a microparticle. For scale, the white box at the upper right is 27 micrometers long. The hole is about 100 micrometers in diameter. *Courtesy of NASA.*

of the tail eventually becomes distributed throughout the comet's orbit. If Earth's orbit intersects the comet's orbit, that material will fall to Earth as meteors. Cometary material is very small; little of it survives the atmospheric heating to reach the ground.

Asteroids are concentrated in the region of space between Mars and Jupiter. However, the orbits of some of them cross

Earth's orbit and enter the inner Solar System. Non-cometary meteoroids are probably related to the asteroid belt.

Hazards to Spaceflight

We have discussed the contents of space at some length. Let us now summarize the aspects of the space environment that need to be considered when planning either manned or unmanned spaceflights. At this point we are concerned mainly with the environment itself. The biological effects will be discussed further in Chapter 9.

An asteroid hitting a spaceship would of course be a catastrophe. Even a pea-sized meteoroid could be disastrous. But what is the probability of such a collision? Statistics on the number of meteoroids and micrometeoroids in space and their size distribution can be calculated from observing meteors in the night sky, from experience with satellites that carry impact counters, and from experiments carried on Gemini, Apollo, Skylab, and LDEF and returned to Earth for examination.

To get an idea of the meteoroid content of near Earth space, consider a square surface 1 yard on a side. How frequently will meteoroids of various sizes impact that square? Table 4.1 shows a statistical estimate. However, this doesn't tell us whether or not the particle will penetrate the spacecraft. That depends on the thickness of the skin. Table 4.2 gives that information for a large aluminum space station. Credit for these calculations goes to P. M. Millman of the National Research Council of Canada.

Using a double-layered hull for the spacecraft increases the protection markedly. When a meteoroid impacts and penetrates the outer skin, it shatters into many small pieces which have a much lower probability of penetrating the inner skin.

The probability of collision with a meteoroid large enough to cause serious damage is never zero, but the hazard is relatively small compared to other hazards of spaceflight.

Effects on People

Electromagnetic radiation from the Sun can harm an unprotected astronaut. For us on Earth the atmosphere absorbs and filters out most of the harmful rays. Even so, you should avoid too much exposure to the Sun on the beach or in the mountains. Sunburns are not only uncomfortable, but excessive exposure to ultraviolet causes skin cancer. An astronaut does not have the protection of an atmosphere and must avoid exposure to the undiluted full blast of solar ultraviolet and x-rays. Such protection is not difficult to achieve. The cabin of a spacecraft or the material of a spacesuit is sufficient.

Particle radiation is another matter. Normally the solar wind does not pose a threat to life. The electrons and protons

TABLE 4.1 Meteoroid Impacts

Mass of Meteoroid	Time Between Impacts on One Square Yard Surface
One-millionth gram*	116 days
One-thousandth gram	63 years
One gram	3,200,000 years
Ten grams	320,000,000 years

*There are about 30 grams in an ounce.

TABLE 4.2 Meteoroid Penetrations

Thickness of Aluminum Skin	Frequency of Penetration
0.015 inch	10 per day
0.2 inch	1 per year
2.75 inches	1 per 10,000 years

are travelling slow enough that they are not harmful for short exposure times. Galactic cosmic rays are of concern on long duration flights. They travel at very high speeds and are very penetrating, but are relatively few in number. It has been estimated that continuous exposure to galactic cosmic rays for a year would exceed a safe dosage by some 60 times, but for short trips into space the dosage is not excessive. Trapped radiation in the Van Allen belts would be a serious threat but these regions can be avoided in manned spaceflight.

Solar cosmic ray protons emitted during a large flare are a very serious problem. At least one solar proton event in recent years was of sufficient intensity that astronauts inside a Space Shuttle in polar orbit would have suffered from severe radiation sickness, perhaps death.

Effects on Equipment

Early in the space age, before much was known about the space environment, equipment aboard satellites sometimes failed from exposure to x-rays or cosmic rays. For example, photographic film is fogged by x-rays and cosmic rays. Computer chips and other microcircuitry are susceptible to damage from x-ray bursts or high energy protons emitted during solar flares. Measuring instruments give false readings when high speed protons or cosmic rays pass through, and long exposure to the contents of space invariably degrades equipment operation.

Electrons or protons impacting a spacecraft can build up a charge, sometimes a charge large enough to cause a spark, similar to the spark that jumps from your finger when you shuffle across the room and touch a light switch. That can seriously damage electronic equipment. As much as 20,000 volts of potential has been observed on research satellites in geosynchronous orbit.

Star sensors are used on many satellites to give the spacecraft a reference point from which it can keep itself pointed in the proper direction. It is possible to lose control of a satellite if protons entering a star sensor cause a flash of light which appears to be a star. Random flashes confuse the sensor because it cannot make a correlation with known star charts.

We have learned to properly construct and shield equipment from most of these environmental hazards. In spite of all precautions taken to protect equipment, inexplicable events sometimes occur. On January 20, 1994, for example, operators lost attitude control of Intelsat K, a communications satellite in geosynchronous orbit on the equator over the Atlantic Ocean, and several hours later lost control of another geosynchronous communications satellite, Anik E-1. Anik E-2 suffered the same problem the following day. Operators recovered the first two by switching to backup circuits, but problems with Anik E-2 continued for months. Several other spacecraft reported problems, but none were as severe as the Anik satellites experienced.

At the same time, satellites equipped with sensors to measure particles in the magnetosphere and in geosynchronous orbit reported an unusually large flow of high energy electrons which were traceable back to the Sun. Apparently the satellites intercepted the electrons which imbedded themselves in the circuit boards. The electric fields emanating from the electrons may have caused the problem. However, such enhanced electron flow is observed from time to time by the sensors without any spacecraft malfunctions. It is difficult to understand why there should be problems with some such events but not with others.

Another problem is the nearly pure oxygen atmosphere. Although there are not many atoms in low Earth orbit, oxygen atoms are extremely reactive and degrade many surfaces. Shuttle astronauts report seeing the leading edge of the tail and the pods on either side of the tail glow in the dark due to the impact and reaction with atomic oxygen. Surfaces facing in the direction of motion, into the wind, intercept more atoms than other surfaces and are more likely to glow.

Space Environment Forecasting

It would be nice to know if and when solar flares and cosmic ray proton events are going to occur so flights are not planned at that time. Unfortunately, forecasting flares is somewhat comparable to forecasting thunderstorms and predicting solar proton events is like forecasting tornadoes. Forecasts are made routinely at a joint civilian-military forecasting center and provided to NASA. They state the probability of occurrence of a large solar flare and a proton event; but a definite "yes or no" forecast giving the exact time and intensity of a proton storm in space is beyond anyone's capability at present. If there are no active regions or sunspots visible on the Sun, the forecaster can issue a near-zero probability forecast. If there are several active regions

with complicated sunspot configurations and large plage areas, then the probability of a flare is high.

Making flare forecasts involves evaluating the flare potential of each active region on the Sun. Some of the more important criteria used in this evaluation are:

a. high magnetic field strength around sunspots and plages,

b. rapidly increasing magnetic fields,

c. magnetic complexity,

d. emergence of new magnetic fields in a stable or decaying region,

e. shearing or twisting of the field,

f. frequent small flares with increasing x-ray emissions,

g. regions moving toward each other.

There are other factors as well, but note that the most important ones are concerned with the history of the active region, how it develops and changes, particularly the magnetic field configuration.

Figure 4.17 shows hydrogen-alpha pictures of the Sun taken 2 days apart. Notice how the plages and filaments appear to move from left to right (east to west; south is at the bottom) across the disk. Actually, this apparent motion is the rotation of the Sun on its axis. If the features seen on the disk persist, they eventually disappear off the right (west) side. About 2 weeks later those active regions that rotated off the west side will come back into view on the east side. Thus their return can be anticipated. However, during the 2 weeks when they are out of view, the regions may increase or decrease in complexity, or even fade away completely. Likewise, new regions are born and develop. All this must be taken into account by the space forecaster when the forecast is prepared.

The forecasts are quite reliable, especially the "no flare" forecast. Forecasters can generally tell whether a region will produce large flares, but predicting the time of occurrence varies from difficult to impossible.

Proton events are even more difficult to predict. However, given that a flare has occurred, the forecaster uses several parameters to state whether a proton event is likely in near Earth space. Some of the characteristics of a proton flare are:

a. a pair of parallel ribbons,

b. coverage of the darkest portion of the sunspot,

c. the location of the event on the Sun's disk,

d. long bursts of radio waves and x-rays,

e. high intensity radio noise at certain wavelengths and less intensity at certain others.

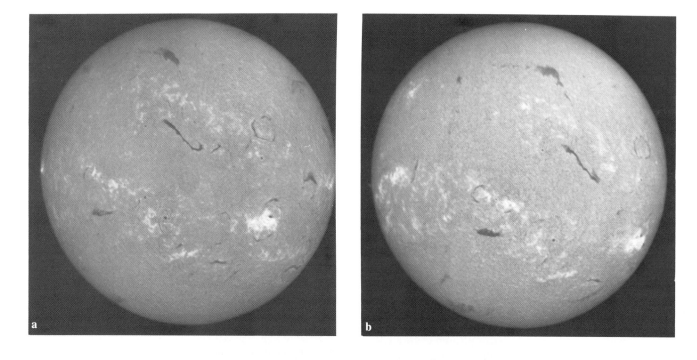

Figure 4.17   The Sun. (a) July 2, 1982; (b) July 4, 1982; (c) July 6, 1982. *Courtesy of the National Solar Observatory, Association of Universities for Research in Astronomy, Inc.*

If the forecaster believes that a proton event is likely, a warning can be issued.

This is similar to the way in which severe weather warnings are issued. A probability forecast is issued first, then an advisory if conditions favorable to severe weather develop, and finally a warning when the severe weather is actually sighted.

To keep track of events on the Sun, a joint civilian-military network of solar observatories has been established around the world in such places as Hawaii, Australia, New Mexico, and Italy. Sites were chosen for their good weather for observing the Sun. In addition, they are so spaced around the world that at least one is always able to see the Sun. Observations are sent from the observatories to the forecasters in the United States who evaluate the information and pass it on to NASA and other interested parties. If a manned flight is in progress, the forecasters keep close contact with mission control, updating it on any solar activity that may prove hazardous.

Contamination by Spacecraft

All spacecraft create a contaminated "atmosphere" which travels with them in orbit. This was first noticed in the Mercury flights. John Glenn and others reported "fireflies" following the capsule. Windows clouded over on Mercury, Gemini, and Apollo vehicles. The density of the induced atmosphere around Skylab was nearly 10,000 times greater than the normal space environment. This type of contamination has many sources:

a. Outgassing of molecules from the body of the vehicle when it is in the vacuum of space

b. Small leaks of fluids and gases from spacecraft systems

c. Exhaust gases from rocket engines and maneuvering thrusters

d. Deterioration of paints and other coatings due to ultraviolet radiation from the Sun

e. Leakage of air from the cabin

f. Water and urine dumped overboard

g. Gases from the waste management systems

This contamination follows along with the spacecraft because those molecules of gas are in the same orbit as the vehicle and, perhaps, because of the gravitational attraction between the molecules and the vehicle. As a result of this contamination several problems arise. The surfaces of windows and optical equipment such as lenses and mirrors become fogged. Deposits on solar panels reduce their electric power output. Deposits on radiators and the outer surface of the vehicle affect their heat-absorbing and radiating properties. At certain angles, sunlight scatters off this induced atmosphere causing a glare and making it difficult for sensitive optical instruments to work properly.

Besides gaseous contamination, an orbiter produces "light pollution." It glows as it moves through the oxygen of the thin atmosphere at 200 miles. The phenomenon is not completely understood, but it may have to do with the electrical and magnetic fields produced by the motion of the orbiter. The oxygen probably becomes ionized, then emits a glow as it recombines into atomic oxygen.

Space Debris

Objects in orbit around the Earth remain in orbit unless some force returns them to Earth. Some satellites and pieces of rocket have been up there for over 30 years. By 1988 there were over four million pounds of material in Earth or-

bit. Almost half of that is debris, including empty propellant tanks, shrouds that covered the top of the satellites, lens caps, and other protective materials discarded on reaching orbit. One-fourth of the debris is pieces of rockets that went into orbit with the payload. Another one-fourth consists of satellites, 80 percent of which do not work. The number of spacecraft approaches 2,000 and large pieces of debris number about 5,000. In addition, there are at least 30,000 marble-sized to baseball-sized pieces of debris and perhaps a hundred million paint chips and smaller sized particles.

With new satellites being launched every month, something will soon have to be done about the increasing amount of artificial space debris. The smaller pieces are already a hazard, probably a greater hazard than meteoroids. During an early flight of the orbiter *Challenger*, a flake of paint hit a window, causing enough damage to necessitate replacing it. The Soviets reported that something struck a Salyut space station window with a loud noise. They believed it to be a micrometeoroid, although it could have been a manmade object.

Not enough is known about these smaller objects because they are difficult to detect with present tracking equipment. The North American Aerospace Defence Command (NORAD) can track and catalog objects larger than 10 centimeters, about the size of a baseball. One of NORAD's tasks is to check out the planned trajectories of satellites and Shuttles and to compare them with the orbits of the trackable objects in space to assure there will be no collision. Occasional "holds" have been necessary during Shuttle launch preparations because of passing objects. It is not possible to track all the small pieces of junk, however, and they pose the greatest hazard. A pea-sized fragment moving at high speed relative to a satellite could easily put it out of commission. During a Shuttle mission, NORAD warns the crew of known objects on a collision course, giving time for an orbital maneuver to avoid the object. When the International Space Station becomes operational, bumper shields may be attached to the habitats and laboratories, because maneuvering a large, massive space station to avoid a collision would be difficult. Firing rockets would disrupt station operations, particularly zero-g experiments.

Many objects return to Earth naturally as atmospheric drag decays the orbit and they burn up on reentry into the atmosphere. As discussed earlier, geomagnetic storms are helpful in this regard. They help to clean up the trash in low Earth orbits because drag increases as the atmospheric density increases during the storm. The largest object to burn in was the Skylab space station, which broke up on reentry in 1979. The pieces fell into the Indian Ocean and the Australian desert. Many Russian spacecraft use nuclear generators for electric power. One of them reentered and fell in Canada, spreading radioactive material over a wide area. Another fell into the Indian Ocean in 1983. A piece of Soviet satellite Cosmos 954 is shown in Figure 4.18. Because

Figure 4.18 A piece of the Soviet satellite Cosmos 954 which fell to Earth. It is probably part of a propellant tank for attitude control. *Courtesy of U.S. Space Command.*

most of the Earth is covered with water and only a small part of the land area is populated, the probability of a falling satellite doing any serious damage is remote.

So how do you cope with the problem? You cannot get rid of a satellite by blowing it up. Instead of one large object, you would have hundreds of smaller ones, each in a different orbit following the explosion. One solution is to boost the satellite to a high "parking orbit" well above any remnant atmosphere where it will remain essentially forever. The Russians now do that to satellites carrying nuclear generators. Still, such action only moves the problem to a higher altitude. Perhaps a better approach is to retrofire the satellite onto a tumbling collision course with the Earth so that it burns up in the atmosphere on the way down. The Space Shuttle has picked up several disabled satellites and returned them to Earth, but what are those few among so many? The real problem of the millions of smaller pieces remains.

DISCUSSION QUESTIONS

1. Why doesn't the solar wind blow satellites out of orbit?

2. The Sun is composed mostly of hydrogen. Why, then, is the solar wind mostly protons and electrons?

3. If the Earth's atmosphere suddenly became transparent to all electromagnetic radiation, what changes, if any, would you expect to take place on Earth?

4. If the Earth's atmosphere had always been transparent to all electromagnetic radiation, how might Earth be different?

5. If you had an important space mission to fly but there was a high probability of a solar proton storm, would you go anyway? What factors would go into your decision?

6. Give some suggestions as to how to reduce the amount of space debris in orbit around Earth.

ADDITIONAL READING

Hunton, Donald E. "Shuttle Glow." *Scientific American*, November 1989. Technical but readable.

Johnson, Nicholas, and Darren McKnight. *Artificial Space Debris.* Updated edition, Krieger Publishing Company, 1991.

Millman, Peter M. "The Meteoritic Hazard of Interplanetary Travel." *American Scientist*, November-December 1971. Technical but readable.

NASA. *Our Prodigal Sun.* NASA Facts, KSC 116-81. Government Printing Office, 1982. Easy reading pamphlet.

Stafford, Edward P. *Sun, Earth and Man.* NASA EP-172. Government Printing Office, 1982. Easy reading booklet.

Tascione, Thomas F. *Introduction to the Space Environment.* Second edition, Krieger Publishing Co., 1994.

Van Allen, James. "Radiation Belts Around the Earth." *Scientific American*, March 1959. Discovery of the Van Allen belts.

White, Oran R. *The Solar Output and Its Variation.* Colorado Associated University Press, 1977. Technical treatise.

NOTES

Chapter 5

Satellites

Earth has one natural satellite: the Moon. Sputnik, the first artificial satellite of Earth, was launched by the Soviet Union on October 4, 1957. It carried a small transmitter that beeped a signal to ground stations following its trek across the sky. That is all it did.

As of mid-1994, over 7,000 man-made objects were being tracked in their orbits around Earth, only one-fourth of which were useful satellites. The rest were pieces of burned out boosters, satellites whose transmitters or sensors had gone dead, and a few small objects such as bolts and an astronaut's glove. Most of the payloads and pieces of debris shot into orbit since Sputnik have fallen out of orbit, and most of them burned up in the atmosphere during their fall. All objects of importance and interest to the United States are catalogued by the North American Aerospace Defense Command (NORAD) at its command post inside Cheyenne Mountain in Colorado, near Colorado Springs, Figure 5.1. Tracking data is fed into NORAD computers from a world-

wide network of radars and telescopes. Orbital parameters for the most important satellites are frequently updated so their locations are known at all times.

What are they all doing up there? Some of the practical uses for satellites include communications relay, weather observing, military surveillance, monitoring ocean wind and waves, geodetic mapping, and monitoring of Earth resources. Satellites are used for search and rescue, locating fishing grounds, observing icebergs, and for scientific research in astronomy, space physics, and planetary science.

This chapter will deal with satellites for communications, navigation, energy production, and defense. Satellites for viewing Earth will be covered in Chapter 6. Spacecraft for astronomical research and exploration will be discussed in Chapter 7.

Communications Satellites

Radio waves propagate in straight lines and cannot travel around the curvature of the Earth. Thus, a radio or television receiver cannot pick up broadcasts from a transmitter that is beyond the horizon; see Figure 5.2. The *ionosphere*, a naturally occurring ionized layer of the atmosphere from about 60 miles to 200 miles altitude, reflects certain frequencies so they can be heard over the horizon. Long distance shortwave broadcasts may be received by ionospheric

Figure 5.1 NORAD Space Surveillance Center. A space analyst studies the ground trace of a spacecraft on the computer display. The center operates 24 hours a day to maintain an up-to-the-minute catalog. *Courtesy of U.S. Space Command.*

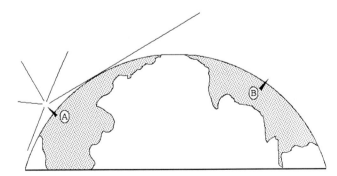

Figure 5.2 Radio signals travel in straight lines; they do not curve around Earth. A radio receiver at B cannot receive the broadcast from the transmitter at A.

reflection, but the ionosphere is irregular and somewhat undependable. If an artificial "mirror" could be placed in orbit, communications could be reliably established between any two points on Earth.

The first U.S. experiment in communicating via satellite was Echo 1, a 100 foot aluminized mylar balloon launched into space folded in a small package which opened and inflated on arriving in orbit. How do you fold a 100 foot balloon into a 28 inch diameter canister? One of the engineers was inspired by his wife's plastic rain cap which folded into a long narrow strip and unfolded into a perfect hemisphere. It worked.

Because it carried no receiver or transmitter, an Echo balloon could be used only passively as a reflecting surface to bounce radio signals beyond the transmitter's horizon (Figure 5.3). The spherical shape scattered the radio waves in many directions; only a small amount arrived back at the ground. High powered transmitters, 10 kilowatts or more, were necessary to assure that a detectable amount of radio signal reached the receiver. Movable large dish antennas on Earth followed the "satelloons" across the sky.

The balloons were launched into elliptical orbits between 600 and 1,100 miles initially but, because of their size, air drag had a major effect on them and the orbits varied substantially from week to week. Echo 1 was launched August 12, 1960, and remained in orbit for nearly 8 years before burning in. Echo 2, 135 feet in diameter, was launched in 1964 and lasted about 5 1/2 years.

Later, active relay satellites replaced the passive balloons. Active relay communications satellites carry receivers and transmitters. The signal is sent to the satellite, picked up by its receiver, and retransmitted on another frequency to a ground station.

Placing relay stations in space was a natural follow-on to the microwave relay stations on Earth. During the 1940s

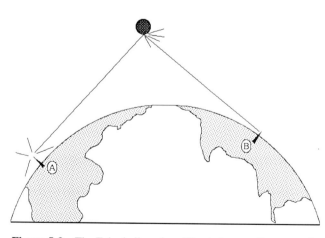

Figure 5.3 The Echo balloon in orbit acted as a reflector so signals from the transmitter at A bounced to the receiver at B. Later, relay satellites received the signal from the transmitter on the ground and re-transmitted it to receivers beyond the horizon.

and 1950s, based on technology developed in World War II, communications companies had set up a network of microwave stations across the country to replace the long wires and to extend telephone and television services to more remote areas. At each relay station the signal was received and retransmitted to the next station in line. Because radio waves travel in straight lines, the microwave towers had to be placed within line of sight of each other. On flat terrain the towers were built about 50 miles apart. This communications technology was well developed, and when satellites came along there was the opportunity to place the relay tower in space where broadcasts could be transmitted to vast areas from a single transmitter in orbit.

The first active repeater satellite was Courier, a Department of Defense experiment which operated from October 4 to October 21, 1960. Spherical in shape, it was covered with solar cells, the first satellite to use solar cells instead of batteries for power.

The first commercial active repeater satellites were Bell Telephone's Telstar 1 and RCA's Relay 1, both launched in 1962. Telstar 1 weighed 170 pounds and was capable of handling one television channel or 600 telephone circuits. By comparison, the newest communications satellites weigh nearly 8,000 pounds and can handle more than 250 television channels or 50,000 voice channels at once. Telstar 1's orbit had an apogee at about 3,500 miles, perigee near 600 miles, and a period of 157 minutes. This allowed it to "hang" at apogee in position to provide communications between two ground terminals for up to 2 hours. Relay 1 was placed into a higher orbit, perigee at about 800 miles, apogee near 4,600 miles, and period of 185 minutes, allowing for longer operating times.

All these early communications satellites were in relatively low orbits which meant that they rose above the horizon, ambled across the sky, and set below the horizon. Of course, they were only useful when they were above the horizon. To provide continuous communications between any two points would require a "constellation" of satellites so that at least one was visible from the ground at all times.

Geosynchronous Satellites

The favored orbit for communications satellites is the geosynchronous orbit, a circular orbit with an inclination of zero degrees, 22,300 miles above the equator. (Refer back to Figures 3.15 and 3.16.) At that altitude the orbital period is 23 hours and 56 minutes, the time it takes the Earth to make one rotation on its axis. Therefore, the satellite's angular speed is the same as the Earth's, 360 degrees per day. To a ground observer, the satellite appears to remain stationary in the same spot in the sky. This has three advantages for communications satellites. First, the satellite can be used continuously; it never goes below the horizon. Second, the large dish antennas do not have to track a moving satellite. Construction is much simpler and cheaper. Third, from a distance of 22,300 miles, a satellite can "see" almost one-third of the Earth's surface. Theoretically, only three satellites are needed to cover the entire Earth except for the high latitude regions around the north and south pole. Figure 5.4 shows the view from geosynchronous orbit at three points on the equator, 120 degrees apart. Notice how their viewing areas overlap at the equator up to midlatitudes.

These ideas originated with science and science fiction writer Arthur C. Clarke, coauthor with Stanley Kubrick of the movie *2001: A Space Odyssey*. In the 1940s and 1950s, Clarke was active with the British Interplanetary Society, a far-sighted group of scientists and engineers that studied and developed many futuristic ideas about space. In the October 1945 issue of *Wireless World*, Clarke published a paper showing the technical feasibility of communications satellites in geosynchronous orbit. Actually, the Russian school-teacher, Constantin Tsiolkovski, recognized the existence of the geostationary orbit around Earth decades earlier, but the idea of using it for communications satellites originated with Clarke. Therefore, the geosynchronous orbit is often called the Clarke orbit in western countries.

The first geosynchronous satellite was Syncom II, which achieved orbit on July 26, 1963. On August 19, 1964, Syncom III was locked into a stationary location near the international date line. It carried the first trans-Pacific television broadcast, the Tokyo Olympic Games. By 1994 there were more than 150 communications satellites in geosynchronous orbit.

Geosynchronous satellites, positioned over the equator as they must be, are difficult to use in high latitude regions because they appear so low on the horizon. At a low angle, the radio waves must traverse a longer path through the atmosphere, making them more subject to atmospheric fluctuations and disturbances. The Soviet Union had a unique solution to this problem with their Molniya communications satellites which they began to launch in 1965. The Molniyas are placed in 63 degree, 12 hour orbits of high eccentricity with perigee in the southern hemisphere at 330 miles and apogee in the northern hemisphere at 24,400 miles as described in Chapter 3. The Russians continue to use the Molniya orbits. ("Molniya" is Russian for "lightning.") Besides dwelling for many hours over the high latitudes, less energy is required to launch into a 64 degree orbit than into a geosynchronous orbit from their northern launch sites at Plesetsk and Tyuratam. Three of them equally spaced around the orbit can provide continuous communications links; as one goes down below the horizon, another is rising. More than 135 Molniya satellites have been put into orbit, many of which have since burned up on reentry back into the atmosphere.

We now find dish antennas popping up in backyards all over town to receive television programs directly from the satellites. In the 1970s it was necessary to have a 30 foot dish antenna to receive signals reliably and clearly; now an 8 foot dish will do the job. Practically all television programming, except for local programs, is relayed around the world by satellite. A dozen or more satellites, each carrying half a dozen or more programs, can be seen from a single location. Even television broadcasts via Molniyas can be picked up from the northern United States, although the color and sound are not compatable with a standard American television set.

A new generation of direct broadcast satellites (DBS) began operating in 1994. Because they transmit with higher power at higher frequencies, smaller receiving antennas, only about 18 inches in diameter, are able to pick up their broadcasts. One satellite has been launched on French Ariane 4 boosters from the equatorial site in French Guiana to geosynchronous orbit at 101° west longitude. A second one will follow soon. They will carry 150 TV channels, capable of digitizing and compressing the image for high definition TV (HDTV). The service operates similarly to a cable system, that is, subscribers need decoders.

To prevent interference between satellites, they are each assigned specific operating frequencies and are spaced at least 4 degrees apart, about 2,000 miles in orbit. This limits the number of satellites that can operate on any given frequency in geosynchronous orbit. The International Telecommunications Union, an organization under the United Nations, is responsible for assigning frequencies and orbit

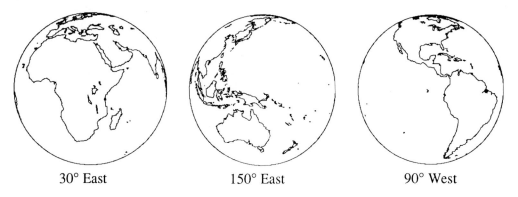

30° East   150° East   90° West

Figure 5.4   View from geosynchronous orbit over spots on the equator at 30° E, 150° E, and 90° W longitude.

positions so there is no mutual interference. By mid-1994 about 150 commercial communications satellites were in geosynchronous orbit, some of them not operational.

Most communications satellites operate in the C-band, 3,700 to 4,200 MHz (3.7 to 4.2 GHz). Recently satellites have been moving to higher frequencies in the Ku-band, 11.7 to 12.2 GHz. This allows the transmitter to carry more channels or conversations on a single channel. However, the higher frequencies are attenuated, absorbed by clouds and rain. Because satellite transmitters have limited power output, any attenuation of the signal can be a problem. Absorption by large size shower-type raindrops is greater than by drizzle or small cloud droplets. A climatological study of the receiver site gives a clue as to the percent of time that the downlink may be wiped out due to heavy precipitation. The engineers' solution to the problem has been to increase the power of the satellite transmitters. However, as higher frequencies are called into use, the problem becomes more acute.

The antennas on the satellites may be designed for maximum area coverage like the whip antenna on an automobile, or they may be narrow beams using large dish reflectors, beaming their signals to relatively small regions on the Earth. A NASA Tracking and Data Relay Satellite, its antenna open for inspection, is shown in Figure 5.5. It is folded

into a compact package for launch (Figure 5.6). Figure 5.7 shows it deployed in space.

As the state of the art advances, it may become possible to transmit to and receive from communications satellites directly using a handheld cellular phone type of radio. Satellites in low Earth orbit (about 200 miles) would not have to carry high power transmitters like those in geosynchronous orbit (22,300 miles up) and they would be able to receive transmissions from low power transmitters on the ground. The main problem with this concept was mentioned before: low orbiting satellites spend much of their time out of sight below the horizon. However, communications companies now plan to launch constellations of several hundred small, simple satellites which would assure that at least one is always in view. As electronic components have become smaller, more reliable, and cheaper, it is now possible to make satellites smaller, lighter weight, cheaper to build assembly line fashion, and cheaper to send into low orbit on small rocket launchers. With hundreds in orbit at a time, failure of a few would not disrupt the entire system, and replacements could be launched quickly. Pocket telephones may become a reality with such a satellite-based system, an extension of the present cellular phone system.

Military Communications Satellites

The defense communications relay satellites are similar in operation to their civilian counterparts. To prevent outsiders

Figure 5.5 A NASA Tracking and Data Relay Satellite (TDRS) antenna open for ground testing. *Courtesy of NASA.*

Figure 5.6 The TDRS satellite being lifted from a Space Shuttle orbiter for dropping off in space. The spacecraft folds into a compact package to fit into the cargo bay. A folded umbrella-like antenna and a dish antenna can be seen inside the folded solar panels. A rocket attached to the bottom will boost it from the Shuttle's orbit to geosynchronous orbit. *Courtesy of NASA.*

Figure 5.7 An artist's conception of TDRS in orbit. Seven antennas (can you find them all?) cover three frequency bands. The two largest ones, over 16 feet in diameter, are steerable. Receivers and transmitters are in the hexagonal box in the center. Arrays of solar cells on the two square paddles provide power. *Courtesy of NASA.*

from listening in, the messages are generally encoded electronically before transmission to the satellite and automatically decoded at the receiving station.

The United States is probably more dependent on satellites for military communications than other countries. American forces are spread around the globe on all continents; positive command and control are essential worldwide. In the 1990s the U.S. military will have a new satellite system called Milstar. Four satellites in geosynchronous orbits and three satellites in highly elliptical (Molniya-type) orbits will provide worldwide command, control, and communications for land, air, sea, and undersea forces. A number of spare satellites may be placed in 100,000 mile orbits. It is easier and quicker to move a satellite from a high orbit to geosynchronous altitude than it is to prepare and launch a replacement from the ground, especially during a war.

The dividing line between civil and military space applications is less clear in Russia than in the United States. We have previously discussed the Russian development and use of the Molniya orbit for communications relay to the far northern regions. In addition, the Russians use numerous communications satellites in circular orbits inclined at 74 degrees, about 930 miles high. Eight satellites are launched on one booster and placed into orbits with periods differing by a few seconds so they slowly drift apart and spread themselves around the Earth. Over 300 such satellites have been launched since 1970 and 30 or more may be operational at any time to assure that at least one is always in sight of the ground stations.

Navigation Satellites

Using satellites it is now possible to find your precise location at any time and place on Earth with an accuracy of about 30 feet. How is it done? Look first at an example of how you determine distances traveled in your car.

Let's take a trip from Colorado Springs to Denver. We leave at 2:30 on the dot and we arrive in Denver at exactly 3:45. Assume we traveled at a constant speed of 60 miles per hour. How far is it from Colorado Springs to Denver? Time elapsed from 2:30 to 3:45 is 1 hour and 15 minutes or 1.25 hours. Multiply that times the speed and we get 75 miles. The accuracy of our answer depends on whether our watch gained or lost time during the trip and how precisely we held to a constant speed.

Using navigation satellites to locate our position on Earth is based on the same principles. A satellite transmits a radio pulse at a given time. The clock in our receiver tells us what time we receive the pulse. Assume that the radio pulse travels at a constant speed, the speed of light, and assume that the clocks are synchronized together so we can determine the elapsed time. We can then calculate the range, the distance to the satellite.

The range by itself is not sufficient information; it is also necessary to know the location of the satellite. The latest ephemeris (orbital data) is also transmitted from the satellite so we can calculate its precise location. Now that we have the satellite's location and its range we know that we are somewhere on a sphere with a radius equal to the range and with the satellite at the center; see Figure 5.8.

This still does not tell us our precise location. Next we obtain the same information from a second satellite to locate ourselves on a second similar sphere. If we are on both spheres at the same time, we must be at the place where the two spheres intersect. See Figure 5.9 (it looks like a snowman). Notice that two spheres intersect in a circle (the waist of the snowman) so we know we must be somewhere on that circle. Getting warmer! Reading out a third satellite pins us down to our precise location (X on Figure 5.10), our latitude, longitude, and altitude.

Errors in Position Finding

There are two difficulties with putting this relatively simple idea into practice.

First, in order for this to work, the clocks in the satellite's transmitter and the user's receiver have to be exactly synchronized. Each satellite carries four atomic clocks which are updated and synchronized daily using atomic clocks on the ground which are accurate within 1 second in 300,000 years. However, atomic clocks are not portable enough and the clock in a mobile receiver cannot be made accurate enough to precisely measure the travel time of the pulse. A small error in the travel time produces a large error in the range calculation. MATHBOX 5.1 shows some sample calculations. An error of only 50 billionths of a second in time produces a range error of 50 feet.

The solution to this problem is to read out a fourth satellite and treat the receiver's clock error as a fourth unknown. Working out our position is a mathematical problem of no small magnitude. It involves solving four equations in four unknowns: latitude, longitude, altitude, and clock error. Re-

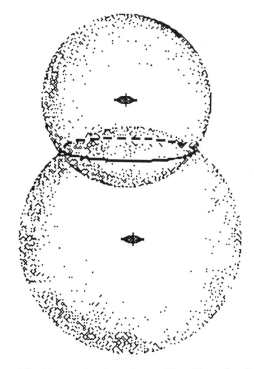

Figure 5.9   Two overlapping sphere, with radii equal to the ranges of the two satellites. They intersect in a circle.

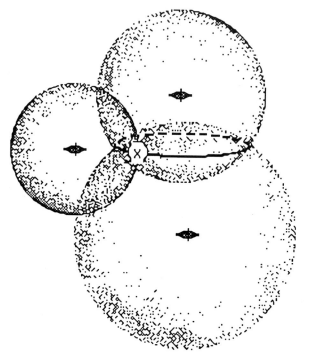

Figure 5.10   Three overlapping spheres with radii equal to the ranges of the three satellites. The position of the observer is marked "x".

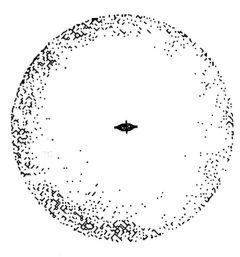

Figure 5.8   A sphere with a satellite at the center. The radius of the sphere is the distance (range) from the observer to the satellite.

ceivers specially designed for use with the navigation satellites incorporate a microcomputer chip preprogrammed to handle the calculations. In actual practice, the user receives transmissions from four satellites and the computer solves the four equations and finds the four unknowns. MATHBOX 5.2 shows the algebra. It is absolutely necessary to read out four satellites to get the 30 foot accuracy mentioned above.

A second problem comes from the assumption that the radio pulse from the satellite travels at 186,282 miles per second, the speed of light in a vacuum. But here the pulse does not travel in a vacuum. It travels through the Earth's atmosphere during part of its trip to the ground, and its speed depends on atmospheric conditions, particularly conditions in the ionosphere. The ionosphere is the upper part of the atmosphere extending from about 50 miles to 250 miles. In this region the air is ionized by the intense solar x-ray and ultraviolet radiation, and the free electrons cause an apparent slowing of the radio pulse. The density of free electrons determines how much the actual speed of the pulse deviates from the speed of light in a vacuum. Electron density in the ionosphere varies in a somewhat predictable fashion through the day, through the seasons of the year, and with the 11 year solar cycle. But minute-by-minute variations can be quite large, especially during periods of high solar activity, and can cause errors in the range calculations.

An additional error in the assumed speed is introduced by the lower atmosphere, the troposphere. However, the error is smaller and less variable. Uncorrected, these ionospheric and tropospheric effects on pulse speed cause range errors amounting to nearly 20 feet.

The problem of pulse speed can be handled in either of two ways, one more accurate than the other. The amount of error depends on the frequency of the transmitted pulse; the higher the frequency, the lower the error. Each satellite carries two transmitters operating on different frequencies. By receiving pulses on both frequencies and comparing the range calculations, it is possible to remove the error. This correction can be applied only by military users; civilian users have access to only one frequency. However, a less accurate correction can be made using information about the ionosphere contained in the satellite transmission. It is based on the average daily and seasonal variations in the ionosphere. For this method of correction the user needs to listen to only one frequency. While the error correction is not exact, it is accurate enough for many purposes.

Navstar Global Positioning System

The Navstar Global Positioning System (GPS) was designed, launched, and funded by the Department of Defense primarily for military use. However, it is available to anyone with a commercially available GPS receiver. The system consists of 21 satellites and 3 spares in six circular 12 hour orbits at 12,500 miles altitude, with an inclination of 55°. All transmit on two frequencies, 1227.60 MHz and 1575.42

MATHBOX 5.1

Range Error

Radio waves travel at the speed of light, about 186,000 miles per second. At a satellite range of 12,000 miles, travel time is

$$\text{time} = \frac{\text{distance}}{\text{velocity}} = \frac{12,000}{186,000} = 0.0645 \text{ second.}$$

An error of 1% in measuring the time would produce an error of 1% in computing the distance. In our example, a 1% timing error is .0006 second; that would produce a range error of 120 miles! A navigator can guess his position better than that!

Suppose we want to reduce the range error to 50 feet. The distance to the satellite is

$$12,000 \text{ miles} \times 5,280 \text{ feet per mile} = 63,360,000 \text{ feet.}$$

What percentage of this distance is 50 feet?

$$\left(\frac{50}{63,360,000} \right) \times 100\% = 0.00008\%.$$

We must reduce the timing error to

$$0.00008\% \text{ of } 0.0645 \text{ second} = 0.00000005 \text{ second!}$$

MATHBOX 5.2

Satellite Navigation Equations

Before we examine the equations used by a satellite navigation system in three dimensional space, let us consider a more manageable problem—determining the position of an object, P, on a two dimensional surface with respect to another object, A.

Look at Figure 5.2.1. The point A is at position $A_x = 4$, $A_y = 3$. This is written (4, 3). Similarly, P is at $P_x = 1$, $P_y = 1.5$, written (1, 1.5). Notice the right triangle formed by a horizontal line from P, a vertical line from A and the line r connecting A to P. Our question is, how far apart are A and P? Because r is the hypotenuse of a right triangle,

$$(A_x - P_x)^2 + (A_y - P_y)^2 = r^2.$$

In this case, $r^2 = (4 - 1)^2 + (3 - 1.5)^2 = 9 + 2.25 = 11.25$. So $r = \sqrt{11.25} = 3.4$.

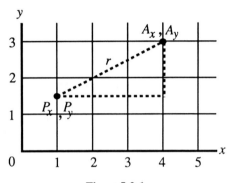

Figure 5.2.1.

When we deal in three-dimensional space, we must include a third term:

$$(A_x - P_x)^2 + (A_y - P_y)^2 + (A_z - P_z)^2 = r^2.$$

Let us apply this to the satellite system. The distance to satellite A is r_A; the distance to the satellite is also the speed of light times the time of travel: $r_A = vt_A$. But we cannot count on our clock to be correct, so we must add a correction factor, e, to the time; $r_A = v(t_A + e) = vt_A + ve$.

Now we are ready to read out four satellites (A, B, C, and D) and write four equations, one for each of them.

$$(A_x - P_x)^2 + (A_y - P_y)^2 + (A_z - P_z)^2 = (vt_A - ve)^2$$
$$(B_x - P_x)^2 + (B_y - P_y)^2 + (B_z - P_z)^2 = (vt_B - ve)^2$$
$$(C_x - P_x)^2 + (C_y - P_y)^2 + (C_z - P_z)^2 = (vt_C - ve)^2$$
$$(D_x - P_x)^2 + (D_y - P_y)^2 + (D_z - P_z)^2 = (vt_D - ve)^2$$

The four unknowns in these equations are our position (P_x, P_y, P_z) and the clock error, e. The positions of the four satellites are given in their transmissions, the travel time of the wave is measured by our receiver, and the speed of light is a constant. These are knowns. Because we have four equations we can solve for the four unknowns. (That is, we can let the computer do it).

MHz. Each is powered by two large panels of solar cells which are kept oriented toward the Sun while the transmitting antennas remain pointed toward Earth. GPS reached an initial operational capability in December 1993 with 24 operating satellites and a guarantee that civilians can rely on its availability.

GPS gives an unlimited number of users the capability of determining their position within 30 feet, their velocity within 0.07 mile per hour, and the time within a millionth of a second. The Department of Defense guarantees an accuracy of at least 100 meters (330 feet) for general civilian use. For even more precise position measurements over a small area, another correction technique is sometimes employed. A GPS receiver is placed at a location whose *exact* coordinates are known. That base receiver then continuously reads out the GPS satellites, processes the data as a portable receiver would, and notes the error in the readout. The error is then sent to the other GPS receivers in the area, often by radio, so they can apply a correction to their data. With some receivers, the error messages are automatically incorporated into the position calculations. By this means, accurate locations can be found within a few inches.

If the user is in a moving vehicle, its speed and direction can be calculated by getting two positions in a short time and finding the change in location during that time. Its velocity can be calculated more precisely by measuring the doppler shift in frequency of the signal coming from the satellite. A doppler shift is an apparent change in frequency of a wave when the source or the receiver or both are moving toward or away from each other. You have perhaps noticed the change in pitch of a siren on an ambulance. When the vehicle is moving toward you, the siren sounds higher in pitch than when it is moving away. The faster the vehicle moves toward you, the higher the pitch. Similarly, the faster it moves away from you, the lower the pitch. In this example we are referring to sound waves, but the same doppler effect applies to electromagnetic waves. The GPS satellite transmitter operates on a very stable frequency. If the receiver can precisely measure the received frequency and detect a shift in that frequency, then the speed of separation or approach between the satellite and the observer can be calculated. Using the orbital elements of the satellite included in the transmitted signal, the receiver can calculate the velocity of the satellite and subtract it from the doppler shift leaving only the observer's velocity with respect to that satellite. Then, by making similar measurements of velocity with respect to four satellites, the absolute velocity of the observer is calculated.

Military controllers at a ground station near Colorado Springs send messages daily to the satellites to update the clocks and the orbital parameters. To prevent an enemy from using the satellites in time of war, the controllers can encode the transmissions and can send a message to deliberately produce an error. Friendly forces would be able to decode and remove the error but an adversary would not. The satellites are protected against damage from nuclear and laser weapons.

The obvious military use is for ground troops, aircraft, ships, and tanks to know where they are and where they are going. In the Persian Gulf War, troops used GPS receivers to find their way day or night across trackless desert terrain without visible landmarks. Some GPS receivers allow the users to enter the coordinates of their destination. The receiver then finds their current location and calculates the direction to head toward their destination. Soldiers who had to make a trip away from home base would enter the location of their camp before leaving. GPS then continually updated the direction back to camp.

Although several thousand GPS receivers were available at the beginning of the Persian Gulf action, the number was insufficient to meet the demand. Many were mounted permanently in tanks and other vehicles and were capable of both receiving and decoding satellite frequencies. When the U.S. forces realized how valuable the small handheld units were, several thousand more were purchased off-the-shelf from companies in the United States and shipped to the Gulf area. Some units were available at aircraft and marine supply stores in the United States and numerous soldiers asked relatives to send them one. Some soldiers simply called suppliers and bought them on their credit cards. Initially, the satellite transmissions were encrypted, but when thousands of commercial GPS receivers were in use in the area, the coding was removed.

GLONASS

The former Soviet Union simultaneously developed a similar system called GLONASS for both military and civilian use. Its theory of operation is identical to the Navstar GPS. GLONASS will have 24 satellites in three orbits at inclinations of 64.8°. The broadcast frequencies are near 1600 and 1250 MHz. The Russians have made the details of their satellite signals available to anyone worldwide. They say that they do not intend to introduce deliberate errors like the U.S. military could do in wartime. Only a few hundred GLONASS receivers were built in the former Soviet Union. Manufacturers in both countries have developed equipment to receive and process both GPS and GLONASS signals.

The most obvious use of GPS and GLONASS is for navigation, both in the air and at sea. The International Civil Aviation Organization, an agency of the United Nations, has endorsed the use of satellite navigation systems for commercial aviation, anticipating that both GPS and GLONASS will become fully operational, thus providing redundancy in case one becomes inoperable. A single channel receiver receives only one satellite at a time; it must switch to four satellites in succession, requiring several minutes to get an accurate position. That may be adequate for some applications, but because of their speed, aircraft must have multichannel receivers which can read out six or more satellites simultaneously. Merchant ships can follow the most favorable routes more precisely taking advantage of ocean currents. The Maritime Administration estimates that a large tanker could save $17,000 in time and fuel in crossing the Atlantic. Sea and air collisions can be reduced by each craft knowing its exact position. A GPS combined with a walkie-talkie radio would be extremely useful to rescue downed aircrews who become separated from their aircraft.

Besides the obvious uses, there have been many innovative applications. For example, treasure hunters may locate a sunken ship but then have difficulty finding it again later, especially in murky water. With precise position information from GPS, it is easy to return to the exact spot. When archaeologists locate prehistoric and historic artifacts, they make precise maps of the sites using compass, topographic map, transit, and measuring tapes. Now, with GPS receivers, an area can be mapped in far less time, with greater accuracy, and sites that become covered or otherwise lost from view by vegetation, fire, flood, drifting sand, etc., can be easily found again at a later date.

On September 16, 1992, five helium-filled balloons from five countries lifted off near Bangor, Maine, on a 3,000 mile race across the Atlantic. Each carried GPS equipment which continuously displayed its position and transmitted this hourly to a command post in Rotterdam via the Inmarsat satellite communications system. Both the crews and the command post knew exactly where the balloons were at all times. Balloons drift with the wind and the crews used the GPS velocity information to find the best winds. By slowly ascending or descending and noting the velocity every few hundred feet, they selected flight altitudes with the most favorable winds, i.e., the maximum speed in the correct direction. Four of the five completed the transatlantic flight. One ran into an icy rainstorm in mid-Atlantic and was forced down. Because the command post knew where the balloon went down, they could direct a nearby ship to the exact spot to pick up the crew. Another balloon was forced down in the English Channel. Its GPS receiver was still working and the rescue team reached it within 6 minutes.

GPS receivers have been miniaturized down to pocket size. Because the U.S. Army is ordering them by the thousands, prices will undoubtedly come down in the next few years. Carry a GPS receiver with you and you will never be lost!

Energy Satellites

Proposals have been made to use satellites to collect solar energy and beam it to our energy-hungry Earth. The prime source of energy consumed by humans now comes from the fossil fuels: coal, oil, and natural gas. But they are not inexhaustible; they will eventually run out. In addition, most of Earth's air pollution problems are traceable to the com-

bustion of fossil fuels. Even nuclear energy is not an unlimited source of power. Nuclear electric generators run on uranium fuel, but uranium is not inexhaustible either. An additional factor is the strong public concern over how to dispose of the radioactive waste from nuclear power plants.

One practically endless source of nonpolluting energy exists: the Sun, which will continue in its present state of energy production for another 5.5 billion years. The problem is how to tap into it efficiently and economically.

There is much popular interest in solar energy. We see solar collectors on roofs around town. In larger scale installations they are designed to heat the building interior; smaller collectors just heat water for household use. In principle, a solar collector is simply a glass-covered box which gets hot when the Sun shines into it. Interior objects absorb the sunlight, thus raising the temperature. The heat produced does not readily escape through the glass. Water is pumped through pipes in the box or air is blown through ducts. The heated water or air is then sent to the building or to a water storage tank to do its job. In the early 1980s, interest was so high in solar energy that the federal government and some state governments gave tax breaks to people who installed solar collectors in homes or businesses.

Photovoltaic Cells

Another way to collect and use solar energy involves *solar cells*. More properly called *photovoltaic cells*, they generate electricity when light shines on them. They are semiconductor devices made of thin wafers of pure silicon, like transistors and computer chips but much larger in diameter. Figure 5.11 shows a typical cell. There are no moving parts and no fuel is needed to produce clean, quiet, nonpolluting, free electricity. As a bonus, another country cannot cut off the source of supply. Photovoltaic cells have been used for low power requirements in rural areas and in isolated locations on Earth where it may be too expensive to run an electric power line (Figure 5.12). They are also used to power pocket calculators and exposure meters in cameras.

You could install an array of solar cells on the roof of your house or in your yard and generate electricity to run your house. A 20 by 30 foot panel would yield a peak of 5,000 watts at midday and an average of more than 1,000 watts over the year. There are a few problems, however. Most obvious, what do you do after sunset and on rainy days? One solution is to charge up storage batteries during the daytime and run the house from them at night. Car batteries are common, reliable, and have the lowest cost of appropriate storage batteries, but even they would be quite expensive. A car battery costs perhaps $50.00 with a 4 year warranty. To assure that you have enough electricity for overnight plus periods of heavy cloudiness when less electricity is generated by the cells, you would need perhaps 40 batteries. If they needed to be replaced every 4 years, the annual cost of your "free" electricity is $500.00 per year.

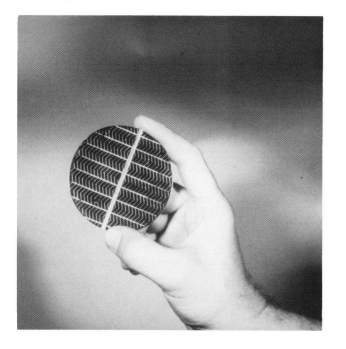

Figure 5.11 Typical photovoltaic cell. *Courtesy of Solar Energy Research Institute.*

Solar cells are quite inefficient. The best they can accomplish is to convert 14 percent of the light falling on them into electricity, although experimental cells have been constructed which have an efficiency of 27.5 percent. Also, sunlight is lost on the way down through the atmosphere by scattering, clouds, and pollution. The solution to such inefficiency is simply to build a larger array.

Photovoltaic cells are expensive at present; buying and operating a solar array to run a house would cost about 35 cents per kilowatt hour. By comparison, electricity purchased from utility companies varies from 6 to 20 cents per kilowatt hour in different parts of the United States. The cost has been high because present-day photovoltaic cells are manufactured from single crystals of silicon which requires much handwork, similar to the processes involved in making semiconductors a few years ago. Techniques for mass production have been under development for several years and there is some hope that prices will come down dramatically just as they have for other mass-produced semiconductor devices. A promising approach is to use a triple layer of cells. Some light is converted to electricity in the top layer of silicon while the two lower layers of silicon-germanium alloys produce electricity from wavelengths that passed through the top layer. Thus, each layer captures different wavelengths of light. Although each layer is inefficient in itself, the combination of the three converts about 10 percent of the sunlight into electricity. While this is not as efficient as pure silicon crystals, the layered cells can be mass-produced inexpensively by depositing hot gases containing silicon and germanium onto a substrate such as stainless steel. This new development could bring the cost

Figure 5.12 Radio relay repeater station on remote island powered by a photovoltaic array. *Courtesy of Solar Energy Research Institute.*

down to 12 to 16 cents per kilowatt hour, competitive with utility companies in high cost or isolated areas.

Solar Cells for Spacecraft

Photovoltaic cells are commonly used as power sources for spacecraft. In fact, the first practical application of solar cells was for powering satellites. You can see them covering the sides of spacecraft or sometimes installed on large wing-like panels attached to the vehicle (Figures 5.13 and 5.14). Skylab, the first U.S. space station, shown in Figure 1.14, was powered by an array that produced 10 kilowatts.

Most solar cells used in space applications have been similar to the one shown in Figure 5.11. They are heavy and expensive to produce, but highly reliable. However, the new methods of depositing thin-film photovoltaic cells on lightweight substrates described above appears to be practical. These cells may not be as efficient as the silicon crystals, but they are lightweight and can be folded into compact packages.

An experiment in deploying a large photovoltaic array

was carried on the Space Shuttle *Discovery*. See Figures 5.15 to 5.17. The accordion-pleated array 13.5 feet wide and 105 feet tall was extended and retracted 17 times successfully. When folded into its box in *Discovery*'s cargo bay, it was only 4 inches thick! For the experiment, only one of the panels had active solar cells. If completely covered with photovoltaics it would have produced 12.5 kilowatts of electrical energy.

Power Stations in Space: 1970s Concept

From favorable experiences with photovoltaic power supplies came the idea to produce power in space for use on Earth. The concept is to erect very large arrays of solar cells in geosynchronous orbit to produce electricity to generate microwaves and beam them to Earth where they are reconverted to electricity and fed into the commercial power grid to customers. The night and atmospheric loss problems are absent in space. A satellite in geosynchronous orbit is in almost continuous sunlight and can harvest solar energy continuously, 24 hours a day.

A study in the 1970s showed the feasibility of solar power satellites. The photovoltaic array would be very large, 20 square miles in area, covered with millions of solar cells. On Earth it would weigh about 75 million pounds. The electricity would be converted to microwaves and transmitted to Earth in a narrow beam using a dish antenna half a mile in diameter (Figure 5.18). The Earth stations receiving the beams would be located in remote areas where there would be no hazard from the microwaves (Figure 5.19). Receiving antennas would cover perhaps a square mile. Because the satellite power station would be in geosynchronous orbit it would appear fixed in the sky and the antennas on both ends would point in fixed directions.

Although the microwave power density in the center of the beam would be more than double the safe limit, the edge would be at very low power. A restricted area would have to be designated to keep airplanes from flying into the beam. (There are many such restricted areas in the world where aircraft are prohibited.) An automatic system on the satellite would shut the beam off if it strayed from its designated path. One concern is the effect of the microwave beam on the ionosphere as it passes through. Microwave beams affect the ionosphere, but the extent and consequences are not well known.

At the ground station the microwaves would be converted back to electricity and fed into the established national electric power grid just like power from a coal-powered generator or a hydroelectric dam generator. One such satellite would produce 5 gigawatts (5 million kilowatts) of electricity. MATHBOX 5.3 shows how to calculate this. Two of them would meet the maximum requirements of Manhattan or Chicago or Houston. Projections showed that 10 percent of the United States energy needs could be met by 300 power satellites.

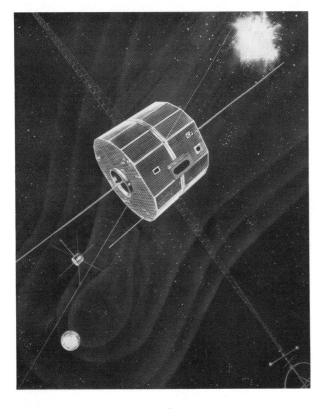

Figure 5.13 Power for the Dynamics Explorer is provided by solar cells covering the main body of the satellite. *Courtesy of NASA.*

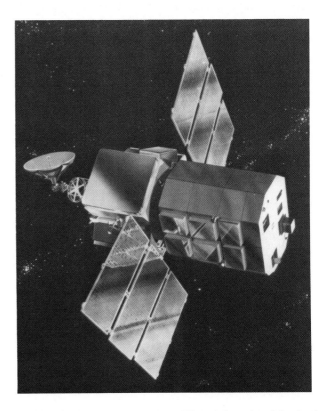

Figure 5.14 Solar cells on six paddles produce electricity for the Solar Maximum Mission satellite. *Courtesy of NASA.*

Figure 5.15 The solar cell array during preflight tests prior to installation in *Discovery. Courtesy of Lockheed.*

Figure 5.16 Photovoltaic array beginning to extend from the cargo bay of the Space Shuttle *Discovery*. Photo was taken from the flight deck looking toward the rear of the Shuttle. *Courtesy of Lockheed and Threshold Corp.*

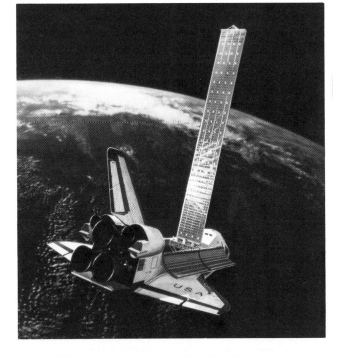

Figure 5.17 Artist's concept of the collapsible solar array carried to orbit on the maiden voyage of the Space Shuttle *Discovery.* *Courtesy of Lockheed; artist Joe Boyer.*

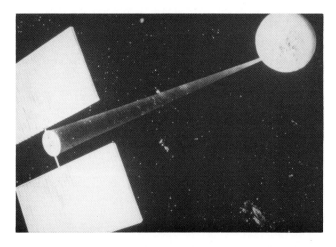

Figure 5.18 Artist's concept of a solar power satellite, transmitting microwave energy to a receiving station on the ground. The photovoltaic array may have an area of 20 square miles. *Courtesy of Solar Energy Research Institute.*

There are, of course, a few problems with this 1970s vintage design. Building such a power station in space would be extremely expensive in time and money. The cost of lifting the required mass into orbit is enormous. There would be three possible approaches. One, carry the pieces into low Earth orbit, assemble the structure there, and boost it to geosynchronous altitude. There are serious problems with this approach. The large booster rockets would require gigantic amounts of fuel which would have to be carried up to the construction site. Also, the structure would have to have

Figure 5.19 Antenna "farm" for receiving microwaves transmitted from a solar power satellite. The antenna elements are arranged in a circle because the beam coming from the satellite has a circular cross section. *Courtesy of Solar Energy Research Institute.*

additional bracing to withstand the stresses of acceleration when the boosters fire. That means additional weight. Another approach is to build the power station in geosynchronous orbit. The problem here is that the Space Shuttle cannot reach geosynchronous altitude. New heavy lift launch vehicles would have to be designed to carry workers and materials to that altitude. Also, the radiation exposure of the workers would be greater. A third approach is to manufacture the parts on the Moon and transport them to Earth orbit. Launching from the Moon to geosynchronous orbit requires less than 8 percent of the energy required to launch from Earth to geosynchronous orbit. A colony on the Moon is a prerequisite which we will discuss in Chapter 12.

Power Stations in Space: 1994 Concept

Research in 1994 by S. D. Potter sponsored by the Space Studies Institute suggests an approach to the construction of solar power satellites that may be more practical. It reduces their mass to less than 10 percent of the 1980 concept. Two recent technological developments contribute to this mass reduction: (1) thin film deposition of photovoltaic cells on a thin plastic material such as Kapton and (2) miniaturization of solid-state microwave transmitters which can be deposited among the solar cells on the same substrate.

As discussed earlier, thin-film solar cells have been made only experimentally and only on heavy substrates such as stainless steel. Deposition on lightweight substrates is still in the research stage. However, results to date look very promising.

In the 1970s power satellite concept, high power bus wires carried the current from the entire array to the microwave transmitter and large dish antenna. The transmitter used klystron tubes. In the new approach, small solid-state microwave transmitters are deposited over the entire surface of the substrate in among the photovoltaic cells. Integrated

---

MATHBOX 5.3

Solar Energy

The intensity of solar radiation decreases as the square of the distance from the Sun. At the Earth, 93 million miles from the Sun, solar energy above the atmosphere is about 128 watts per square foot. Consider an array of solar cells covering 20 square miles:

20 square miles $\times$ 5,280^2 square feet per square mile = 558 million square feet.

The array would intercept

558 million square feet $\times$ 128 watts per square foot = 71 billion watts

= 71 gigawatts of solar energy.

Solar cells at best are only about 14 percent efficient. The array would produce

71 gigawatts $\times$ 0.14 = 10 gigawatts of electricity.

Converting the electricity to microwaves and transmitting them to the ground is about 50 percent efficient, so 5 gigawatts would reach the receiving site on Earth.

"wiring" runs to the transmitters only from nearby solar cells. Each miniature transmitter has its own small antenna; thus the transmitting antenna is an array of small antennas as large as the entire photovoltaic array. The larger the antenna, the more concentrated is the beam reaching Earth. This fact limits the size of the array in order not to exceed the safe intensity level of the microwave beam.

Two possible designs have been investigated: a bicycle wheel configuration and an inflatable balloon. The bicycle wheel support structure is shown in Figure 5.20. For clarity, the size of the hub is grossly exaggerated in the diagram. The wheel is about 1 1/4 mile in diameter while the hub is about 6 feet in diameter and 4 inches thick. Eight spokes on each side provide rigidity. The spokes are made of a very thin but strong silicon carbide fiber. The rim is made of the same material, four times as thick. The structure is then covered by the plastic material with the solar cells and microwave transmitters deposited on it.

The logical location for power satellites is, of course, geosynchronous orbit so they always appear to be in the same spot in the sky. As the bicycle wheel power station revolves around the Earth, though, it is impossible for it to always face both the Sun and the Earth at the same time. At midnight it is on the side of Earth directly opposite the Sun, sunlight falls fully on it, and its illuminated side faces the Earth like the full Moon. At noon, however, it passes between Earth and the Sun, like the new Moon, so the side facing the Earth is not illuminated and not producing electricity. There are several possible ways to overcome this problem. One is to build a large mirror in orbit near the power satellite oriented so that the sunlight is reflected onto

the dark side. Such a mirror could be made out of aluminized plastic film and would weigh about as much as the power satellite itself. A second solution is to cover the dark side with microwave transmitters deposited on a lightweight substrate and supply them with power from the sunlit side by connections through the interior space between the two substrates.

Another configuration for a power satellite is an inflatable balloon made of lightweight material with photovoltaic cells and microwave transmitters deposited over the entire surface. As it moved in its geosynchronous orbit, one side would always be illuminated and collect solar energy. The energy collected on the Sun-facing side would have to be

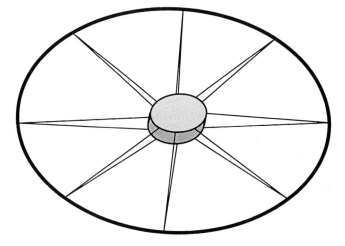

Figure 5.20 Bicycle wheel solar power satellite concept. For clarity, the central hub is shown oversized.

distributed to the transmitters on the Earth-facing side via wiring in the interior of the balloon. Comparing a spherical balloon with a circular wheel of the same radius, the surface area of the sphere is four times the area of the circle so the cells, transmitters, and substrate would weigh four times as much. However, the collecting area is the cross-sectional area of the balloon which is equal to the area of a wheel of the same diameter. The advantage of a balloon is that it needs no support structure. When it reaches orbit, it is inflated with a compressed inert gas such as nitrogen.

If we divide the total weight of the satellite by the power it delivers to the consumer, we get the number of pounds of satellite required to deliver a kilowatt of energy. Potter calculated this number for various configurations of thin-film solar power satellites. A 1.25 mile diameter bicycle wheel won hands down. It would yield 109 megawatts into the electric power grid and would weigh about 1 pound for each kilowatt. By comparison, the original 1980 design weighed about 22 pounds per kilowatt of useful energy.

The 1980 design produced 5,000 megawatts from one satellite weighing about 75 million pounds. It would take 46 bicycle wheels at 109 megawatts each to equal that. However, their total weight would be only 5 million pounds, one-fifteenth as much. Recall that the cost of delivering material to orbit is the limiting factor in space exploration and exploitation. This new concept for building power satellites may soon make them economically feasible.

Meteoroid impacts, even micrometeoroids, will slowly degrade an array of photovoltaic cells and their electrical output will gradually decrease. The intense sunlight above Earth's protective atmosphere will also cause some degradation. Even so, it should be possible to build a power station that will last 30 years. They are expensive to build, but cheap to operate.

Space Defense

Space defense operations have been conducted since the beginning of the space age. We often hear slogans opposing the militarization of space; but it is naive to think that space is not already militarized. Nuclear-tipped intercontinental ballistic missiles were operational before the first satellite went into orbit. In fact, it was the development of those heavy-lift missiles in both the United States and the Soviet Union that made it possible to launch satellites. When the Russians successfully orbited their first Sputnik, President Eisenhower announced that we had a satellite in preparation, but our highest priority was the continued development of the ICBM, and other space activities had a secondary role.

In addition to ballistic missiles, both countries immediately recognized the advantages of satellites for many purposes. By far the greatest number of satellites in orbit today are for defense. Indeed, much of the spacecraft technology now in use was developed for defense purposes and had its origin in military satellites. Communications relay and navigation satellites were discussed earlier in this chapter. Other defense satellites such as photo reconnaissance, electronic intelligence gathering, weather observing, and warning of missile launches are in the category of remote sensing which will be covered in Chapter 6.

The defense activities in space listed above are obvious, contribute substantially to national security, and are accepted by all adversaries. International agreements on the uses of space do not preclude any of these defense satellites. All sides take advantage of their capabilities. Some things have been prohibited by international agreement, however. Nuclear weapons in orbit are prohibited but nuclear power generators are not. Establishing military bases on the Moon or other celestial bodies is prohibited, but military personnel may be used for exploration and scientific pursuits.

Defense Against Ballistic Missiles

The 1972 antiballistic missile (ABM) treaty with the U.S.S.R. allowed for the deployment of a single system to protect an ICBM missile complex and one to protect its seat of government. The Soviets elected to deploy an ABM system around Moscow.

The United States began construction of a system to protect some of its missile silos. It used large northward pointing radars which could see warheads (and decoys) rising over the horizon. Their detection information was passed to terminal guidance radars which tracked the objects. Spartan antimissile missiles would be fired to intercept the incoming warheads above the atmosphere using x-rays from a nuclear explosion to disable as many of the incoming objects as possible. The lethal range for x-rays from a nuclear detonation in space is a few kilometers. As the remaining objects entered the atmosphere, the decoys would burn up while the heat-shielded warheads continued on toward their targets. Sprint antimissile missiles would then be fired and steered by radar to intercept and destroy the warheads before they reached their targets. By the time the real warheads were identified, the Sprints had less than 30 seconds to do their job.

One major problem that apparently was never overcome is that x-rays produced in a nuclear weapon explosion above the atmosphere will cause increased ionization in the ionosphere which the ground-based radar waves could not penetrate, blacking them out. Another problem is that unless the destroyed warheads vaporized completely, their radioactive residue would scatter over the northern United States and Canada.

Strategic Defense Initiative

A topic of great controversy has been the Strategic Defense Initiative (SDI), sometimes called "Star Wars." President Reagan announced in 1984 that it appeared to be technologically possible to destroy ICBMs during their launch phase and that the United States would undertake a research program to investigate the feasibility of developing a defense against nuclear missiles.

Even though the ICBM threat seems to have ended, new scientific research and technological developments under SDI are worthy of discussion here. An ABM system stationed in space might be able to destroy an ICBM over enemy territory while it is still in the boost phase with engines running. Figure 5.21 shows the various stages of a missile flight launched from a land site. A submarine-launched ballistic missile or a short range missile like the Scud has a similar but much shorter flight. For an ICBM, the boost phase lasts 2 minutes, warheads separate 2 minutes later and coast for 7 minutes, reentry phase lasts only a minute.

There are numerous advantages to attacking the missile while it is still under power. First, it is easier to detect and track using infrared sensors. Second, the missile and its bus carrying the payload is only one target; later, up to 10 warheads and 100 decoys may separate from the bus, becoming a multitude of targets. Third, a rising missile under power and full of fuel is particularly vulnerable.

A space-based early warning detection system now in operation detects missile launches and gives some information on their trajectories, but it does not track the missiles with the precision required to aim and fire a weapon against

them. The task of SDI was to detect, track, and destroy hundreds of ICBMs—each carrying several warheads and decoys, each launched within a few minutes, and each having a slightly different trajectory from its launch site to its target a third of the way around the globe, that is, a mass raid on the United States.

The heart of the command-control system would have to be an automated computer which would receive information from the detection system to aim and fire the weapons. A mass raid would take place too rapidly for human decision making. A computer program would have to be written which would completely and automatically manage such a battle. It could never be tested under actual battle conditions; it would be written for a one time use and would have to be 100 percent reliable.

Some computer scientists doubt that a 100 percent reliable computer program of such complexity could ever be written. It must be noted, however, that computers are now doing things reliably that were undreamed of only 20 years ago. For example, extremely complex programs handle our charge cards and our banking; we can withdraw money from our bank account via a computer-driven machine anywhere in the country. Industrial computer programs control machines that build everything from automobiles to computer chips. The NORAD computers have never had to operate under real battle conditions; yet simulations are routinely done which give credibility to their ability to operate prop-

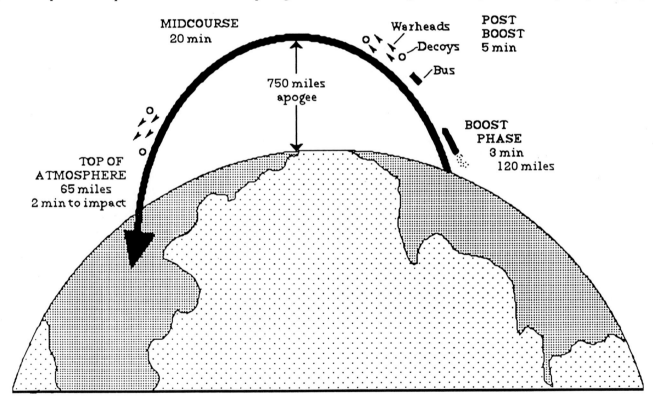

Figure 5.21 Typical ballistic missile trajectory. For about 10 minutes after booster burnout, the bus uses thrusters to aim and release warheads and decoys. On reentry into the atmosphere, shielded warheads glow with heat; lighter decoys slow and fall behind. *After Office of Technology Assessment, A. Carter.*

erly should a mass raid occur. In addition, rapid advances are being made in artificial intelligence techniques for programming computers and the construction of reliable programs is a major area of research in computer science.

So how do you shoot down an ICBM? Two types of anti-missile weapons were being considered for SDI: *directed energy weapons* and *kinetic energy weapons*. Directed energy weapons send a beam of electromagnetic waves to heat up and burn through the outer skin of the missile, while kinetic energy weapons fire projectiles into the path of the missile, destroying it by collision.

Laser Weapons

One type of directed energy weapon is a powerful laser which produces a narrow intense beam of light with a single wavelength and focuses it on the target. Lasers are now used for microsurgery, spot welding, cutting metal, and other applications. Most of them involve heating the area or shocking it with a short, high intensity pulse. To attack a missile, the laser beam must be focused on the surface long enough to burn through or at least to weaken the skin and cause it to rupture. A laser weapon has a very large advantage: the destructive beam travels at the speed of light. The nuclear weapon aboard the missile would not be triggered, but would be destroyed in the explosion.

Lasers with sufficient power to destroy a missile pour out tremendous quantities of power, tens of millions of watts, and require very large power supplies. At least two concepts were being considered. First, the laser and its power supply could be stationed in orbit where it would shoot at rising missiles. An orbiting laser weapon would be about 80 feet long and weigh 50 tons with its large power supply. Second, the laser could remain on the ground with mirrors in orbit to reflect and focus the laser beam. This concept saves placing the power supply in orbit, but requires several orbiting mirrors. One mirror would be positioned at geosynchronous altitude, stationary like a communications satellite, and other mirrors would be in low orbit to redirect the laser beam to the targets. The laser could be located on United States territory and the constellation of mirrors could reflect the energy to any spot on Earth.

Particle Beams

Another type of directed energy weapon uses a particle beam accelerator. Electrically charged negative hydrogen ions, that is, hydrogen atoms with an extra electron, are accelerated by injecting them into the field of an electromagnet. Because charged particles in motion also have a magnetic field, the two fields interact which causes them to accelerate. Other magnets focus and steer the beam in the desired direction.

The electron gun in a television picture tube works the same way. Electrons are accelerated through the neck of the tube by an electromagnet wrapped around it. When the electrons strike the fluorescent coating on the face of the tube, they cause it to glow. Varying the electrical current in the electromagnet deflects the beam back and forth and up and down the face of the tube, creating the picture.

A particle accelerator in space will not hit a missile by firing charged particles directly at it, however. The particle beam would be deflected from its intended path by Earth's magnetic field. Recall from Chapter 4 how charged particles in the solar wind are affected by the Earth's magnetic field. Therefore, the particles must be neutralized by removing the extra electron from each hydrogen ion. This is done by passing the beam through a thin gas. The electrons are stripped off by the collisions with the gas atoms. The beam then consists of neutral hydrogen atoms.

The beam diverges and loses its intensity if the particles collide with other atoms such as atmospheric gases. Therefore it is effective only when the target missile is well above the atmosphere. When a high speed hydrogen atom hits the skin of the missile, it splits into a proton and an electron. The proton penetrates deeply into the metal and does the damage. Electronics would be disrupted by only a small dose; very heavy doses would be needed to melt the aluminum skin.

Kinetic Energy Weapons

Exotic laser weapons get a lot of attention, but as with any new technology, they would require a great deal of research and development before they become practical. Kinetic energy weapons, on the other hand, are "old fashioned rifle bullets" that damage a missile by hitting it. There are two general types of kinetic energy weapons: those which accelerate the interceptor "bullets" with rockets and those which use electromagnetic devices to fire the projectiles.

The first type operates rather like a retrorocket by taking the projectile out of a higher storage orbit onto a collision course with the rising missile or the warheads on their ballistic path. The interceptor carries sensors so it can home in on its target and may be able to differentiate between warheads and decoy. It does not carry any explosive; it simply collides with the target (see Figure 5.22). Rockets have been miniaturized so an orbiting platform could carry a large number of such interceptors.

One type of electromagnetic accelerator is the rail gun. It has two parallel rails carrying a high electric current. When a piece of metal shorts across the rails, the high current flowing through it causes it to suddenly evaporate to a high pressure gas which accelerates a projectile out the barrel. Velocities of over 50,000 miles per hour may be achievable in space. A barrage of small plastic projectiles fired into the path of a missile could destroy it by collision. Rail guns are a difficult technology. Problems with handling the high current and with damage to the rails and gun barrel during rapid repeated firing would have to be overcome.

Figure 5.22 Artist's concept of a kinetic energy weapon in space. A number of interceptor rockets are carried in the platform at the center. The spacecraft at the right is a radar which tracks the rising missile and directs the interceptor on a collision course. *Courtesy of Lockheed; artist Louis De La Torre.*

Countermeasures

Whenever a new military system is developed, opponents immediately try to find ways to counter it. Several countermeasures against a directed energy weapon have been suggested. The attacking beam of energy must dwell on the missile outer shell for a period of time to weaken the metal sufficiently to cause it to rupture. Spinning the missile as it rises would spread the energy of the weapon over the missile so it would not dwell on one spot. Since the beam width is similar in size to the missile, however, spinning is probably ineffective as a countermeasure.

A shiny or glossy coating on the missile may reflect rather than absorb the energy. However, a shiny surface which we usually think of as reflecting light may not reflect x-rays or infrared delivered by a laser weapon. Perhaps a coating could be found which does.

A skirt over the rocket engine's nozzle to hide the heat of the flame from the infrared detectors may prevent the tracking of the rising missile. However, the hot exhaust plume is larger than it appears and the added weight of a large skirt would reduce the effective payload of the missile.

Because the missile is best attacked when it is in the boost phase before the warheads and decoys separate from the bus, rapidly burning boosters would be an effective countermeasure. They would be more difficult to attack if they burned out in, say, 90 seconds, before they left the dense part of the atmosphere. However, replacing an entire fleet of missiles with fast burn boosters would be extremely expensive.

Antisatellite Weapons

With the high value placed on observation, detection, and communications systems in space, it is not surprising that both superpowers developed means of disabling an opponent's spacecraft. Compared to missile defense, it is relatively easy to "shoot down" a satellite. Once its orbit is determined, its exact position in space can be predicted with great accuracy, unless it is maneuvered into another orbit.

For many years the Soviet Union had the capability of disabling or destroying a satellite in orbit. A rocket interceptor was designed to rendezvous with the target spacecraft and, using a high explosive warhead, blast it with pellets. If rendezvous is accomplished during the interceptor's first orbit, the target spacecraft does not have much warning time to make an evasive maneuver. The Soviets also had a powerful ground-based laser which may have been capable of damaging optical sensors and electronics on orbiting satellites.

The United States tested a satellite interceptor. A two stage rocket was carried to high altitude by an F-15 airplane and launched to rendezvous with the target spacecraft. It carried a homing device and destroyed the satellite by collision. Congress forbade any further tests.

Conclusion

The problems in developing a space-based antimissile system are enormous, as with any other new undertaking. The objective of SDI has been to complete enough research and development to determine the feasibility and possible architecture of a space defense system.

Cost estimates of such a system run into the trillions of dollars, depending on whom you talk to and whether you are talking about development or deployment, space-based or ground-based, conventional or nuclear, directed energy or kinetic energy weapons. Even if each of the various components prove to be technologically feasible, many notable scientists doubt that a reliable working system could be built at any price. A system that would guarantee protection of the entire United States against all attacking missiles, they argue, would be entirely too complex and could not be realistically tested to see if it worked. And, they add, anything less than complete protection would not be justifiable. However, scientists are not invariably correct. Such criticisms have frequently been voiced against just about every new state-of-the-art undertaking. As late as the 1920s one prominent scientist said that a rocket ship would not work in space because there was no air for the exhaust to push against. A few years ago, a friend of Dr. Edward Teller, father of the hydrogen bomb, said, "When Edward says something can be done, it can be done. When he says something cannot be done, he is frequently wrong."

The dissolution of the Soviet Union has reduced the danger of a massive exchange of nuclear-tipped intercontinental ballistic missiles. Deployment of an SDI type of system has become a moot question as the two major world powers dismantle their stockpile of those weapons. That does not mean that a defense against such missiles will never be needed. Several other countries are developing both nuclear weapons and short-to-medium range rockets. Iraq's Scud missiles fired at Israel and Saudi Arabia during the Gulf War are a case in point. Their launches were detected by early warning satellites. In fact, in 1992, Russia and the United States,

formerly sworn enemies, jointly announced a concept for a global missile defense including a joint early warning center. If a danger to the world arises in the future, the research and development done by SDI will be useful in building a defense against those missiles.

DISCUSSION QUESTIONS

1. What has been the social impact of international television broadcasts in less developed countries and countries with closed borders?

2. Name possible users of a precision positioning system other than those mentioned in this chapter.

3. If you could build your own satellite, what would you use it for?

4. Should the United States continue to develop a defense against nuclear missiles?

5. What factors might be considered in determining whether or not to deploy an antimissile weapons system in orbit?

ADDITIONAL READING

Caprara, Giovanni. *The Complete Encyclopedia of Space Satellites*. Portland House, 1986. Describes over 1,000 civil and military satellites of all nations launched or planned as of 1986.

Garwin, Richard L., et al. "Antisatellite Weapons." *Scientific American*, June 1984. Argues against them.

Hansen, James R. "The Big Balloon." *Air & Space, Smithsonian*, April/May 1994. History of Project Echo.

Hudson, Heather E. *Communications Satellites, Their Development and Impact*. The Free Press, 1990. Political, economic, military, cultural, and social impact of communications satellites.

Kaplan, Marshall H. *Modern Spacecraft Dynamics and Control*. John Wiley and Sons, 1976. Calculus level math.

King-Hele, Desmond. *Observing Earth Satellites*. Van Nostrand Reinhold, 1983. Do-it-yourself visual observing of satellites.

Logsdon, Tom. *The Navstar Global Positioning System*. Van Nostrand Reinhold, 1992. Comprehensive description of GPS and its applications for both novices and professionals.

Patel, C., and Nicolaas Bloembergen. "Strategic Defense and Directed Energy Weapons." *Scientific American*, September 1987. Findings from American Physical Society study.

Potter, Seth D. *Low Mass Solar Power Satellites Built from Lunar or Terrestrial Materials: Final Report*. Space Studies Institute, 1994. Research report on use of thin-film materials for photovoltaic solar power satellites.

Strategic Defense Initiative Organization. *Report to the*

Congress on the Strategic Defense Initiative. Government Printing Office, April 1987. Summary of SDI concepts and research.

U.S. Congress, Office of Technology Assessment, *Ballistic Missile Defense Technologies*, Government Printing Office, September 1985. Nonjudgmental survey of SDI.

Williamson, Mark. *Dictionary of Space Technology.* Adam Hilger, 1990. Comprehensive.

Wilson, Andrew, ed. *Interavia Space Directory, 94–95.* Jane's Information Group Inc., 1994. Excellent reference on space programs and aerospace industry of all nations, and more; updated regularly.

PERIODICALS

Air & Space/Smithsonian. P.O. Box 53261, Boulder, CO 80322-3261. Bimonthly magazine with two or three articles about space in each issue.

Aerospace America. American Institute of Aeronautics and Astronautics, 370 L'Enfant Promenade, S.W., Washington, DC 20024. The monthly popular journal of the AIAA.

Aviation Week & Space Technology. P.O. Box 1505, Neptune, NJ 07754-1505. Weekly magazine, current events.

GPS World. P.O. Box 7677, Riverton, NJ 08077-9177. Monthly magazine covering news and applications of GPS.

Space News. 6883 Commercial Drive, Springfield, VA 22158-5803. Weekly newspaper covering space business, policy, and some technology.

SSI Update. Space Studies Institute, P.O. Box 82, Princeton, NJ 08542. Bimonthly newsletter devoted mostly to space manufacturing, lunar mining, power satellites, and space colonization.

Via Satellite. 7811 Montrose Road, Rockville, MD 20850. Devoted to communications satellites.

NOTES

Chapter 6

Remote Sensing

When you look across the room at an object you are "sensing remotely." Your eyes are sensors, sensitive to light waves, and your brain interprets their message. When you use a camera to make a photograph you are also sensing remotely, using a machine to sense the object and make a permanent record on the light sensitive film. The camera becomes an extension of your eyes.

Hearing is also remote sensing. Ears and their associated anatomy are sensitive to sound waves. Microphones, amplifiers, and speakers are machines which extend the capabilities of the human ear.

Even human touch sensors can operate remotely. Consider that you can feel a hot stove without actually touching it. The touch sensors are sensitive to the infrared heat waves emitted by the hot object.

Remote sensing, then, is learning something about an object at a distance, without coming into physical contact with it.

Remote Sensing from Space

Many spacecraft are remote sensors, acting as extensions of human senses. They look up into the universe and down to Earth. One of the earliest and most practical applications of remote sensing has been for weather observing using both visible and infrared waves. Mapping and cataloging the resources of Earth are another application of remote sensing with widespread benefits. Military early warning satellites detect launches of missiles and spacecraft by means of infrared sensors which respond to the heat of the rocket exhaust. Spy-in-the-sky satellites looking and listening from their orbits keep each side informed in remarkable detail what the other side is up to. The news media use satellite images to cover news events from space. A satellite may soon carry a sensor designed expressly for the media, capable of seeing things as small as a truck, in full color, available within hours, with global coverage.

The science of astronomy is based on remote sensing, first using earthbound telescopes, now using spacecraft carrying sensors above the atmosphere for a better view. They observe the skies not only with sensors of visible light but also radio waves, infrared, ultraviolet, x-rays, and gamma rays. Many of these emanations from stars and galaxies are not visible at earthbound observatories because they are absorbed in the atmosphere before they reach the ground.

This chapter will cover the principles of remote sensing and their applications to Earth observing. The next chapter will deal with applications of remote sensing to astronomy.

Electromagnetic Waves

Radio, infrared, visible light, ultraviolet, x-rays, and gamma rays are all electromagnetic waves, differing only in their wavelengths. The electromagnetic spectrum was described in Chapter 4 and Figure 4.3. The length of a light wave, measured from crest to crest or from trough to trough, is 0.4 to 0.7 micrometer. A micrometer, also called a micron, is one-millionth of a meter. Therefore, visible light waves are very short electromagnetic waves. Shorter still are the ultraviolet, x-rays, and gamma rays, ranging down to a trillionth of a micrometer. On the other hand, infrared, microwaves, television, and radio waves are all longer than visible light, ranging up to many miles for AM radio.

Anything which has a temperature above absolute zero emits electromagnetic radiation. An object of high temperature radiates shorter waves; an object at a lower temperature emits radiation of longer wavelengths. The photosphere of the sun at 11,000 °F emits most of its radiant energy in the visible light portion of the spectrum. On the other hand, a human body at 98.6 °F emits longer infrared waves. When you turn on the burner of an electric stove you can feel the heat, that is, the infrared waves, before you see the glow. When the temperature goes higher, the burner becomes red hot and emits visible light as well as infrared. Since almost everything emits some amount of infrared, it is a favored region for remote sensing. The infrared portion of the electromagnetic spectrum is sandwiched between visible light and radio waves, with wavelengths from 0.8 micrometer to about 1 millimeter (1/25 inch). MATHBOX 6.1 shows how to calculate the wavelength at which an object radiates the

MATHBOX 6.1

Wavelength of Maximum Emission

In the late 1800s Wilhelm Wien found a simple relationship between the temperature of a body and the wavelength at which it emits the maximum amount of electromagnetic radiation:

$$L = \frac{2900}{T}$$

where T is the temperature in Kelvins (degrees above absolute zero) and L is the wavelength in micrometers. (To find the temperature in Kelvins, just add 273 to the Celsius temperature.)

Wien's law says that the higher the temperature of an object, the shorter the wavelength of maximum emission. Very hot objects emit ultraviolet; cold objects emit radio waves.

For example, the Sun's surface temperature is about 5,700 K. So

$$L = \frac{2900}{5700} = 0.51 \text{ micrometer}$$

in the middle of the visible light part of the spectrum.

A human body is at 98.6 °F which is 37 °C or 310 K. It emits radiation at

$$L = \frac{2900}{310} = 9.4 \text{ micrometers}$$

in the near infrared. People glow at that wavelength.

maximum amount of energy when you know its temperature.

Objects also reflect electromagnetic waves which fall on them from other sources. A green leaf reflects green light; a red flower reflects red light. Most objects also reflect other wavelengths of the electromagnetic spectrum—ultraviolet, infrared, for example—but human eyes are sensitive only to visible light so we cannot see these other wavelengths.

We are accustomed to viewing the world around us by the light emitted and reflected by objects surrounding us. See Figure 6.1. But that is not all there is. Using imaging devices which are sensitive to other wavelengths we can get quite a different view of the world. Doctors use x-rays striking a piece of film to create an image of the inside of the body. X-rays pass through a body but are absorbed variably by tissue and bone on the way. When they strike a piece of film they create a shadowgram of the interior. Figure 6.2 shows an x-ray image of a broken leg.

Film sensitive to infrared is available at camera stores. Photographs made on infrared film look quite different from those made on regular film. With proper equipment it is possible to create an image of an object solely by the infrared heat waves that it gives off. For cooler objects which emit longer infrared waves, a mirror imaging system is used instead of a lens because the longer wavelengths are absorbed by glass lenses and do not reach the film.

Landsat

The exploration of space has led to an unprecedented and remarkable new exploration and understanding of Earth. For more than 15 years, a series of Landsat satellites have been recording pictures of the Earth's surface. Landsat 1 was launched in 1972. The newest spacecraft, Landsat 5, is shown in Figure 6.3 being checked out prior to launch. Figure 6.4 is an artist's rendition of how it appears in space.

Landsat's orbit is shown in Figure 6.5. It is *Sun-synchronous* which means that it reaches the equator at the same local time on every pass, approximately 9:37 a.m. It makes images only on the north-to-south part of the orbit. The south-to-north part on the other side of the Earth is in darkness.

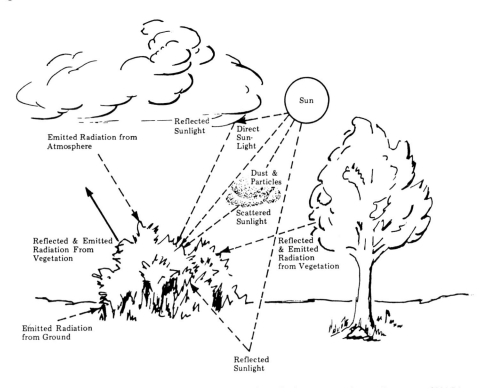

Figure 6.1 Emitted and reflected electromagnetic radiation surrounds us. *Courtesy of NASA.*

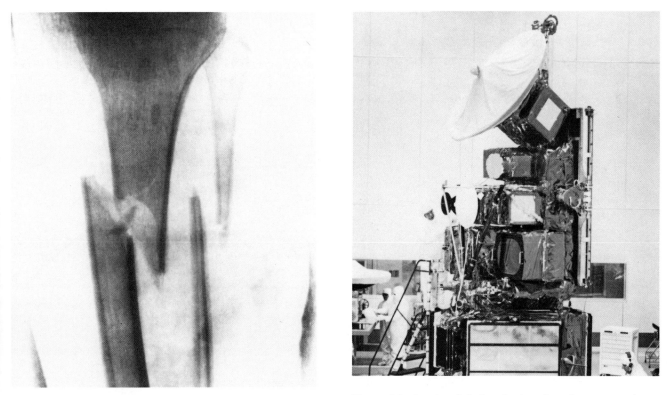

Figure 6.2 X-ray photo of a broken leg. Both the large bone (tibia) and the small bone (fibula) are broken through. The knee joint can be seen at the top. *Courtesy of C. Hudson.*

Figure 6.3 Landsat 5 during checkout in a clean room prior to launch. The communications antenna is covered with a shroud, and the solar cell panels are folded against the right side of the spacecraft. *Courtesy of EOSAT.*

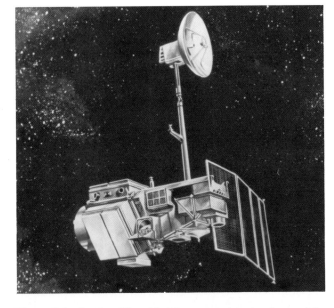

Figure 6.4 Landsat 5 in orbit. The antenna is pointed toward the TDRS communications satellite for relay of the data to receiving stations on the ground. Solar cell arrays are extended for power. *Courtesy of EOSAT; artist P. A. Bertolino.*

A rotating mirror produces east-west scan lines covering a swath 115 miles (185 kilometers) wide as shown in Figure 6.6, making a continuous strip picture. Electromagnetic waves reflected from the mirror pass through a device which divides them into the six different *spectral wavelength bands* shown in Table 6.1. The six beams then pass into six detectors which measure their intensities. A seventh detector gives a broader heat measurement.

Scan lines are broken into short bits called picture elements or *pixels*, a 100 by 100 foot square, approximately a quarter of an acre, on the surface of Earth. This 100 foot

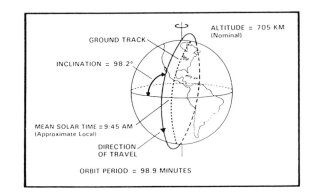

Figure 6.5 Orbit of Landsat 4. *Courtesy of EOSAT.*

Scanning Arrangement

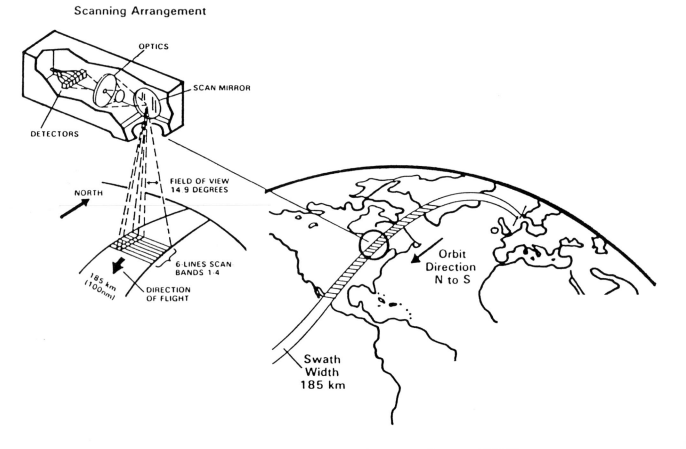

Figure 6.6 Scanning geometry of Landsat sensors. *Courtesy of EOSAT.*

TABLE 6.1 Spectral Bands of Landsat Sensors

Band	Spectral Range	Color	Principal Applications
1	0.45–0.52 micron	Blue-Green	Coastal water mapping. Soil/vegetation differentiation. Deciduous/conifer tree differentiation.
2	0.52–0.60 micron	Green-Yellow	Reflected by healthy vegetation.
3	0.63–0.69 micron	Red	Chlorophyl absorption for plant species differentiation.
4	0.76–0.90 micron	IR	Biomass surveys. Water body delineation.
5	1.55–1.75 microns	IR	Vegetation moisture measurement. Snow/cloud differentiation.
6	10.4–12.5 microns	IR	Plant heat stress measurement. Thermal mapping.
7	2.08–2.35 microns	IR	Hydrothermal mapping.

(Note: A micron is a micrometer, one-millionth of a meter.)

resolution is less than one-third the area of a football playing field. Each pixel is represented by six numbers which tell the intensity of the waves in each spectral band, reflected and emitted from the pixel. The satellite then transmits the stream of numbers to a ground station below.

Figure 6.7 shows the strips covered by three consecutive orbits. Because of the rotation of the Earth, different areas are covered on each orbit with gaps between them. On the following days the strips will fall in between until after 16 days the pattern begins to repeat itself (Figure 6.8). Notice that on any pass, the strip covers an area just west of the strip produced one week earlier, and the next strip west will be covered one week later. Except for small regions around the poles, the entire Earth is completely photographed every 16 days.

Landsat 4 has been in orbit since 1982, Landsat 5 since 1984. Both were designed for 3 year lifetimes, but were still working at the end of 1993. Landsat 5 remains healthy with enough fuel for maintaining its orbit and attitude control until the turn of the century. A new Landsat 6 was launched

from Vandenberg AFB, California, on October 5, 1993, but never made it to orbit. Apparently the apogee kick motor, which was supposed to circularize the orbit, failed and the satellite fell back to Earth. It carried a 50-foot resolution grey-scale sensor as well as the sensors carried by Landsat 5. Meanwhile, Congress cancelled funding for the follow-on satellite, Landsat 7.

Satellite Pour l'Observation de la Terre (SPOT)

Similar to Landsat, the French Earth observation satellites, SPOT, have been operating since 1986. The current spacecraft, SPOT 3, was launched on September 25, 1993, into a Sun-synchronous orbit with an inclination of 98.7 degrees at an altitude of 516 miles, crossing the equator from north to south at about 10:30 a.m. local time. Segments of orbits of SPOT 3 and Landsat 5 are shown in Figure 6.9. They are almost identical, but cross the equator at different times with respect to the Sun.

SPOT pictures have a higher resolution than Landsat. A pixel is 33 feet square for grey-scale images and 66 feet square for multispectral images. SPOT images a swath 73 miles wide so it covers the entire Earth in 26 days as compared to Landsat's 16 days. However, a moveable mirror allows the satellite to create images up to 27 degrees either side of the point directly beneath it. This capability permits it to view a selected area on 7 successive passes in equatorial regions or on 11 successive passes at midlatitudes.

SPOT carries only three sensors for multispectral imaging: the green band from 0.50 to 0.59 micrometer; the red band from 0.61 to 0.68 micrometer; and a near infrared band from 0.79 to 0.89 micrometer. It also carries a single sensor that covers almost the entire visible range 0.51 to 0.73 micrometer for highest resolution in a grey-scale (panchromatic) image.

Russian Remote Sensing Images

The Russians have just begun releasing high resolution images taken from Cosmos satellites. Originally made for military purposes, the once super-secret grey-scale images have a resolution of just 6 feet, the best resolution of any photos available to the public. The image archive dates back to

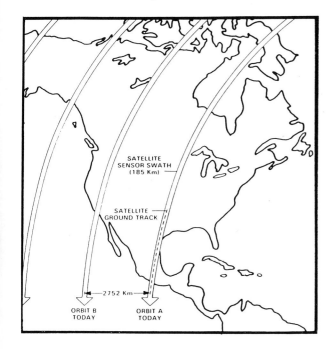

Figure 6.7 Landsat coverage on each orbital pass. *Courtesy of EOSAT.*

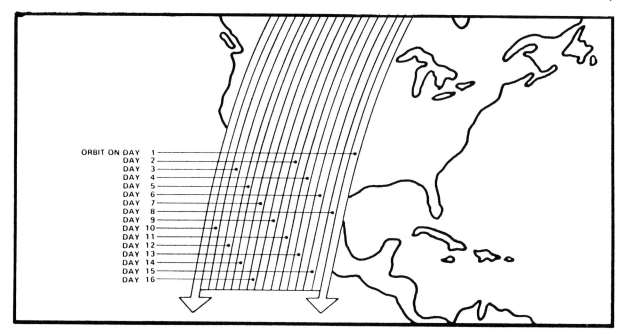

ORBIT ON DAY 1
DAY 2
DAY 3
DAY 4
DAY 5
DAY 6
DAY 7
DAY 8
DAY 9
DAY 10
DAY 11
DAY 12
DAY 13
DAY 14
DAY 15
DAY 16

Figure 6.8 Landsat 4 and 5 coverage pattern. The pattern repeats itself over the entire Earth every 16 days. *Courtesy of EOSAT.*

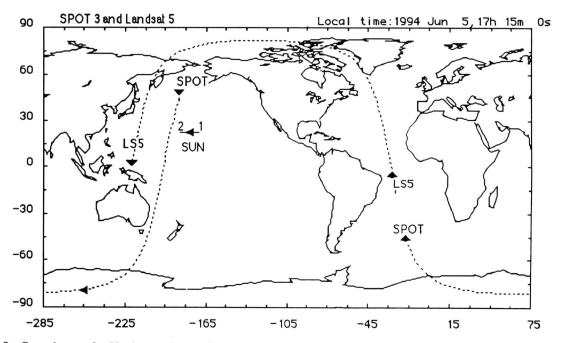

Figure 6.9 Ground traces for 55 minutes of travel for SPOT 3 and Landsat 5 from 4:20 to 5:15 p.m. mountain daylight time on June 5, 1994. At the start time SPOT is over the Bering Sea, Landsat is off the east coast of Brazil, and the Sun is over the Pacific at the position marked 1. SPOT crosses the equator in the Pacific at about 10:30 a.m. before the Sun is overhead at noon. Landsat reaches the Pacific equator later when the Sun is at position 2, but at 9:37 a.m., farther ahead of the Sun than SPOT. At the end of the time period, SPOT is approaching the point in the Atlantic where Landsat began SPOT's next orbit will trace almost precisely over Landsat's previous one.

1984. Stereo images were also made at a resolution of 33 feet.

Image Processing

The streams of numbers transmitted to Earth from both Landsat and SPOT are processed by a computer into grey shades or colored images showing a 115 by 105 mile section of the Earth's surface made up of 35 million pixels. An infrared image of an object is usually reproduced in false colors. Since our eyes are not sensitive to infrared, the "color" of infrared waves has no meaning for us. So the technician who creates the image arbitrarily assigns visible colors to the different infrared wavelengths and prints the picture that way. The first infrared images were printed using various shades of grey to represent the various wavelengths of infrared. However, it is easier to view and analyze an image with colors than it is with grey shades. Therefore, false color images have become popular, not only for infrared images, but for images made in other wavelength bands as well.

Figure 6.10 in the color section, page 115, is a Landsat picture of Copenhagen, Denmark, in nearly natural colors. Compare this with the false color images of Figures 6.11, 6.12, and 6.13 in the color section, pages 116, 117, and 118. Colors can be adjusted to emphasize any particular land use: urban, cropland, vegetation, wetland, sediment, etc.

The data can be manipulated in a number of different ways. Unique patterns of light intensity across the seven bands can be used to identify features on the Earth which may not be apparent to the human eye. See Figure 6.14. For example, a field of soybeans, once its spectral intensity pattern has been identified, can be located on a satellite image. The light intensity levels of the various spectral bands of the soybean field are as unique as a person's handwriting. Other areas on the same image that have the same *signature* can be interpreted as being the same type of soybean fields. Finding the right combination of spectral intensities for a

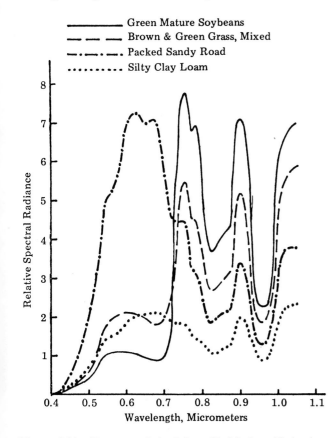

Figure 6.14 Signatures derived from Skylab data. Notice how each of the four different materials emits and reflects light and infrared differently. Although the soybeans and the grass have similar patterns, the intensities are different. The signatures for sand and loam are completely different. *Courtesy of EOSAT.*

soybean field in an image is done by computer evaluation of the pixel data.

Fine details can be recognized in these spectral signatures, even to determining the health of crops. Healthy vegetation looks green because the chlorophyll absorbs blue and red wavelengths, Landsat bands 1 and 3, and strongly reflects green, band 2. Water reflects very little visible light and absorbs infrared, so Landsat bands 5 and 7 are indicative of water content. Thus crops suffering from a drought have a different signature from normal healthy crops. Using this data, estimates of crop production can be quickly assessed. Sediment and pollution in rivers give a different signature from clear water. Trees and lawns in a residential area, asphalt and cement areas, different kinds of exposed rocks, meadows, and fields—all can be recognized.

To accomplish this kind of mapping of Earth resources by the older method of aerial photography would require 100 photographs and months or years to assemble and analyze a 100 mile square scene. Landsat or SPOT can cover the same area in a single picture and the computer analysis can be done in a few hours. The timeliness of the data makes it especially useful in agricultural and environmental studies.

Radar Mapping

Radar transmitters and receivers were carried aboard several Shuttle flights to make images of selected locations on Earth. One spectacular result was the discovery of dry river channels buried under the sands of the Sahara Desert. Radar pulses can penetrate dry soil, but reflect from water droplets in wet ground; in effect, they can see through the sand. Radar waves reflect differently from various surfaces. Interfaces between dissimilar kinds of rock, diverse types of vegetation, sea ice, and paved surfaces, for example, all produce different radar images. The travel time of the signal is measured so the distance to objects in the picture can be mapped. Also, because the antenna points at an angle to the ground, vertical objects have a three-dimensional look in the picture. Anthropologists have recommended further use of radar imaging to search for ancient cities buried under sand and debris.

Images created by a radar operating on only one frequency are like photographs on black and white film. In April 1994, Space Shuttle *Endeavour* carried imaging radar which operated on two frequencies. Recall our discussion above about signatures obtainable from the multiband Landsat or SPOT data. Combining the two radar frequencies into one image gives more information than either frequency by itself.

Canada will launch its Radarsat in 1995 with a pledge to provide data to commercial customers within 4 hours of acquisition. A primary customer is expected to be the shipping industry. Its service will be uninterrupted because, unlike visual sensors, radar can map sea ice through clouds and at night. Its images should be most useful to ships plying the Arctic.

The Magellan spacecaft completely mapped Venus by radar. With its perpetual overcast of heavy clouds, the surface features are not visible by any other means. The details are in Chapter 7.

Military Reconnaissance

Spy satellites operate somewhat like the Landsat and SPOT satellites, but with much higher resolution. The capabilities of intelligence-gathering satellites are, of course, highly classified. However, rumors circulate about spy satellites with sufficient resolution to read licence plates and identify people from a 100 mile orbit. Current U.S. photoreconnaissance spacecraft reportedly have a resolution of about 4 inches.

The camera on one U.S. satellite, called Big Bird, photographs the area of interest, the film is processed like a Polaroid snapshot, and the picture is transmitted to the ground. If still higher resolution is needed, Big Bird can eject a capsule containing the film itself. The capsule deorbits like a miniature Apollo spacecraft. An airplane flies out to the landing point and catches its parachute in midair before it reaches the water. Using a different approach, the Russians recover their entire satellite and its load of film when a mission is completed.

Film returned to Earth yields higher resolution images than pictures transmitted to the ground by radio. The difficulty in transmitting very high resolution images is the quantity of data that must be sent. A picture covering only 1 mile square with 1 foot resolution requires nearly 3 million pixels (and a very high pointing accuracy to photograph the area desired). Saving images from a large area or many smaller areas for subsequent transmission to a ground station may strain the storage capability of the spacecraft. With greater storage capacity in computer memory chips, newer reconnaissance satellites will undoubtedly overcome these problems and will be able to transmit images as fast as they are taken, with higher resolution, in several spectral bands, and perhaps with radar to make images through clouds.

For best resolution it is necessary, of course, to operate as low as possible, down to 90 miles, which means rapid decay of the orbit due to atmospheric drag. Therefore, spy satellites are equipped with thrusters and fuel to periodically boost them back up to higher altitudes. Big Bird has a lifetime of 6 to 7 months; Russian satellites operate at lower altitudes and remain in orbit for about 2 weeks. That explains why the United States launches about three per year while the Russians usually launch more than 30. Newer Russian photoreconnaissance satellites are highly maneuverable, changing orbit frequently, dropping to a low perigee

over a place of interest, taking their photos, then returning to a higher orbit before atmospheric friction does them in. They have also been seen to change apogee which changes the local time over the target for better lighting for their photography.

Electronic intelligence satellites have listening devices which tune in on telephone conversations, radio messages, radar signals, and data transmissions during missile launches. The United States keeps several such satellites in low orbit and at least two in geosynchronous orbit. The Russians use a constellation of satellites in six different orbital planes, 400 miles high, at about 82 degrees inclination.

Ocean surveillance is another major effort by intelligence agencies. Watching the activities of naval surface fleets is relatively easy by radar and ELINT satellites. Keeping track of nuclear submarines with their load of ballistic missiles is more difficult, but several factors make it possible. Although the submarines remain under water for weeks at a time, when they are under power they leave a trail of hot water which may be followed by infrared sensors. They also leave a small wake which may be detectable by radar or by the glint of sunlight from the surface.

Cosmonauts in space stations probably have had intelligence-gathering duties, also. Salyuts carried telescope-cameras with 33 foot focal length lenses. While solar observing was their announced purpose, such a camera in low orbit is more useful for Earth observing.

Whatever their capabilities, governments have become highly dependent on the information satellites gather. Just knowing that others are probably monitoring your every move is a stabilizing influence. It is unlikely that any nation could mount large scale military activities without detection. The fact that other sides are watching and listening is a strong deterrent against any actions which may appear offensive.

Early Warning

Several U.S. early warning satellites located in geosynchronous orbit were designed to watch the Soviet ICBM launch complexes, their space launch cosmodromes, and the Atlantic and Pacific Ocean areas from which submarines could launch missiles. These early warning satellites carry infrared sensors which respond instantaneously to the enormous quantity of heat emitted when a rocket engine is burning (Figure 6.15). Within a few seconds the detection is reported to the North American Aerospace Defense Command (NORAD) inside its Cheyenne Mountain Complex near Colorado Springs (Figure 6.16). Computer analysis and human interpretation of the intensity, duration, and location of the infrared signal make it possible to determine what kind of launch is taking place.

These satellites may seem to be of little use now that Rus-

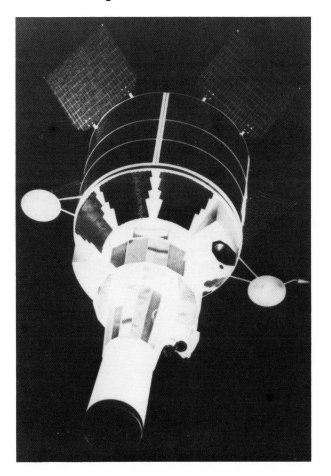

Figure 6.15 Early warning satellite for detection of missile and space launches. Several of these satellites are parked at strategic locations in geostationary orbit. The telescope is pointed slightly off axis so that as it rotates it scans a wide area with its infrared sensors. *Courtesy of U.S. Space Command.*

Figure 6.16 NORAD Command Post inside Cheyenne Mountain near Colorado Springs. The Space Surveillance Center and Missile Warning Center receive data from satellites and ground sensors 24 hours a day to keep track of every object launched into space by any country worldwide. *Courtesy of U.S. Space Command.*

sia and the United States are disarming and disposing of their nuclear weapons arsenals. However, they have proven their value in smaller regional conflicts such as the war with Iraq. The Iraqi Scud missiles were observed igniting and were tracked long enough to give a short advance warning of their probable targets. The satellites operated near the limit of their capabilities, because they were designed to detect much larger rocket boosters.

The infrared sensors on early warning satellites also respond to the heat from nuclear explosions worldwide for the purpose of monitoring treaty compliance and for damage assessment.

Weather Satellites

Tiros, the first true weather satellite, was launched on April 1, 1960, and carried television cameras that viewed Earth's cloud cover by visible and infrared light from a 450 mile orbit. A total of 10 Tiros satellites were launched over a 5 year period. Tiros 8 carried automatic picture transmission (APT) equipment so that anyone with a simple receiver and a facsimile recorder could receive the pictures. Nimbus was a second-generation weather satellite with improved cameras, an APT system, and an infrared sensor that allowed pictures to be made at night. Seven Nimbus satellites were put into orbit.

At least three bands of the visible and infrared spectrum are aboard most of the weather satellites: (1) the visible band from 0.4 to 0.7 micrometer, (2) a band in the range from 5 to 7.5 micrometers which is sensitive to water vapor, and (3) a thermal infrared band in the 8 to 15 micrometer range which indicates temperature. Visual pictures of clouds, storms, and weather fronts are interesting and useful, but the infrared bands show the water vapor in the air and the temperatures of clouds. From this, their altitudes can be estimated, most important information for the weather forecaster.

Weather satellites were placed in geosynchronous orbit beginning in 1974 to transmit images almost continuously of selected areas. Earthbound meteorologists have a near real-time view of the broad weather patterns as they emerge and change. The satellite can also provide an enlarged image of a small area in which storms may be developing. A series of several dozen pictures taken over a period of several hours can be assembled into a movie showing the speeded-up motion of the clouds, a display often shown on television weather programs.

Because the National Weather Service geosynchronous satellites are designed and read out with civilian needs in mind, the Defense Department has its own dedicated weather satellites. Figures 6.17 and 6.18 are visual images from such a military satellite. The spacecraft fly in polar orbits at lower altitude for greater resolution and can observe the

Figure 6.17 Defense Meteorological Satellite Program (DMSP) picture of central Europe. France is in the center, Spain is at the lower left; Great Britain is under the band of clouds at the left. Snow can be seen in the Pyrenees and the Alps as well as in the mountains of Norway at the top of the picture. *Courtesy of U.S. Space Command.*

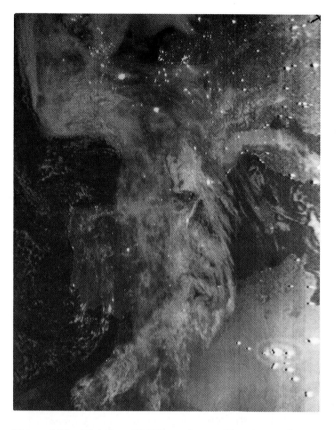

Figure 6.18 Nighttime DMSP weather satellite picture of western Europe and North Africa. Clouds are illuminated by moonlight. Bright spots are city lights. Eastern Spain is covered by clouds, western Spain is clear. Madrid is at the edge of the clouds. The two brightest spots at the upper center are London and Paris. The "cities" in the desert of north Africa are oil well fires (bottom right). *Courtesy of U.S. Space Command.*

weather anywhere on Earth. Pictures may be read out immediately as the satellite passes overhead, or the images may be recorded and played back when the satellite is in sight of a ground station. Readout equipment in mobile vans can be quickly moved by aircraft to any point where military action requires weather information.

It must be emphasized that the satellites are observing tools only. They do not make forecasts, but they give the human forecaster a much broader, more complete, and more timely view of weather systems. Newly developing storms can be observed without delay and their motion monitored. Before the advent of satellites, meteorologists had to rely on ground-based observations from weather stations. New storms developed and old storms intensified or weakened, often between ground stations where the changing pattern was not observed. Also, information concerning the patterns of wind flow in the higher atmosphere, in particular the jet stream, came only from weather balloons sent aloft by a widely scattered network of observing sites. Data from mountainous areas, polar regions, and oceans was almost nonexistent. Weather satellites have changed all that. Now, forecasters, armed with more complete and timely informa-

tion about the state of the atmosphere, are able to make significantly more accurate forecasts.

Mission to Planet Earth

The views of Earth returned by the Apollo astronuats caught the attention of the people everywhere and placed a new perspective on our home planet. See Figure 6.19 in the color section, page 119. People began to view the Earth as a limited, closed particle, isolated from the rest of the universe by vast nothingness. The Moon, our nearest neighbor, appeared to be a desolate, forbidding place. Earth seemed to be all we had, and we had better take good care of it. The exploitation of natural resources, decline and extinction of plant and animal species, potential for global warming, depletion of ozone in the stratosphere, and pollution of water resources have increased the concern over human guardianship of the Earth. We have suddenly realized that we have the power to dramatically change the environment, not always for the better, and perhaps irreversibly. We also realize that we are woefully ignorant of how the Earth's systems work and interact: the biosphere, the atmosphere, the oceans, and the land surfaces. Especially significant is that we do not understand the changes produced by human activities.

Studies of the Earth's environment from the ground have been severely limited because only small areas could be examined for short periods of time, making it difficult to see the big picture. What is needed is comprehensive and continuous observation of everything we can measure about the Earth and its environment. Remote sensing from space provides a unique capability to examine the Earth systems individually and as a whole, to study and to better understand the Earth as a single, unified living system.

Mission to Planet Earth is the name given to an international program getting underway to measure just about everything there is to measure about Earth's environment during the next 15 years From these global observations scientists will try to understand the links between the oceans and the atmosphere, the ozone layer and the troposphere, the rain forests and global climate, greenhouse gases and global warming, Pacific ocean currents and rainfall in southern United States, and many other complex interactions among the Earth's systems.

Measurements of ocean currents, waves, and topography are particularly important in understanding climate. Heat from the Sun is most intense in the tropics where it is absorbed mostly by ocean water and then transported toward the poles by ocean currents. The warm Gulf Stream flowing from south to north in the Atlantic Ocean along the coast of North America is a prime example. Perhaps surprisingly, some regions of the ocean surface are higher in altitude than other regions. The difference is not much, about 7 feet from peaks to valleys, but it is significant in determining ocean circulation patterns. Although the circulation of ocean water is a most important factor in determining weather and climate, it not well understood and will be a research topic.

Chemical reactions between atmospheric gases and pollutants are not well understood, either. For example, we are just beginning to realize that chlorofluorocarbons from man-made sources on the ground react with ozone in the stratosphere to deplete the ozone concentration. Ozone in the stratosphere is essential to life on Earth because it absorbs ultraviolet rays from the Sun and prevents them from reaching the ground. Earth as a habitat for humans could be jeopardized if the ozone depletion continues.

Periodically, the waters of the eastern Pacific near South Amercia become abnormally warm. Named "El Nino" for the Christ Child because they usually happen near Christmas, these events cause dramatic changes in the weather, including torrential rains in South America. The effects maybe felt as far away as the southern United States. What causes the onset of El Nino and what causes its eventual demise? Perhaps these "abnormal" events should be considered "normal." Somehow they fit into the regular global ocean-atmosphere circulation pattern, but no one knows how. These and many other questions will be addressed by Mission to Planet Earth.

Earth Observing Satellites

Remote sensing satellites like SPOT and Landsat have been making their rounds for over 10 years along with the weather satellites of the United States, Russia, and the European Space Agency. Now, exploiting this experience, space agencies of 20 countries are orbiting a fleet of new Earth-observing spacecraft for Mission to Planet Earth.

The United States and Russia will, of course, continue to operate their existing collection of remote sensing satellites as well as develop new more capable sensors and spacecraft. One such instrument built for the U.S. Earth Observing Satellite (EOS) will carry 36 sensors observing in 36 different wavelength bands from 0.4 to 14.4 micrometers. They are designed to measure land, atmosphere, cloud, and ocean temperatures as well as ocean color, atmospheric water vapor, cloud tops, and ozone.

The European Remote Sensing satellite (ERS-1) was launched to orbit in July 1991. In addition to infrared sensors, it carries a radar imaging instrument which can see the ground through clouds, day or night. It monitors the ocean temperatures, waves, currents, drifting ice floes, and even oil spills. Follow-on satellites called Envisat and Metop carrying advanced radar and weather-monitoring equipment will become part of the Mission to Planet Earth. Topex/Poseidon is a joint U.S.-French spacecraft for measuring ocean currents and surface topography. Canada's Radarsat will gather global data on crops, forests, ocean ice conditions, and geological structures.

Japan will operate two new Earth-observing satellites, ADEOS and JERS. ADEOS will carry ocean wind and wave measuring instruments and a variety of atmospheric sensors, including an ozone mapping device. JERS has both radar and optical sensors. India also operates a series of remote sensing satellites observing in four bands similar to the Landsat bands. Earlier sensors had a resolution of about 125 to 225 feet; new ones will have a 65 foot resolution. China and Brazil are cooperating in building a SPOT type of Earth resources satellite.

The vast quantity of data anticipated from all these satellites in the next 15 years would be useless without some means of acquiring it, archiving it, and making it easily accessible to all researchers. Formerly, prime investigators on a project had exclusive use of data from their experiments for a year or more. Not so with Mission to Planet Earth. A data center is expected to make raw data available to users within 48 hours of collection and processed data within 96 hours. This speed of handling is unprecedented in the research community.

Not only does Mission to Planet Earth offer some hope

of discovering how the Earth works as a closed system, it also is a rare opportunity for international cooperation on an extraordinary scale for a common purpose.

DISCUSSION QUESTIONS

1. What is the value of remote sensing of the Earth's resources?

2. Could very high resolution photo reconnaissance satellites of the future be considered an invasion of your privacy?

3. Do you see any objection to the use of high resolution satellite photographs by the news media?

4. What other uses could be made of satellites other than those described so far in this book?

ADDITIONAL READING

EOSAT. *EOSAT Notes*. Earth Observing Satellite Co., 4300 Forbes Blvd., Lanham, MD 20706-9954. Quarterly publication about applications of Landsat imagery.

Hafemeister, David, et al. "The Verification of Compliance with Arms-Control Agreements." *Scientific American*, March 1985. Includes satellite photography.

NASA. *Landsat Tutorial Workbook, Basics of Satellite Remote Sensing*. Government Printing Office, 1982. Comprehensive reference book.

NASA. *Planet Earth Through the Eyes of Landsat 4*. NASA Facts NF-138, Government Printing Office, March 1983. Pamphlet, select example images in color.

Weaver, Kenneth F. "Remote Sensing: New Eyes to See the World." *National Geographic*, January 1969. Somewhat dated, but excellent fundamentals.

NOTES

Chapter 7

Astronomy from Space

Astronomers learn about the universe by remote sensing of the remotest kind, at first with naked eyes, then with telescopes, now with spacecraft. The people of bygone centuries probably knew more about the location and arrangement of the stars in the sky and the motions of the Sun, Moon, and planets than their modern descendants. Without electric lights and TV, they spent much time outside at night just looking and wondering what it was all about. We have come a long way since the ancient astrologers watched the motions of the celestial bodies with the unaided eye and tried to relate heavenly events to human affairs. As the twentieth century and the second millennium draw to a close, astronomers have incredible new tools on spacecraft above Earth. Not since Galileo pointed his primitive telescope at Jupiter has there been such a transformation in the way astronomers go about their business. Spaceflight has aided astronomers in two ways: first, by transporting sensors and men to other celestial bodies, moving up close for a better look, and second, by carrying the instruments above the "muddy" and turbulent atmosphere of Earth for a clear view in all wavelengths of the electromagnetic spectrum.

Exploring the Solar System

The greatest space spectacular of all human experience was undoubtedly the Apollo Moon landings. For the first time, men left Earth, landed on another celestial body, and returned home. Astronauts walked on the Moon, poked holes in its surface, photographed its landscape, installed instruments for measuring moonquakes, and brought back hundreds of pounds of rocks. (See Figure 7.1.) From their work we have a clearer understanding of the origin of the Moon and the Solar System. The Apollo program was summarized in Chapter 1.

We have learned far more about our Solar System by sending robotic spacecraft to distant places than we have by sending people, simply because we have sent many more robots than people. To survive in space, people require oxygen, food, water, and a secure environment, which add weight, require more space, and must meet stricter safety standards. Robots have needs, too, but not as demanding.

Pre-Apollo Lunar Exploration

Thirteen unmanned flights preceded Apollo to the Moon. Three Rangers transmitted closeup pictures of the lunar surface as they approached and crash landed. Five Surveyors went to the Moon, three soft-landed and two not-so-soft. They tested the composition of the soil and transmitted more then 17,000 pictures from the lunar surface. Apollo 12 landed so close to Surveyor 3 that the astronauts could just walk to it, look it over, and see how it had survived the trip and the lunar environment for 2 1/2 years. See Figure 7.2. Finally, five lunar orbiters made medium resolution photographs of 99 percent of the lunar surface to aid in selecting Apollo landing sites. When the orbiters finished their job, they fired retrorockets and crash landed on the moon so they would not interfere with future Apollo activities.

When the Apollo program ended, lunar exploration was in hiatus for many years while much discussion and debate ensued as to the value in returning to the Moon. (Chapter 12 will examine this question in detail.) Finally, on January 25, 1994, a remote sensing spacecraft named Clementine

Figure 7.1 An Apollo scientist-astronaut scoops up rock and soil from the surface of the Moon. *Courtesy of NASA.*

Figure 7.2   Apollo 12 Astronaut Charles Conrad examines the TV camera on Surveyor 3 which had been on the Moon more than 2 1/2 years before his arrival in the lunar lander seen in the background. *Courtesy of NASA.*

was launched toward the Moon, the first lunar mission in almost 22 years.

Clementine to the Moon

Clementine was injected into a highly elliptical polar orbit around the Moon with a near point periselene (or perilune) at about 270 miles and a far point at aposelene of 1,840 miles. (Selene is the Greek goddess of the Moon, comparable to the Roman goddess Luna.) It carried two cameras, each with CCD imaging devices similar to those found in camcorders. A strip of data was taken from the south pole to the north pole on the periselene side of the orbit when the spacecraft was closest to the surface. Then Clementine turned toward Earth as it approached aposelene and the data was transmitted. Because of the elliptical orbit, pixel size varies from 20 to 30 feet. Data was taken at each pixel through 11 filters to make images of the Moon in 11 different "colors" in the visible light and infrared wavelength bands. The spacecraft also carried a Laser Imaging Detection and Ranging (LIDAR) system which is similar to radar except that it operates with laser light instead of microwaves and yields much higher resolution measurements of the heights of lunar surface features.

Clementine sent back 25,000 pictures per day during more than 2 months in lunar orbit, a total of more than 1.5 million images covering the entire 15 million square mile lunar surface. Like Landsat and SPOT, these multispectral images can be analyzed by the signature at each pixel. Data through the 11 filters at each pixel is combined to determine rock, mineral, and soil types and to produce geological maps. For the first time we have a complete high resolution digitized map of the Moon showing surface features, composition, and topography.

After completing its lunar mapping mission, Clementine is on its way to Geographos, an asteroid whose orbit crosses the orbit of the Earth, to determine as much about its surface features and composition as possible in a quick flyby.

Clementine, also called the Deep Space Program Science Experiment, is a unique mission for two reasons. First, it was designed and built by the Ballistic Missile Defense Organization and the U.S. Naval Research Laboratory to test the durability of new military sensors and electronics in the radiation environment in space and the ability of a spacecraft to operate autonomously. With the end of the cold war, much defense research has been more open. The Navy engineers got together with space scientists, decided they could combine missions, and sent Clementine to the Moon. The scientists got their long-sought-after Moon map and the military tested its equipment. Second, it is a small spacecraft built with new lightweight components: star tracker, reaction wheels for attitude control, nickel-hydrogen battery, and solar panel. After the failure of several very large, very expensive spacecraft in recent years, NASA is considering designs which are smaller, cheaper, and more quickly built. The 300 pound Clementine was built in 2 years for $55 million and was launched on a $20 million Titan 2 rocket. Compare that with the 25,000 pound, $1.5 billion Hubble Space Telescope which took nearly 20 years to design, build, and place in orbit. Such a comparison may be unfair because the Space Telescope is a far more capable and versatile piece of equipment. Nonetheless, 20 Clementine-class spacecraft could be built for the price of one Hubble.

Advances in robotics, telecommunications, and virtual reality now make it possible to send an unmanned vehicle to another planet and operate it from Earth. The former Soviet Union had experience with roving vehicles on the Moon and NASA has proposed sending a more advanced Russian rover back to the Moon and using American "telepresence" technology to operate it. The rover would carry two cameras which would transmit stereo pictures back to Earth. The images would be computer processed and displayed in a virtual reality kind of helmet. The operator wearing the helmet and controlling the rover would seem to be sitting on the vehicle and driving it like an automobile. When the controller looks around, left or right, up or down, the cameras on the rover follow the action to give a view in that direction. In tests of telepresence technology, researchers in California have operated an underwater vehicle in Antarctica and a rover in the Mojave Desert.

Missions to Mars

Two Viking spacecraft began their yearlong flight to Mars in 1975. Both of them had orbiters and landers that examined the red planet, studied its geology, and searched for living organisms. They sent back spectacular and surprising information. Mars is no longer a point of light in the sky; it is a real place, a world that is both very much like Earth and very different from Earth. Chapters 12 and 13 will examine the Viking results, the planet itself, the search for life, and future missions.

The Mars Observer was the next U.S. mission to the red planet, departing Earth in September 1992 and arriving at Mars in August 1993. Unfortunately, controllers lost contact with the billion dollar spacecraft just 3 days before it was to enter orbit around Mars. Commands had been sent to pressurize the propellant tanks in preparation for firing the thrusters which would insert the spacecraft into a long elliptical orbit. The transmitters were shut down during that time to avoid damage to their tubes when explosive devices were fired to open the propellant valves. Ten minutes later, when the tank pressurization was completed, the transmitters were supposed to turn back on. Mars Observer was never heard from again. There have been a number of suggestions as to what happened, but the real cause of its demise may never be known. It was a very complex and expensive spacecraft. Some scientists and engineers had been working on it for their entire careers.

In a program named Mars Surveyor, NASA has sent out a request for innovative, faster, better, cheaper ways to explore Mars. During the past decade, computers, electronic equipment, sensors, data storage, and power supplies have become dramatically smaller while their capability and reliability have dramatically increased. As a consequence, it is now possible to build lighter weight spacecraft with as much capability as the larger ones of the past. A small remotely operated vehicle, called a microrover, is planned for a 1996 launch to Mars. Duplicates of the Mars Observer instruments will be sent on smaller vehicles, some in 1996 and others in 1998. Subsequently, it has been proposed to send a series of spacecraft, each with an orbiter and a lander, every 2 years when Earth and Mars are properly aligned. The United States and Russia have entered into an agreement to develop a long term joint Mars exploration program. Russia plans flights every 2 years, although financial problems in that country have delayed some of the missions. Mars 94 has slipped to 1996 and the Mars 98 expedition will carry a remote controlled roving vehicle and a balloon. Japan will also send a spacecraft to orbit Mars in 1998.

Voyager

Perhaps the most successful planetary probe was the pair of Voyager spacecraft that made closeup inspections of the outer planets, Jupiter, Saturn, Uranus, Neptune, their moons, and their rings.

Twelve Mariner and Pioneer spacecraft had flown past Mercury, Venus, Mars, Jupiter, and Saturn from 1962 to 1978. They photographed the surfaces and cloud tops, detected magnetic fields, studied the atmospheres, and took measurements of the interplanetary solar wind. They were the forerunners of the more sophisticated and technically capable Voyager spacecraft. Figure 7.3 is a diagram of Voyager. Notice that it did not carry solar panels for electric power. Because it was to travel so far from the Sun, sunlight would not be sufficient to power the spacecraft. Therefore, it carried electric generators powered by the heat from radioactive decay of plutonium.

Once every 175 years the outer planets line up on the same side of the Sun, making it possible to send one spacecraft to all of them on one flight. This "Grand Tour" was conceived and carried out by the Jet Propulsion Laboratory. The plan was that as Voyager fell toward Jupiter, the gravitational force would increase the spacecraft's speed suf-

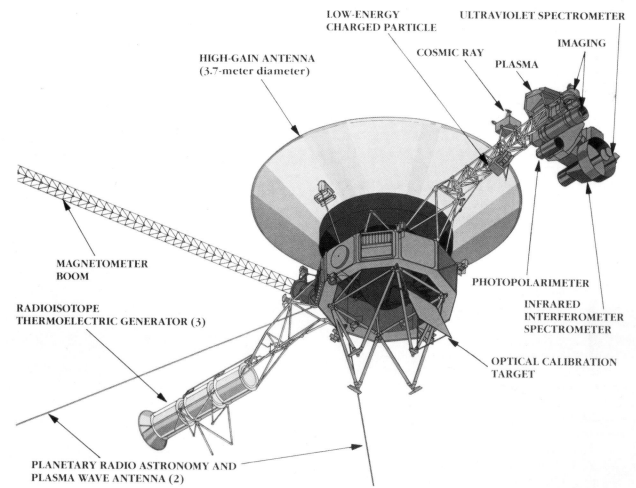

LOW-ENERGY
CHARGED PARTICLE

ULTRAVIOLET SPECTROMETER

COSMIC RAY

IMAGING

HIGH-GAIN ANTENNA
(3.7-meter diameter)

PLASMA

MAGNETOMETER
BOOM

RADIOISOTOPE
THERMOELECTRIC GENERATOR (3)

PHOTOPOLARIMETER

INFRARED
INTERFEROMETER
SPECTROMETER

OPTICAL CALIBRATION
TARGET

PLANETARY RADIO ASTRONOMY AND
PLASMA WAVE ANTENNA (2)

Figure 7.3 Voyager spacecraft. The spacecraft was powered by nuclear electric generators because the intensity of solar energy is so low at the outer fringes of the Solar System that solar cells cannot be used. THe communications antenna is over 12 feet in diameter. Other components are sensors which measure particles and electromagnetic radiation in space and in the vicinity of the target planets. The imaging devices created spectacular photos of the planets and their moons. *Courtesy of NASA.*

ficiently and curve its trajectory around the giant planet just properly so it would go on to Saturn, the next planet in line. Similarly, at Saturn Voyager would pick up enough speed and change course for Uranus, and so on to Neptune. These trajectories are shown in Figure 7.4.

Two Voyager spacecraft were sent on their way in 1977. Voyager 2 was launched August 20; Voyager 1 was launched into a faster trajectory on September 5 and reached Jupiter first on March 5, 1979. Voyager 2 arrived 4 months later on July 9.

Voyager at Jupiter

The Voyager 1 trajectory past Jupiter is shown in Figure 7.5. What it found was awesome. Three of the 33,000 photographs returned are shown in Figures 7.6 to 7.8 in the color section, page 120.

The giant red spot in Jupiter's atmosphere, seen in exaggerated color in Figure 7.6, had been observed for three hun-

dred years. It was found to rotate counterclockwise, like a hurricane on Earth, once in 6 days at the outer edge. Lightning bolts and aurora were seen and a magnetic field and magnetosphere were discovered. In addition, Jupiter was found to have a thin ring which could only be seen looking back from the opposite side into the glare of the Sun.

The moons of Jupiter were spectacular. Planetary scientists had expected cold, rocky bodies riddled with craters. The surprises were many. Io (Figure 7.7) is the most geologically active body yet found in the Solar System, the only place other than Earth with active volcanoes. Europa (Figure 7.8) is as smooth as a billiard ball covered with water ice but with dark linear markings. Standing on Europa, you would see Jupiter in the sky 20 times larger than our Moon. You could watch it rotate once in 10 hours and see it go through phases like our Moon (full, crescent, new) in 3.5 days. Ganymede, 1,640 miles in diameter, is the largest moon in the Solar System and is covered with both craters and grooved ice terrain. Callisto has wall-to-wall craters, proba-

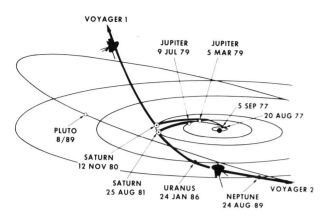

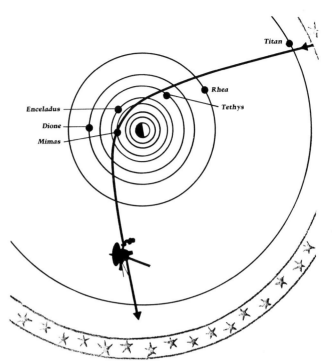

Figure 7.4 Voyager spacecraft "Grand Tour" of the outer Solar System. *Courtesy of NASA.*

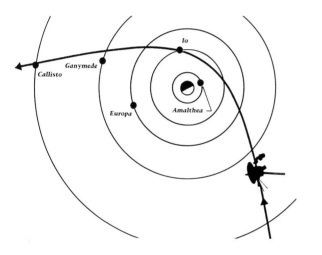

Figure 7.5 Voyager 1 track past Jupiter and its moons. The curvature of the trajectory is due to the gravitational force as Voyager approached Jupiter, which deviated the spacecraft from its elliptical orbit around the Sun. *Courtesy of NASA.*

bly the most intensely cratered body in the Solar System. Other lesser moons were photographed as well and three new ones were discovered.

Voyager at Saturn

Voyager 1 arrived at Saturn on November 12, 1980, only 12 miles off course; Voyager 2 followed on August 25, 1981, 2.7 seconds early and 30 miles from the aim point. The Voyager 2 trajectory through the Saturnian system is shown in Figure 7.9. Photos are shown in Figures 7.10 and 7.11 in the color section, pages 120–121. Again the discoveries were many.

Figure 7.10 in the color pages shows a view of Saturn and its rings from the back side looking toward the Sun. Saturn's rings are very thin, composed mainly of small chunks of water ice, and so numerous they look like grooves in a phonograph record. Spoke-like features were observed in the rings. Small moons were found on the edge of some of the rings. These moons act to keep the ring particles in

Figure 7.9 Voyager 2 trajectory through the Saturnian system.

place and the edge of the ring sharply defined, and consequently are called "shepherding moons." Cloud features are similar to Jupiter but not as pronounced in color. The wind at the equator was measured at 1,100 miles per hour. Saturn has a magnetic field and a magnetosphere.

Titan, the largest moon, has a dense nitrogen atmosphere, with haze layers formed from organic compounds such as methane, propane, and acetylene. The surface temperature on Titan is −288 °F. At that temperature methane could play much the same role as water does on Earth. The innermost moons are composed mostly of water ice. Figure 7.11 in the color section (p. 121) is Enceladus. Notice the strange distribution of craters.

Voyager at Uranus

Voyager 1 was steered under Saturn and allowed to accelerate to escape velocity from the Solar System. It is headed upward out of the plane of the ecliptic where the orbits of the planets lie. Voyager 2 took a trajectory around Saturn and on to Uranus, arriving there on January 24, 1986. In the Voyager images, Uranus is a blank bluish colored ball and clouds are not as outstanding as the clouds of Jupiter and Saturn. Strangely, Uranus lies on its side as compared to the other planets, that is, its axis of rotation lies close to the plane of its orbit. Consequently, as the planet moves in its orbit, the north pole is pointed toward the Sun for about 20 years, then the south pole for the same time. The equatorial regions receive less sunshine than the poles. Even so, the winds were clocked at 375 miles per hour and cloud top temperatures were quite uniform over the entire planet. The surprise of the encounter was the moon Miranda, shown in Figure 7.12. It has geological features and patterns never before seen in planetary exploration. One possible explanation is that it collided with another moon-sized object, broke apart, and the pieces rejoined a bit out of order.

Voyager at Neptune

The final planet visited by Voyager 2 was Neptune on August 25, 1989. Although it is the smallest of the large outer gas planets, Neptune is about 31,000 miles in diameter, almost four times the diameter of Earth. Voyager was then nearly 3 billion miles from the Sun, 30 times farther than the Earth. At first Neptune, too, appeared as blank as Uranus, but with some processing and enhancing of the images, clouds and a giant blue spot can be seen. See Figure 7.13 in the color section, page 122. These features all move at different speeds. The highest wind speed at cloud tops was over 700 miles per hour.

The largest of Neptune's moons is Triton. seen in Figure 7.14 in the color section, page 187. Its surface temperature is down to −400 °F and it has a thin nitrogen atmosphere with some methane mixed in. The only planet left unvisited is Pluto. Planetary scientists believe that Pluto is probably

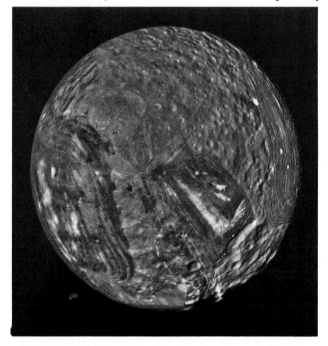

Figure 7.12 Uranus's moon Miranda has features never before seen on Voyager's trek through the outer planets. The left and right edges show "racetrack" grooves. A large chunk seems to be missing at the bottom. V-shaped grooves are at the center. Only the top part looks like Earth's cratered Moon. Resolution is less than half a mile. Because of the low light level at Uranus (it is 19 times farther from the Sun than Earth is) long time exposures were necessary. To avoid blurred pictures, Voyager controllers instructed the spacecraft to fire thrusters causing it to spin in synchronization with its 43,000 miles per hour motion past Miranda. Without that trick, resolution would have been no better than about 16 miles. *Courtesy of NASA.*

very much like Triton. They are about the same size and density, and Pluto's atmosphere, as observed from Earth, is similar to Triton's.

One of the significant discoveries of Voyager is that, although they may look superficially alike, each body in the Solar System has unique features all its own, different from every other body.

Pioneer and Voyager Beyond the Solar System

In 1990, Pioneer 6 celebrated its 25th birthday, still orbiting the Sun with aphelion near Earth's orbit and perihelion halfway between Earth and Venus. The 160 pound spacecraft was launched in 1965. Three of its six instruments were still operating and, although its solar panel power supply had degraded over the years, it continued to send data to whomever was listening. Interplanetary spacecraft in orbit around the Sun continue on their rounds forever, unless they collide with something.

In that same year, Pioneer 10 was 4.6 billion miles from the Sun, 50 times farther than Earth is from the Sun. Its 8 watt transmitter was still sending data 18 years after its 1972 launch and will probably do so beyond the year 2000. Its speed is greater than escape speed from the Solar System so that, even after its power supply dies, it will travel on forever in interstellar space, unless it collides with something.

Three other spacecraft are also on their way out of the Solar System. The trajectories of these four spacecraft are shown in Figure 7.15. Pioneer 11 is going in the opposite direction to Pioneer 10. Voyager 1 is headed up out of plane of the ecliptic, Earth's orbital plane, and Voyager 2 is headed downward. "Up" and "down" are roughly the north and south directions from Earth's point of view.

How long the Voyagers will be able to communicate with Earth depends on several factors. They are powered by electricity produced by heat from the decay of plutonium and do not depend on the Sun for power. The radioactive material eventually will be used up as will the hydrazine fuel needed for the thrusters to keep the large antenna pointed toward Earth. Actually, the Voyagers cannot sense the Earth but

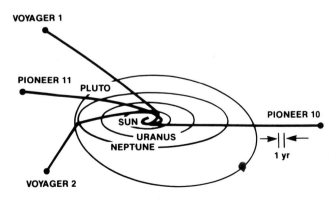

Figure 7.15 The trajectories of four spacecraft leaving the Solar System for interstellar space. *Courtesy of NASA.*

their sensors can detect the Sun and use it as a reference to find Earth. Some of the instruments and communications equipment have failed and other equipment may stop functioning also. Unless a catastrophic failure occurs, we will continue to hear from the Voyagers until perhaps 2030. Where will they be then? Barely out of the Solar System. Alpha Centauri, the nearest star (after the Sun) is about 4 light years away. At Voyager speed, about 19,000 years to travel 1 light year, it would take about 76,000 years to travel that distance. Neither Voyager is headed in that direction, however.

Unless they collide with something, the spacecraft themselves will remain intact essentially forever. Perhaps some alien species will find them someday. Science fiction has it that V—ger will be picked up by the crew of the starship *Enterprise* sometime in the twenty-third century. Messages from Earth to whomever may find them are attached to both Voyager spacecraft. They are sounds and pictures recorded on a gold-plated record carried in an aluminum case. Instructions on how to play the record are included. It is assumed that anyone smart enough to intercept Voyager will be smart enough to decode the instructions. We will have more about this in Chapter 13.

Galileo to Jupiter

Voyager results at Jupiter whet the appetite of planetary scientists for more. Everything discovered by Voyager was so completely beyond expectations that the mission raised many new questions. Galileo is a spacecraft designed to take a more thorough look at the Jovian system, visiting Venus, two asteroids, and the Earth-Moon system twice, before heading for Jupiter. It is shown in an artist's rendering in Figure 7.16 in the color section, page 121.

Taking the long way around to Jupiter by way of Venus was necessitated because of a change in the rocket which would boost the spacecraft out of Earth orbit onto an escape trajectory. Following the *Challenger* accident, NASA decided it was not safe to carry a liquid-fueled Centaur rocket as an upper stage for interplanetary spacecraft in the Shuttle orbiter's cargo bay. The solid propellant Inertial Upper Stage (IUS) was to be used instead, although the IUS does not have the power of a Centaur and cannot send a large spacecraft like Galileo on a direct path to the outer planets. Therefore, it was necessary to devise a flight plan which used Venus and Earth for gravity boosts. The planets had to be located in the correct positions at launch so each would be at the correct spot in space when Galileo arrived. The schedule is shown in Table 7.1. At each encounter with Venus and Earth, the spacecraft fell toward the planet to increase its speed and extend its apogee until it could reach Jupiter. Figure 7.17 shows this complex trajectory. Follow the path and notice the change in aphelion after each planetary encounter.

Unfortunately, the Galileo spacecraft has a stuck antenna.

TABLE 7.1 Galileo Spacecraft Mission Events and Gravity Assisted Orbital Changes

Date	Event	Perihelion	Aphelion
October 18, 1989	Launch from Space Shuttle *Atlantis*	—	Earth
February 10, 1990	Venus flyby, gravity boost	Venus	Beyond Earth orbit
December 8, 1990	First Earth flyby, gravity boost	Earth	Asteroid belt
October 29, 1991	Asteroid Gaspra flyby	No change	No change
December 8, 1992	Second Earth flyby, gravity boost	Earth	Jupiter
August 28, 1993	Asteroid Ida flyby	No change	No change
December 7, 1995	Jupiter arrival	Jupiter capture	

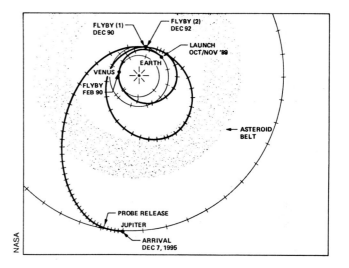

Figure 7.17 Galileo spacecraft trajectory to Jupiter via Venus and the asteroid belt.

The 16 foot antenna used for communicating with Earth was designed to open like an umbrella, but for some unknown reason it did not open all the way. This severely limits the capacity for speedy transmission of the data collected by the sensors. Using the smaller antennas, data transmissions are very slow: 10 bits per second instead of the 131,000 bits per second possible if the balky antenna were fully open. At this leisurely rate it takes days to send a single picture. Two things can be done to improve the situation. First, reprogram the spacecraft's computer to compress the data before sending it. Second, improve the sensitivity of the ground receivers. These actions could increase the data rate to 100 bits per second.

At Venus, Galileo used its infrared sensors to map the turbulent middle level cloud layer about 30 miles above the surface where winds blow more than 150 miles an hour.

Closeup photos of asteroid Gaspra showed not only numerous small craters, but several crisscrossing grooves 30 to 60 feet deep, suggesting that it had collided with other objects sometime in its history. Scientists were taken by surprise to find that asteroid Ida has a moon. See Figure 7.18. The possibility had been suggested previously by amateur astronomers who observed stars "blink out" momentarily when asteroids moved in front of them. On some occasions the star would blink off and on more than once suggesting that there was more than a single asteroid in the vicinity. The picture of Ida was the first confirmation that asteroids may indeed have their own moons.

Earth was photographed in 500 frames to make a movie of its 24 hour rotation. Both Earth flybys were made in December during the southern hemisphere summer. South America, Australia, Africa, and Antarctica are featured in the images. Figure 7.19 in the color section, page 188, is one of these images. From Galileo's vantage point, about 1.2 million miles away, we see Earth as a distant planet with white wispy clouds, blue oceans, brown land, and white ice cap. These features, which are important to life, distinguish it from all the other planets we have seen. The spacecraft also made close passes to the Moon and photographed, among other things, the south polar region and a large impact crater on the side which never faces Earth.

A probe to investigate Jupiter's atmosphere will be released several months before it arrives at the planet. The probe will enter the atmosphere at the same time the spacecraft brakes into an orbit around Jupiter. For the next 2 years Galileo will make side trips to some of the many Jovian moons for closer looks than were made by the Voyagers in their quick flybys.

Figure 7.18 Asteroid Ida and its moon. Ida is about 35 miles long and 15 miles wide. By preliminary estimates its moon is about a mile in diameter and roughly 60 miles away from the asteroid. Both Ida and its moon appear to be made of silicate rock. *Courtesy of NASA.*

Cassini to Saturn

Cassini is a spacecraft under construction to investigate Saturn, its surrounding environment, and its 22 moons. The objective is to gather sufficient data to develop an understanding of the Saturnian system as a whole: the interactions among the planet itself, the moons, the ring system, the particle environment, and the magnetic fields. Also, scientists are very interested in comparing the atmospheres of the giant gas planets, Saturn and Jupiter, to discover their similarities and differences. Remote sensors on Cassini will examine the atmosphere of Saturn in great detail so the data can be compared with the data on Jupiter gathered by the Galileo spacecraft.

Of particular interest is Saturn's largest moon, Titan, because, like Earth, it has an atmosphere composed mostly of nitrogen, although its surface pressure is greater than Earth's surface pressure. Smog has been detected in Titan's upper atmosphere similar to the smog found around the large cities on Earth. Even though the temperatures on Titan are much colder than on Earth, *photochemical processes* are going on which manufacture hydrocarbon compounds. The reaction is probably similar to that which occurred in Earth's atmosphere billions of years ago before oxygen and living organisms existed here. It is likely that there are lakes or oceans of liquid nitrogen and hydrocarbons. Carl Sagan suggests that Titan is a planet just waiting for a global warming so biological reactions can develop from the chemical reactions that have been going on for millenia.

Galileo first viewed Saturn through a telescope in the 1600s and noticed bumps on both sides of the planet He referred to Saturn as the planet with ears; his simple telescope could not resolve the bumps into rings. Nearly 50 years later, Huygens, with an improved telescope, was able to discern the rings as a disk, and correctly surmised that the rings change appearance from time to time because we see them from different points of view as Earth and Saturn revolve around the Sun. Huygens also discovered Titan, the largest moon of Saturn. In the 1670s Cassini, the first director of the Paris Observatory, saw that the ring was not solid but there was a dark region that looked like a blank space dividing it in two. That dark region is now referred to as the Cassini division. Cassini also discovered several of Saturn's moons and was the first to report the giant red spot on Jupiter.

The United States previously sent three spacecraft to Saturn: Pioneer and two Voyagers. Each mission answered many questions about the Saturnian system and raised more new ones. Cassini, is another in the series of large, complex, billion dollar spacecraft for planetary exploration. It stands about 30 feet tall and has a large antenna for communicating with Earth. See Figure 7.20.

The spacecraft will carry a dozen experiments including

Figure 7.20 The Cassini spacecraft. At the top is the communications antenna. Instrumentation is attached to brackets and booms sticking out in all directions. The disk on the left is the European Space Agency's Huygens probe which will be ejected and descend to Saturn's moon, Titan. *Courtesy of NASA.*

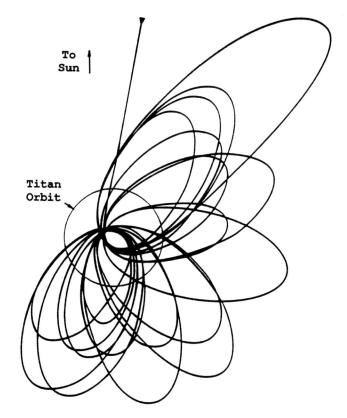

Figure 7.21 Cassini's trajectory after arrival at Saturn. It passes Titan on each orbit. *Courtesy of NASA.*

a radar to map Titan, spectrometers to measure various bands of electromagnetic radiation and a magnetometer to measure magnetic fields. Many of the experiments are attached to two movable rotating platforms on long booms so they may be pointed at objects of interest. The magnetometer is also mounted on a long boom to avoid interference from magnetic fields in and on the spacecraft body. Cassini will also release a probe to investigate the atmosphere of Titan. Named Huygens and built by the European Space Agency, the probe is designed to land on the surface or to float if it lands in a liquid nitrogen lake or ocean. A variety of cameras will make images of the various objects in Saturn's neighborhood.

According to the current schedule, the Cassini spacecraft is to launch on a Titan 4 rocket on October 12, 1997, fly by Venus in 1998 and again in 1999 to receive gravity boosts from that planet, then fly past Earth and on to a gravity boost from Jupiter in 2001. These gravity boosts save about 2 years travel time by putting the spacecraft on a faster trajectory. It will reach Saturn June 25, 2004, fly very close to the rings, and pass the rings before the rockets are retrofired to inject it into a highly elliptical orbit around the planet. Periapsis lies close to the planet and apoapsis lies out beyond the rings and moons. Each time the Cassini orbiter passes close to Titan its direction changes, a sort of gravity boost, and the orbit changes inclination, rotates, or changes plane. Some orbits pass by the other moons for closeup study; some move up out of the equatorial plane for a different perspective of the rings and magnetic fields. Some of these orbits are seen in Figure 7.21. It is expected to maneuver through 40 orbits in 48 months, each maneuver the result of a close encounter with Titan. No thruster

firings are needed. During each pass by Titan, the radar makes an image of the strip of the surface beneath the spacecraft at a resolution of several kilometers. Some selected areas of the moon can be mapped with a 1,000 foot resolution. As with Magellan at Venus, the Cassini radar can "see" through the clouds of Titan.

On the first flyby of Titan, Cassini will be steered on a collision course with the moon so the Huygens probe can be released from the spacecraft for a landing on Titan. This is shown in Figure 7.22 in the color section, page 189. Once the probe is released, the Cassini orbiter monitors its motion for several hours, then fires a thruster so the orbiter doesn't hit Titan too! A shield protects the probe from overheating and slows its fall. A parachute opens and slowly carries it through the atmosphere, striking the surface at about 1 mile per hour. Most of the useful data will be taken during the 2 to 3 hour descent through the atmosphere.

The Huygens probe contains instruments to measure the atmospheric pressure, temperature, density, and composition; a collector to study aerosol particles; a camera that looks down to take pictures of the surface, then up to measure the light transmission through the atmosphere; a lightning and radio wave detector; and surface-measuring instruments. A small transmitter on the probe sends its data to the orbiter which then relays the information to Earth.

That way the probe doesn't need a high power transmitter, power supply, or large antenna. The Huygens probe will last only last a minute or so on the surface before the battery runs out. The next pass of the orbiter is 120 days later and it would add too much weight to supply batteries to last that long.

Magellan to Venus

Venus has a very dense carbon dioxide atmosphere. It is completely and perpetually shrouded in cloud cover. Images through telescopes show only a bright white disk. Venus became the target of intense effort by the Soviet Union's planetary space program. At least 25 spacecraft were launched with the intention of studying that planet. Although many failed, many were phenomenally successful. Venera 1, launched February 12, 1961, was the first spacecraft to escape Earth orbit and reach another planet as it flew by Venus in May of that year. On February 27, 1966, Venera 3 was the first man-made object to land on another planet. In total, Soviet spacecraft made seven successful soft landings on Venus between 1965 and 1985. They found the surface temperature to be about 700 °F and the atmospheric pressure to be 90 times that at the Earth's surface. At least two of the Venera spacecraft survived the 700 °F heat long enough to send pictures of the dry, rocky surface.

The Venusian cloud cover is so dense that it is impossible to photograph its surface from orbit in the usual way with television type cameras. In 1978 a U.S. Pioneer spacecraft arrived to begin mapping 92 percent of the planet by radar with a resolution of 60 miles. Like Earth, Venus has high continents and low basins. The basins contain no water, of course. Water cannot exist in liquid form because of the extreme heat at the surface. Pioneer Venus finally ran out of gas for the thrusters and fell out of orbit into the dense Venusian atmosphere in October 1992 after sending its messages for 14 years. A better radar map of Venus came from Venera 15 and 16 which arrived at Venus in 1983. Their radar had a resolution of 6,500 feet and covered about 30 percent of the planet.

Magellan, launched from the Space Shuttle *Atlantis* on May 4, 1989, carried an even higher resolution imaging radar to map Venus through the clouds. It could distinguish objects as small as 350 feet across and measure altitudes with an accuracy of 100 feet. From its highly eccentric orbit with apoapsis of 5,255 miles, periapsis of 180 miles, and period of 3.26 hours, the radar recorded strips of the ground 12 to 15 miles wide from pole to pole on the periapsis side of the orbit. Then, as it slowed down approaching apoapsis, it turned and transmitted its findings to Earth. Mapping data was taken for 37 minutes of the orbit and played back to Earth in 114 minutes, leaving about 45 minutes of each orbit for turning, scanning the stars for navigating, and spacecraft "housekeeping." As Venus rotated on its axis, Magellan scanned a different strip of the surface on each orbit. One Venusian day (243 Earth days) was required to cover the entire planet except for small regions at the poles. After three complete mapping cycles from mid-September 1990 to mid-September 1992, 98 percent of the planet had been covered.

The radar was not aimed straight down, but looked sideways at angles from 14° to 52° which gives a more pronounced perspective to the images. Combining the radar data with the altimeter data produces spectacular three-dimensional pictures. As with every other spacecraft that has visited another planet, Magellan found some surprising terrain on Venus. Three of the many interesting images are shown here. Figure 7.23 shows some highly unusual circular dome-shaped features that are thought to be lava flows.

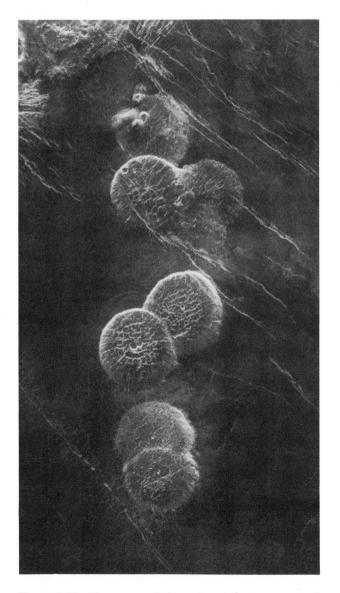

Figure 7.23 These unusual dome-shaped features apparently formed when very thick lava flowed over flat terrain on the hot surface of Venus like pancake batter on a hot griddle. They are about 15 miles in diameter and 2,500 feet high. *Courtesy of NASA.*

Three large impact craters are shown in Figure 7.24. In the color pages (page 190), Figure 7.25 shows a Venusian mountain, similar in structure to the Hawaiian shield volcano, Mauna Loa. It rises 5 miles above the surface and the lava flows extend in the foreground for more than 100 miles.

Ulysses to the Sun

The Solar System explorers discussed so far have all traveled in or near the *ecliptic*, the plane of the Earth's orbit projected to the celestial sphere. At the end of their missions, two Pioneer spacecraft and the two Voyagers were steered out of the ecliptic plane and are leaving the Solar System. (Refer back to Figure 7.15.) But until now, no spacecraft were launched out of the ecliptic plane toward the Sun. Ulysses is a spacecraft designed to fly out of the plane of Earth's orbit and over the poles of the Sun, a region of space never before explored.

To leave the plane of Earth's orbit requires more energy than a booster rocket can provide. Therefore, Ulysses was sent in the direction of Jupiter, steered over Jupiter's northern hemisphere, and with a gravity assist from the giant planet, headed south below the ecliptic plane toward the south pole of the Sun. It was launched on October 6, 1990,

reached Jupiter in February 1992, and passed beneath the Sun in mid-1994. In 1995 it will move back up through the plane of Earth's orbit and over the Sun's north pole.

Ulysses was built and is operated by the European Space Agency. It was launched by Space Shuttle *Discovery*. The experiments it carries include detectors to measure solar x-rays, gamma rays, solar wind, high and low energy particles, solar magnetic fields, interstellar gas and dust, and radio waves.

Pluto

The only planet that has never been visited by a space probe is Pluto and its moon Charon. Pluto's 248-year orbit is the most eccentric of any of the planets. Its distance from the Sun varies between 2.8 billion miles at perihelion and 4.6 billion miles at aphelion. In the 1990s it is near perihelion, closest to the Sun. In fact, its orbit crosses Neptune's orbit, so at perihelion it is closer to the Sun than Neptune. Pluto has an atmosphere which will freeze up and fall to the surface as it moves away from perihelion. The 1990s is a good time to send a spacecraft.

The Pluto Fast Flyby mission being planned calls for miniaturized electronics in a very small and lightweight space-

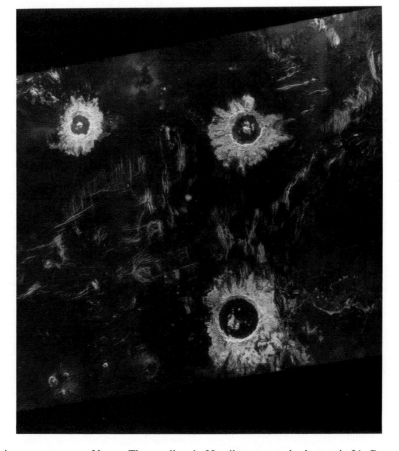

Figure 7.24   Three large impact craters on Venus. The smallest is 23 miles across, the largest is 31. Because of Venus's deep, dense atmosphere, the meteoroids that hit the surface must have been quite large or they would have burned up before reaching the ground. Several volcano vents are located in the lower left corner. *Courtesy of NASA/JPL.*

craft. By launching it on a high powered booster, perhaps a Russian Proton rocket, the craft can be given a higher initial speed and would make it to Pluto in 7 to 10 years instead of a 30 year minimum energy Hohmann transfer orbit. The flyby past Pluto will last only a couple of hours. Data will be stored in memory and sent back to Earth at very slow speed during the following 6 months so that only a small transmitter and antenna need to be carried along. "Minimal weight" is the key phrase for planning this mission. To minimize the chance of failure, project planners recommend sending two separate spacecraft on two separate rockets.

Beyond the Solar System

One of the problems that has always plagued astronomers is having to look through Earth's atmosphere. Light coming from the universe is refracted, that is, bent by the air. That is why stars twinkle; above the atmosphere, they shine with a steady light. Air currents and bubbles of warmer or cooler air cause the light waves to refract as they pass through. Because the air is in constant motion, the refraction varies from moment to moment and the image in the telescope appears to shimmer and dance around.

In addition, atmospheric gases and particulate matter cause light to scatter away from a straight line. Sunlight is a mixture of all colors. As it penetrates the air, the longer wavelengths pass straight through while the shorter wavelength blue light scatters in all directions. That is why the sky is blue. The blue part of the sunlight comes at us from all directions. We cannot see the stars during the daytime because the sky is too bright from the scattered sunlight. Scattering is also the reason that the night sky is often rather bright. Starlight, moonlight, and city lights scatter and reflect from the gas and dust in the air, giving the sky a background glow which interferes with astronomical observations.

Another problem is that the atmosphere selectively absorbs many wavelengths of electromagnetic energy so they never reach the ground. This was pointed out in Chapter 4 and is shown in Figure 7.26. Ultraviolet, gamma rays, and x-rays are absorbed by the ionosphere and ozone layer. Certain infrared wavelengths are absorbed by molecules of water vapor and carbon dioxide. Some radio waves are absorbed by the ionosphere or reflected off its top. Some infrared and ultraviolet observations are possible from the tops of mountains or from balloons above the densest part of the atmosphere. But until the coming of the space age, astronomers have had to be content with looking at the sky primarily by visible light, radio waves, and a few certain wavelengths of infrared, "windows" in which the atmosphere is partially transparent.

A final problem: as pointed out earlier in this chapter, everything emits electromagnetic waves and the atmosphere is no exception. It emits infrared. This fact is a boon to meteorologists whose satellites can detect variations in the temperature of the ground and atmosphere from the infrared emissions. But this same fact is the bane of astronomers who would like to see the universe, not the atmosphere, with infrared sensors on the ground.

Before the days of spaceflight, astronomers used rockets and balloons to send their instruments above the atmosphere for short looks. Putting their observing equipment above the atmosphere allows astronomers to examine all wavelengths of electromagnetic energy coming from the universe. Many astronomy satellites were launched into orbit during the 1960s and 1970s. Uhuru discovered many x-ray sources in its 4 year lifetime. Explorer 53 discovered, among other things, stars that produce enormous bursts of x-rays every few hours. Two Orbiting Astronomical Observatories and the International Ultraviolet Explorer searched the universe for ultraviolet sources. Three High Energy Astronomy Observatories further studied x-ray sources. One of them searched out cosmic ray particles and gamma ray sources. Advanced spacecraft for detecting everything from gamma rays to microwaves are operating or are planned. We will discuss only a few of them here.

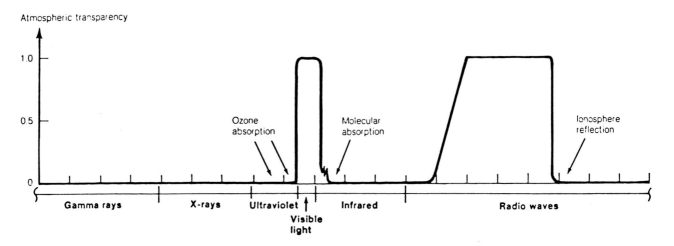

Figure 7.26 Transparency of Earth's atmosphere. *Courtesy of NASA.*

X-ray Telescopes

X-rays bursts are often observed coming from the Sun during a large solar flare. However, some early rocket experiments carried sensors above the atmosphere and found x-rays coming not only from the Sun, but from all directions in the sky. Earth seemed to be bathing in an x-ray bath. Further experiments on rockets, balloons and finally on spacecraft showed that the sources include remnants of supernovas, clusters of stars, neutron stars, pulsars, quasars, superheated gas between galaxies, and perhaps black holes. Detecting and studying x-rays can give a better insight into the workings of these "exotic" objects.

Because x-rays penetrate most substances, the normal telescope design cannot be used for x-rays. They will, however, reflect from a surface if they strike it at a very shallow angle, a glancing blow, so to speak. See Figure 7.27. To accommodate this phenomenon, the reflecting surfaces are not flat, but are curved concentric rings made of fused quartz and coated with a thin layer of nickel.

Several x-ray telescopes have been launched into orbit. The next one will be AXAF, the Advanced X-ray Astrophysics Facility, which should be ready before the end of the century.

Infrared Astronomy from Orbit

One of the most successful of the astronomical satellites, both operationally and scientifically, was the Infrared Astronomical Satellite (IRAS), a joint undertaking by American, British, and Dutch scientists.

A diagram of the satellite is shown in Figure 7.28. It is 12 feet long, 7 feet in diameter, and weighed 2,365 pounds on Earth. It uses a 22 inch diameter concave mirror, shaped like a shaving or makeup mirror, to focus the infrared waves. The mirror is made of beryllium, a very lightweight but strong metal. A secondary mirror reflects the waves to the focal plane where the 62 silicon arsenide and germanium gallium detectors are positioned. The detectors are sensitive to four wavelength bands from 119 micrometers to 8.5 micrometers, designed to observe objects as cold as 27 Fahrenheit degrees above absolute zero. The short wavelength sensors detect hot sources such as stars, while the longer wavelength sensors observe cooler or extended objects such as dust clouds.

As we have pointed out several times before, everything at a temperature above absolute zero emits infrared, even the telescope itself. A cold object emits less infrared at a longer wavelength than a warm object. Therefore, to keep interference to a minimum, the telescope was kept cold with liquid helium. This *cryogenic* system kept the optics and barrel of the telescope and the infrared sensors at just a few degrees above absolute zero. All the liquid helium needed for the lifetime of the spacecraft had to be put aboard before launch. The 125 gallons carried in a "thermos jug" was ex-

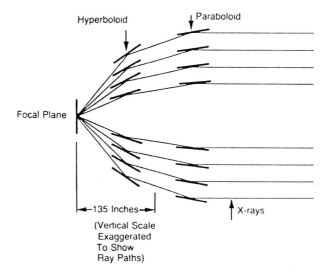

Figure 7.27 Diagram of an x-ray telescope. X-rays enter the telescope from the right, skip off two reflecting surfaces, and converge at the focal plane. "Paraboloid" and "hyperboloid" refer to the shape of the surface. The reflectors are actually rings; we see them edgewise in the diagram. The telescope is about 2 feet in diameter at the entrance. *Courtesy of NASA.*

pected to be a 6 months supply, but it evaporated more slowly than expected and the satellite continued to operate for 10 months.

IRAS was launched on January 25, 1983, from Vandenberg Air Force Base, California, into a 560 mile polar orbit with a period of 100 minutes. That altitude is well above the atmosphere, but below the Van Allen belts except in the South Atlantic Anomaly region. There the data becomes contaminated by many false readings from protons in the radiation belts.

The orbit was oriented over the *terminator*, the sunrise-sunset line between daylight and darkness on Earth's surface (Figure 7.29). The orbit rotated about 1 degree per day so that as the Earth moved around the Sun the satellite remained over the terminator. It was important to keep the orbit oriented in that direction because the spacecraft had to point more than 60 degrees away from the Sun and more than 88 degrees from the Earth's horizon so the sensors would not pick up the Sun or Earth. A sunshade and baffle aided in keeping stray infrared from reaching the sensors.

The spacecraft scanned a one-half degree wide strip on each orbit which overlapped the previous strip by one-fourth of a degree. It did a complete survey of 95 percent of the sky. IRAS catalogued 250,000 individual sources within the limits of sensitivity of its detectors. Most of them are stars in our Milky Way Galaxy, some surrounded by cooler material. The nucleus of the Milky Way is readily discernible in the data, giving further proof that we live in a spiral galaxy. At least 10,000 galaxies outside the Milky Way were also observed. A few sources were seen to move and therefore had to be nearby. Some turned out to be asteroids; five were comets. A ring of dust particles was discovered around the star Vega, the first evidence of cool solid matter around a star other than the Sun. Some newly born stars were also seen, hot bodies that had not turned on as stars yet. From launch of the vehicle to analysis of the data, IRAS was an unqualified success.

IRAS did a complete all-sky survey for infrared sources. A new Infrared Space Observatory (ISO) being developed by the European Space Agency will study individual objects. Similar in design to IRAS, but larger and more capable, ISO will be capable of detecting infrared sources as faint as an ice cube 60 miles away, and will be able to lock onto objects the size of a person seen from 60 miles. With more than 550 gallons of liquid helium, ISO is expected to have an 18 month lifetime. To avoid interference to the detectors from the Earth's trapped protons and electrons, the spacecraft will be inserted into a highly elliptical 24 hour orbit with perigee at 600 miles and apogee at nearly 43,000

miles. It will spend about 16 hours per day near the apogee distance, outside the radiation belts where the detectors can operate at their highest sensitivity. Launch is expected in 1995.

Cosmic Background Explorer (COBE)

The Cosmic Background Explorer searched for microwave noise left over from the big bang. The existence of such radiation was predicted more than 40 years ago and was first detected on the Earth as a hiss of radio noise in 1965. The COBE spacecraft carried a radio receiver to listen for noise from space in a broad range of wavelengths. According to theory, the noise should be most intense at a wavelength of about 2 mm and be less intense at both longer and shorter wavelengths. COBE found exactly that. In addition, COBE made maps of the microwaves over the entire sky. After known sources of microwaves, such as the Sun and Earth, were removed, the maps were extremely smooth. No evi-

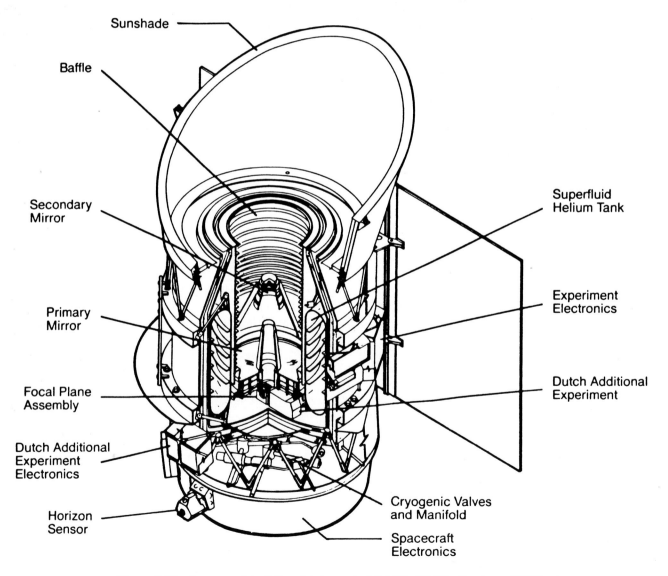

Figure 7.28 Anatomy of the Infrared Astronomical Satellite (IRAS). *Courtesy of NASA.*

Figure 6.10 Landsat 5 image of Copenhagen, Denmark. Bands 1, 2, and 3 were used to make this image in approximately natural color. The image was enhanced somewhat to increase the contrast between rural and urban areas. Water appears black; surf can be seen along the upper right beach. City streets, harbor docks, and the airport are clearly visible. *Courtesy of EOSAT.*

Figure 6.11 Landsat infrared image of Denver, Colorado. Vegetation is printed in various shades of red; urban and commercial areas are light blue or grey; water is black. The mountains on the left are covered with evergreen trees. To the upper right are agricultural fields, some with growing vegetation, others lying fallow. The Platte River runs through the center of the city. Streets and grassy residential areas can be clearly identified. *Courtesy of EOSAT.*

Figure 6.12 Winter wheat crops in Kansas. This infrared image was created from Landsat 5 bands 2, 3, and 4; red is used for vegetation. The circles are irrigated fields, using a sprinkler system which is pivoted at the center of the field and moves in a circle on wheels. Red fields have actively growing crops; white and grey fields lie fallow. The Arkansas River runs from the bottom to the top on the left side of the picture. Garden City is the dark colored area left of the river. *Courtesy of EOSAT.*

117

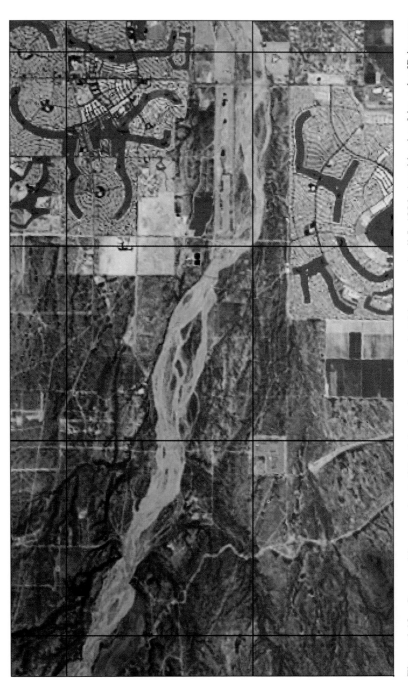

Figure 6.13 Communities carved out of the Arizona desert, a Landsat 5 image on April 5, 1986. As with other false color IR images, vegetation is red, water areas are black, and housing, streets, and bare soil are grey or blue. The Agua Fria dry river bed runs through the center of the image. Sun City developments with parks and golf courses are obvious. Interstate 89 and the city of Surprise are in the lower right corner. This picture has been computer processed to a Mercator map projection and coincides with the U.S. Geological Survey's Calderwood Butte Quadrangle map. *Courtesy of EOSAT.*

Figure 6.19 Earthrise over the Moon with the Apollo 11 lunar module in the foreground. Views like this emphasize the isolation of our small planet with its closed ecosystem. *Courtesy of NASA.*

Figure 7.6 Jupiter as seen by Voyager from 12.4 million miles away. Swirls in the clouds are caused by wind and the rapid rotation of the planet, one rotation in only 10 hours. The giant red spot in the southern hemisphere has been observed for over 300 years. It spins counterclockwise like a low pressure area in Earth's atmosphere. Two moons can be seen: red Io over the giant red spot and Europa just right of center. Sunlight comes from the right, so the left sides of the planet and the two moons are shadowed. *Courtesy of NASA.*

Figure 7.7 Jupiter's Moon Io, sometimes called the "pizza planet" for obvious reasons. The plume rising over the surface in the lower right comes from a volcano. No impact craters are seen because lava flows constantly over the surface. The red-orange color is due to sulfur compounds in the lava, not tomato sauce. *Courtesy of NASA.*

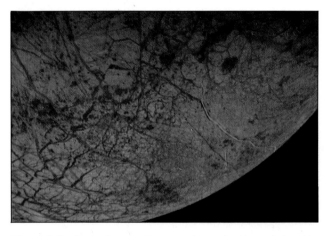

Figure 7.8 Jupiter's moon Europa. Covered with water ice, Europa is the smoothest body yet seen in the Solar System. Dark and light lines on the surface are probably cracks in the ice where material from below squeezed up and froze. Europa's orbital period around Jupiter is about 3.5 days. *Courtesy of NASA.*

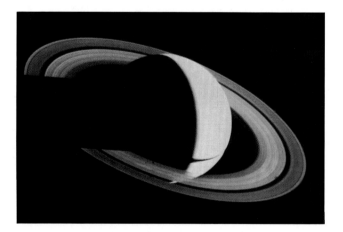

Figure 7.10 A view of Saturn that cannot be seen from Earth. Voyager 1 had passed the planet 4 days before and looked back toward the Sun to make this picture. The rings are composed mainly of small chunks of water ice. They are quite thin and the edge of the planet can be seen through them at the bottom. Notice the shadow of the planet on the rings at the left and the shadow of the rings on the planet near the center of the picture. *Courtesy of NASA.*

Figure 7.11 Saturn's moon Enceladus. The lower part of the picture shows many craters while the upper part has none at all. The grooves and ridges at the center and top are probably faults through which water from the interior flowed to wipe out the craters on that part of the moon. *Courtesy of NASA.*

Figure 7.16 Galileo spacecraft passing Jupiter. *Courtesy of Harris Corporation; artist J. Harrelson.*

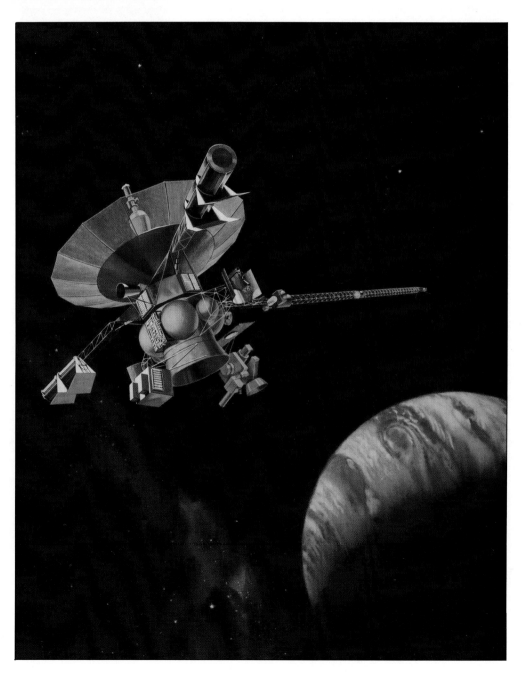

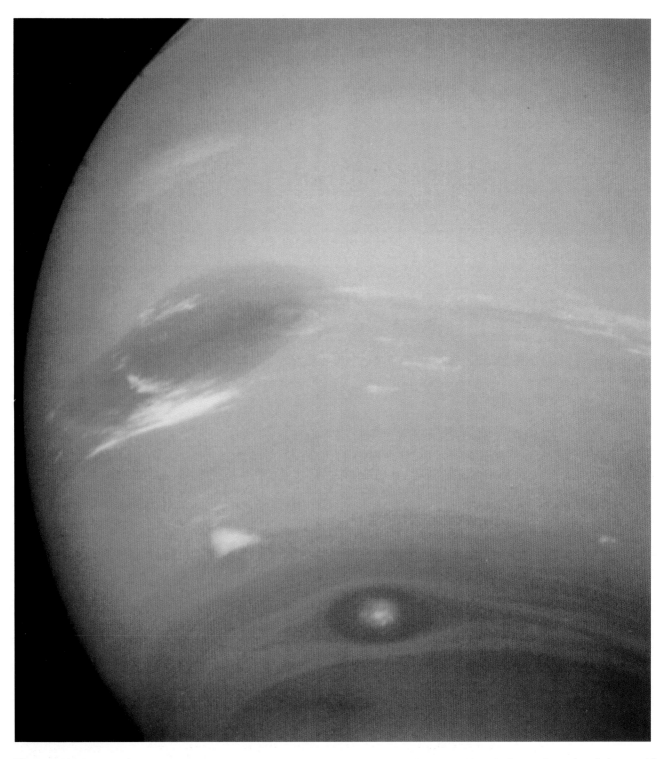

Figure 7.13   Neptune. At the left-center is the great dark spot attended by wispy white clouds. Near the bottom is another dark spot with a bright center. Between the two spots is a white feature nicknamed "Scooter" because it moved faster than the other features. *Courtesy of NASA.*

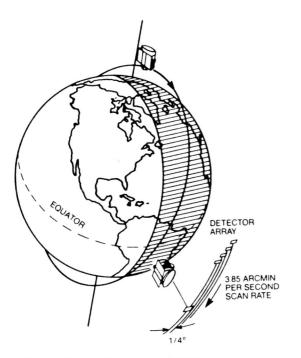

Figure 7.29 IRAS orbit. *Courtesy of NASA.*

dence of early galaxies was found and the origin of galaxies is still uncertain.

Hubble Space Telescope

The Space Telescope is named after Edwin Hubble, the American astronomer whose research on galaxies early in the twentieth century led to the realization that the universe is expanding. The telescope is 43 feet long and weighs (on Earth) about 25,000 pounds. See Figure 7.30. The working end of the telescope is an 8 foot diameter mirror (Figure 7.31) which brings electromagnetic waves from space into focus. The size of this mirror makes it one of the largest instruments ever built; its location in orbit above the atmosphere makes it unique and more powerful than any other telescope. The Space Telescope can observe for longer periods of time unhindered by clouds and scattered sunlight, up to 4,500 hours per year compared with perhaps 2,000 hours per year at the best ground observatories.

Figure 7.32 is a diagram of the Space Telescope and Figure 7.33 shows the light path through the instrument. The telescope is a Cassegrain design, a common design for Earth-based telescopes. The aperture door is opened and

Figure 7.30 Artist's concept of how the Hubble Space Telescope appears in orbit. *Courtesy of Lockheed.*

Figure 7.31 Technicians check the 8 foot diameter mirror of the Hubble Space Telescope. It is made of a lightweight honeycomb glass core fused between two plates of glass. A reflective aluminum coating topped with magnesium fluoride is applied later. Earth weight is 1,827 pounds. *Courtesy of NASA.*

light from the object under study enters. A system of baffles keeps stray light out. The light rays reflect from the large primary mirror to a 12 inch secondary mirror mounted in the center of the tube, 16 feet from the primary. The light beam is reflected back from the secondary mirror and passes through a 24 inch hole in the center of primary mirror to the instruments enclosed in the aft shroud.

Five different instruments for analysis of the light are mounted behind the primary mirror. Two of them are spectrographs which break up white light into its rainbow of colors, two are cameras for wide angle and planetary photography and for faint object photography, and the fifth is a photometer which measures light intensity. Star trackers and fine guidance sensors keep the telescope pointed with unprecedented precision. It could remain locked onto a dime at 450 miles.

The Space Telescope is powered by six nickel-cadmium (Nicad) batteries kept charged by two arrays of solar cells, each providing up to 2,000 watts of power when the instrument is in the sunlight. This operates much like an automobile electrical system where the battery is kept charged by an alternator attached to the engine. The batteries charge

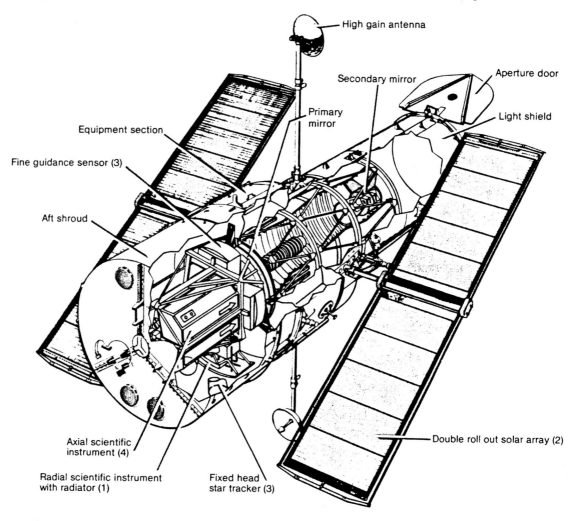

Figure 7.32 Main components of the Hubble Space Telescope. *Courtesy of NASA.*

when the telescope is in sunlight and provide power to the equipment when it is on the dark side of Earth. This charge-discharge cycle recurs 16 times each day as the telescope orbits Earth every 90 minutes. An average of 2,400 watts of electrical energy is required by the telescope and its associated instrumentation.

The telescope is remotely controlled from the ground. Earthbound astronomers at the Space Telescope Science Institute at The Johns Hopkins University in Baltimore evaluate proposed research projects from astronomers worldwide and set the schedule for observing time.

The instrument was expected to detect objects only one-fiftieth as bright as can be seen from the ground. Thus it would be able to see objects seven times farther away than ground-based telescopes. (The square root of 50 is about 7. See **MATHBOX 7.1**.) Ground-based telescopes can detect objects 2 billion light years away; the Space Telescope should be able to observe objects as far away as 14 billion light years. (A light year is the distance light travels in a year, about 6 trillion miles.) Since the universe is thought to be about 15 billion years old, the most distant objects seen by the Space Telescope are observed as they were at the beginning of time, shortly after the big bang. This expands the observable volume of space to 350 times that which is observable from the ground. (Seven cubed is about 350. See **MATHBOX 7.2**.)

Trouble with Hubble

In April 1990, the Hubble Space Telescope was carried by the Space Shuttle *Discovery* to a 320 mile circular orbit with an inclination of 28.5°. That is about the highest altitude the Shuttle can attain with it. Almost immediately, operators on the ground found difficulty focusing the telescope. Within 2 months they came to the awful conclusion that the mirror had been ground to the wrong specifications. For light rays to come to a sharp focus, the curved mirror must be in the shape of a paraboloid. If it is shaped like a section of a sphere, then the light rays reflecting from the mirror do not all merge to the same point and the image is fuzzy. This defect, known to every amateur astronomer who has tried to grind a mirror, is called spherical aberration and that is exactly the problem with the Space Telescope's mirror. The company that manufactured the mirror had used a wrong measuring instrument when grinding the curved surface. For an 8 foot mirror, there is not much difference between a spherical surface and a parabolic one. The edge was too flat by less than the width of a human hair. The great promise of sharp images of faint objects could not be realized.

Even in its flawed condition, much good work was done with the instrument in the first 3 years of operation. The telescope operators designed computer programs which took out some of the fuzziness. Most astronomical obser-

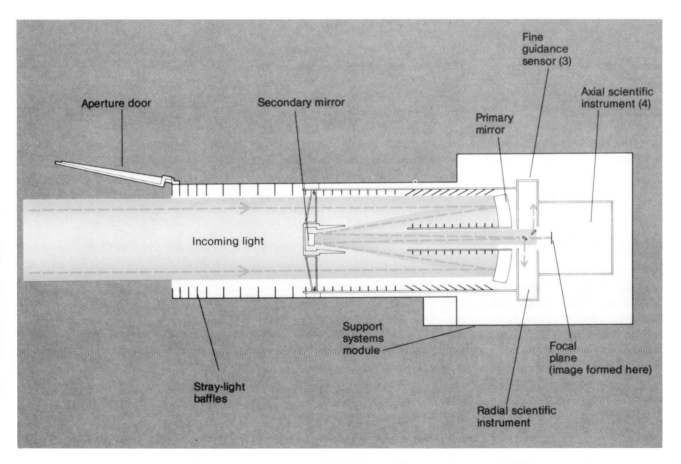

Figure 7.33 Light path through the Hubble Space Telescope. *Courtesy of NASA.*

MATHBOX 7.1

Light Intensity

Light waves expand outward from their source spherically in all directions. The area of a sphere is $a = 4\pi R^2$ where R is the sphere's radius.

Refer to Figure 7.1.1. Sphere A has a radius of 1 mile and an area of 4π. Sphere B has a radius of 3 miles and an area of 36π.

Although the radius of sphere B is three times as large, its area is nine times larger: $3^2 = 9$.

As it expands from sphere A to sphere B, light energy from a source at the center would spread out over nine times the area and would therefore be only one-ninth as intense by the time it reached sphere B.

Thus, light intensity decreases as the square of the distance from the source. If a light source is 7 times farther away it is only 1/49 as bright. If two lights are actually of the same intensity but one appears 49 times fainter, then it must be 7 times farther away.

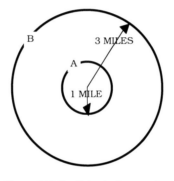

Figure 7.1.1   Spherical expansion.

vations do not require an image; they analyze the light with instruments which break it into its colors. The Space Telescope was able to do that. However, because of the spherical aberration, only about 12 percent of the light came to the focus point; the other 88 percent was spread over a large fuzzy halo, so faint objects could not be analyzed.

Could it be fixed? The Space Telescope was originally designed to be serviced and repaired in orbit by astronauts in space suits. The first servicing mission had been planned for 1993, anyway. Original plans included bringing the instrument back to Earth for complete overhaul every few years. That idea has been generally abandoned because of possible damage and contamination on a return to Earth. The general rule now is to fix it in orbit, and only when it needs fixing. Exchanging the big mirror and realigning it in space were out of the question, but flawed optical devices can often be corrected by adding another optical device. Defective human eyes are corrected with glasses or contact lenses. A very small mirror ground to correct the spherical aberration and placed in the light path of the Space Telescope would remove the spherical aberration, and that is what was done.

A second severe problem was found in the first few months of operation. The two solar panels heated up in the sunlit side of the orbit and cooled down on the dark side as expected. The sudden change in temperature each time the telescope passed from sunlight into darkness or from darkness to daylight also caused them to jitter unacceptably and vibrate the entire instrument. Computer commands corrected some of the jitter, but eventually the panels would have to be replaced.

In December 1993, the Shuttle *Endeavour* came within 30 feet of the Space Telescope and pulled it into the cargo bay with the manipulator arm. Figure 7.34 is an artist's concept of the operation. In five space walks, astronauts inserted the corrective optics, replaced the jittering solar panels, added some computer memory, replaced two pairs of faulty gyros, and replaced the magnetometers with improved ones.

After repairs were complete, the telescope was raised to a higher orbit and released. Atmospheric drag gradually lowers its orbit as with any other satellite. When it gets down below 300 miles, the pointing accuracy is degraded by the air drag and it must be boosted to a higher orbit again.

MATHBOX 7.2

Volumes of Spheres

The volume of a sphere is $V = 4\pi R^3/3$ where R is the radius of the sphere.

Refer to Figure 7.2.1.

Sphere A has a radius of 1 mile and a volume of $4\pi/3$. Sphere B has a radius of 2 miles and a volume of $32\pi/3$.

Although the radius of sphere B is twice as large, its volume is eight times larger: $2^3 = 8$.

If the Space Telescope can see objects 49 times fainter than are now visible from Earth, then it can see seven times farther and can examine a volume of space $7^3 = 343$ times greater.

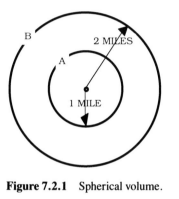

Figure 7.2.1 Spherical volume.

The repair job was a complete success. Now more than 70 percent of the light is concentrated at the focal point, approaching the theoretical limit of what a telescope of that size is able to do. According to astronomers, it is performing as it was designed.

A superb example of its capabilities is shown in Figure 7.35 in the color section, page 191. The image is of a small region in the core of the Orion Nebula, about 1,500 light years from Earth. It is a region in which new stars are forming, at least 500 of them in the past few million years. The hot, bright young star in Figure 7.35 is surrounded by about a dozen smaller stars. These young stars are still surrounded by disks of dust and gas from the clouds in which they formed. It is believed that planets could form from the matter in these disks, which are called protoplanetary disks or proplyds. Intense radiation from θ^1C-Orionis is blowing material from the proplyds into "comet tails" which are clearly seen in Figure 7.35. This remarkable image is a sample of what we can expect from the Hubble Space Telescope in the future.

DISCUSSION QUESTIONS

1. Why is there so much interest in astronomy?

2. Would it be better to build a few very expensive, very capable astronomy spacecraft or many smaller less capable ones? What are the arguments for each?

3. Referring to Figure 7.4, what forces are acting on Voyager at each point along the trajectory?

4. Why was IRAS oriented carefully so it was never pointed toward the Earth or the Sun?

5. What limits the useful life of an infrared telescope in space?

ADDITIONAL READING

Bowyer, Stuart. "Extreme Ultraviolet Astronomy." *Scientific American*, August 1994. Results from the EUV Explorer spacecraft.

Burrows, William E. *Exploring Space, Voyages in the Solar System and Beyond*. Random House, 1990. The human side as well as the technical side of space exploration.

Davies, John K. *Satellite Astronomy*. Ellis Horwood Ltd., 1988. Technical, but readable.

Dunne, James A. *The Voyage of Mariner 10*. NASA SP-424. Government Printing Office, 1978. Detailed results of mission to Venus and Mercury.

Fimmel, Richard O., et al. *Pioneer Venus*. NASA SP-461, Government Printing Office, 1983. Complete document of results

Figure 7.34 Hubble Space Telescope being repaired on orbit. Two astronauts can be seen in the picture, one on the manipulator arm and one in the cargo bay. *Courtesy of Lockheed; artist Gordon Raney.*

Giacconi, Riccardo. "The Einstein X-Ray Observatory." *Scientific American*, February 1980. Results from HEAO-2 spacecraft.

Habing, Harm J., and Gerry Neugebauer. "The Infrared Sky." *Scientific American*, November 1984. IRAS results.

Hart, Douglas. *The Encyclopedia of Soviet Spacecraft*. Exeter Books, 1987. Comprehensive and well illustrated.

Kohlhase, Charles. *The Voyager Neptune Travel Guide*. JPL Publication 89-24, Government Printing Office, 1989. Detailed description of Voyager spacecraft and their mission in anticipation of Voyager 2's encounter with Neptune.

Laeser, Richard P., et al, "Engineering Voyager 2's Encounter with Uranus." *Scientific American*, November 1986. How spacecraft problems were solved.

Luhmann, Janet G., James B. Pollack, and Lawrence Colin. "The Pioneer Mission to Venus." *Scientific American*, April 1994. The 14-year history of the spacecraft.

Saxena, S. K., ed. *Chemistry and Physics of Terrestrial Planets*. Springer-Verlag, 1986. Prior understanding of chemistry, physics, and geology is necessary.

Yenne, Bill, et al. *Interplanetary Spacecraft*. Exeter Books, 1988. Descriptions of U.S. spacecraft, components and subsystems, trajectories and communications; well illustrated and readable.

PERIODICALS

Aerospace America. A monthly magazine of the American Institute of Aeronautics and Astronautics. Technical, but readable.

Astronomy. A popular monthly magazine with frequent articles about astronomy from space.

The Planetary Report. Bimonthly popular magazine published by the Planetary Society for its members.

Sky and Telescope. Monthly magazine for amateur astronomers with frequent articles on spacecraft.

Space News. A weekly newspaper devoted to the business, politics, and technology of space activities.

NOTES

Chapter 8

Space Shuttle

The Space Shuttle is probably the most complex machine ever built. It is the first reusable vehicle for carrying people and cargo into orbit and returning them to Earth. Part of the Space Transportation System, the Shuttle takes off like a rocket, orbits Earth as a spacecraft, and lands like an airplane. It is the outgrowth and merging of two technologies: the Mercury-Gemini-Apollo "man-in-the-can" approach to spaceflight combined with the experimental rocket-powered aircraft. Instead of a capsule sitting atop a heavy-lift booster rocket, this new approach has a rocket-powered airplane, the orbiter, attached to a large external fuel tank with two solid propellent booster rockets attached to it, as shown in Figure 8.1. All except the external tank are recovered, reconditioned, and reused. The orbiter was originally designed for 100 missions; the solid rocket boosters were designed for 20 flights.

The Space Shuttle is the backbone and workhorse of the U.S. manned space program. It was expected to be a low cost bus-truck service to space for 7 to 14 day missions. It can carry satellites into orbit and drop them off; pick up satellites and repair them on site or return them to Earth; conduct scientific, technological, and industrial research; or act as a platform for construction of large objects in space. All of these tasks were done in the first 24 successful flights from April 1981 to January 1986 when the *Challenger* accident occurred. These flights are cataloged in Table 8.1. Flying did not resume until September 1988 while NASA made changes in hardware, procedures, and management. Then 34 more flights were completed successfully by the end of 1993. These are listed in Table 8.2. Eight flights per year are planned through the end of the century.

The Shuttle carries a crew of up to seven. It was designed to open spaceflight to men and women who do not have the test pilot background and training that previous astronauts had. Any person in reasonably good health with the need to go into space for scientific or commercial ventures could do so.

Maximum altitude for a Shuttle orbit is nearly 700 miles. If a satellite must be delivered to a higher orbit, it is equipped with a two stage solid rocket booster called the inertial upper

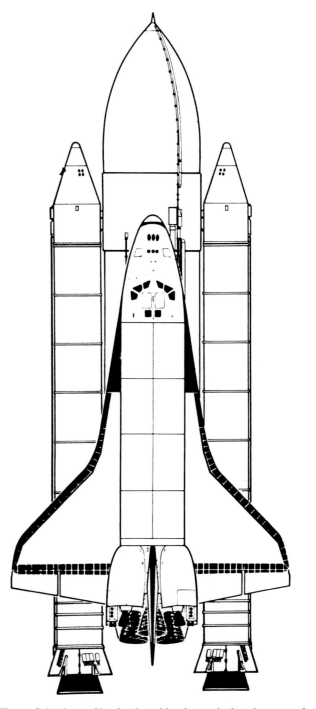

Figure 8.1 Space Shuttle: the orbiter is attached to the external fuel tank and flanked on each side by a solid propellant rocket booster. *Courtesy of NASA.*

TABLE 8.1 Space Shuttle Flights: 1981–1986

Flight	Dates	Orbiter	Missions
STS-1	Apr 12–14, 1981	*Columbia*	First test flight
STS-2	Nov 12–14, 1981	*Columbia*	Test remote manipulator system; Earth survey instruments
STS-3	Mar 22–30, 1982	*Columbia*	Test manipulator arm; materials processing
STS-4	Jun 27–Jul 4, 1982	*Columbia*	Test manipulator arm with scientific payload; materials processing; first DOD payload
STS-5	Nov 11–16, 1982	*Columbia*	Launch two communications satellites
STS-6	Apr 4–9, 1983	*Challenger*	First EVA spacewalk; deploy NASA TDRS satellite
STS-7	Jun 18–24, 1983	*Challenger*	Launch two communications satellites; release and recapture tests
STS-8	Aug 30–Sep 6, 1983	*Challenger*	Deploy comm satellite; test remote manipulator arm
STS-9	Nov 28–Dec 8, 1983	*Columbia*	First Spacelab
41-B	Feb 3–11, 1984	*Challenger*	First test of manned maneuvering unit, backpack propulsion system; launch two comm satellites (boosters failed to send them to synchronous altitude)
41-C	Apr 6–13, 1984	*Challenger*	Retrieve, repair, redeploy Solar Maximum Mission spacecraft using MMU; deploy Long Duration Exposure Facility
41-D	Aug 30–Sep 5, 1984	*Discovery*	Launch three comm satellites; extend and retract large solar array
41-G	Oct 5–13, 1984	*Challenger*	Launch Earth Radiation Budget Explorer; test Earth imaging radar
51-A	Nov 8–16, 1984	*Discovery*	Launch two comm satellites; retrieve two failed satellites from flight 41-B
51-C	Jan 24–27, 1985	*Discovery*	DOD payload
51-D	Apr 12–19, 1985	*Discovery*	Deploy two commercial satellites (one booster failed)
51-B	Apr 29–May 6, 1985	*Challenger*	Spacelab; materials processing
51-G	Jun 17–24, 1985	*Discovery*	Deploy three communications satellites; deploy and retrieve Spartan science experiment
51-F	Jul 29–Aug 6, 1985	*Challenger*	Spacelab, solar physics, astrophysics, plasma physics, materials processing
51-I	Aug 27–Sep 3, 1985	*Discovery*	Deploy three comm satellites; retrieve, repair, redeploy failed satellite from 51-D
51-J	Oct 3–10, 1985	*Atlantis*	DOD mission
61-A	Oct 30–Nov 6, 1985	*Challenger*	Spacelab, materials processing
61-B	Nov 26–Dec 3, 1985	*Atlantis*	Deploy three comm satellites; assemble large structures; study EVA dynamics
61-C	Jan 12–18, 1986	*Columbia*	Deploy comm satellite; deploy hitchiker satellite; IR imaging experiment; images of Comet Halley
51-L	Jan 28, 1986	*Challenger*	The accident

Adapted from the Presidential Commission on the Space Shuttle *Challenger* Accident, *Report to the President*.

stage (IUS). When the Shuttle reaches orbit, the satellite with its IUS is released from the cargo bay, checked out, and left. At the proper time a command from the ground ignites the IUS and the satellite is carried to its higher orbit, often to geosynchronous altitude.

A robotic mechanical arm designed in Canada is used to move things in and out of the cargo bay and to assist in other work. Astronauts in space suits leave the cabin through an airlock to work outside. A rocket-powered backpack called the manned maneuvering unit is used to fly away from the Shuttle to repair satellites or other jobs. Spacelab is a laboratory which is carried in the cargo bay to provide a shirt-sleeve environment for scientific research. All of this and more will be discussed in detail in Chapter 10, Working in Space. This chapter is concerned with the structure and flight profile of the Shuttle.

The Space Shuttle has three major components: the *orbiter*, a pair of *solid rocket boosters* (SRBs), and the *external tank* (ET). See Figure 8.2.

Solid Rocket Boosters

The two solid propellant booster rockets burn for about 2 minutes, providing most of the initial thrust, along with the three engines in the orbiter, to lift the 4.4 million pound vehicle from the launch pad and boost it to an altitude of about 28 miles. While resting on the launch pad, the entire weight of the vehicle is supported by the two booster rockets.

The solid rocket boosters (SRBs) of the Space Shuttle are the largest ever flown, the first designed for reuse and the first to power a manned spacecraft. Figure 8.3 shows SRB structure and Table 8.3 lists SRB statistics. Overall each booster is nearly 150 feet long and 12 feet in diameter, constructed of half-inch thick steel. Because of their size, it is not possible to manufacture the motors and ship them to Kennedy Space Center in one piece. They are made in eleven pieces at a factory in Utah, then partially assembled, filled with propellant, and shipped in four segments on special flatbed railway cars to Florida for final assembly.

The propellant is composed of aluminum powder fuel, ammonium perchlorate oxidizer, a polymer binder, an epoxy curing agent, and an iron oxide catalyst. When hardened into a grain it looks and feels like a hard rubber typewriter eraser. The perforation, the hollow core that runs up the center of the grain, is an 11-point star in cross section. (Refer back to Chapter 2 for a discussion of perforations.) To ignite the propellant, a small rocket motor is fixed at the upper end of the perforation. Its flames ignite the entire exposed inner surface of the perforation and the booster comes up to full thrust in less than half a second.

Each booster contains 1.1 million pounds of propellant and develops a thrust at liftoff of 3.3 million pounds. As the fuel burns, the weight of the booster decreases. As it rises into less dense atmosphere, the drag decreases, and as it leaves the surface of the Earth, the gravitational force decreases. All these contribute to an increase in acceleration.

TABLE 8.2 Space Shuttle Flights Since the *Challenger* Accident: 1988–1994

STS	Launch	Duration	Orbiter	Primary Mission
26	Sep 29, 1988	4	*Discovery*	Launch TDRS communications satellite
27	Dec 2, 1988	4	*Atlantis*	DOD payload
29	Mar 13, 1989	5	*Discovery*	Launch TDRS communications satellite
30	May 4, 1989	4	*Atlantis*	Launch Magellan to Venus
28	Aug 8, 1989	5	*Columbia*	DOD payload
34	Oct 18, 1989	5	*Atlantis*	Launch Galileo to Jupiter
33	Nov 22, 1989	5	*Discovery*	DOD payload
32	Jan 9, 1990	11	*Columbia*	Launch Syncom satellite; Retrieve LDEF
36	Feb 28, 1990	5	*Atlantis*	DOD payload
31	Apr 24, 1990	5	*Discovery*	Deploy Hubble Space Telescope
41	Oct 6, 1990	4	*Discovery*	Launch Ulysses to Sun
38	Nov 15, 1990	5	*Atlantis*	DOD payload
35	Dec 2, 1990	9	*Columbia*	Astrophysics experiments
37	Apr 5, 1991	6	*Atlantis*	Deploy Gamma Ray Observatory
39	Apr 28, 1991	8	*Discovery*	Space science
40	Jun 5, 1991	9	*Columbia*	Space science
43	Aug 2, 1991	9	*Atlantis*	Launch TDRS communications satellite
48	Sep 12, 1991	6	*Discovery*	Deploy UARS atmospheric research satellite
44	Nov 24, 1991	7	*Atlantis*	Launch DSP early warning satellite
42	Jan 22, 1992	8	*Discovery*	Microgravity research
45	Mar 24, 1992	9	*Atlantis*	Space environment research lab
49	May 7, 1992	9	*Endeavour*	Retrieve, repair, release Intelsat
50	Jun 25, 1992	14	*Columbia*	Microgravity laboratory
46	Jul 31, 1992	8	*Atlantis*	European retrievable satellite; tethered satellite
47	Sep 12, 1992	8	*Endeavour*	Spacelab materials and life sciences
52	Oct 22, 1992	10	*Columbia*	Deploy laser satellite; microgravity lab
53	Dec 2, 1992	7	*Discovery*	DOD payload
54	Jan 13, 1993	6	*Endeavour*	Launch TDRS communications satellite
56	Apr 8, 1993	9	*Discovery*	Space environment research lab
55	Apr 26, 1993	10	*Columbia*	Spacelab life sciences
57	Jun 21, 1993	10	*Endeavour*	Retrieve EURECA satellite; Spacehab
51	Sep 12, 1993	10	*Discovery*	Deploy research satellite
58	Oct 18, 1993	14	*Columbia*	Spacelab life sciences
61	Dec 2, 1993	10	*Endeavour*	Repair Hubble Space Telescope
60	Feb 3, 1994	8	*Discovery*	Spacehab microgravity experiments
62	Mar 4, 1994	14	*Columbia*	Microgravity research
59	Apr 9, 1994	11	*Endeavour*	Space Radar Laboratory
65	Jul 8, 1994	15	*Columbia*	International Microgravity Lab
64	Sep 9, 1994	11	*Discovery*	Spaceflight research
68	Sep 30, 1994	11	*Endeavour*	Space Radar Lab

The Shuttle is designed to withstand accelerations of up to 3 g. Therefore, grains are cast in such a way as to produce a regressive burn 55 seconds into the flight to keep the acceleration from exceeding the 3 g design limit. This also keeps stress on the astronauts and the payload within tolerable limits. Figure 8.3 shows the grain structure.

The exhaust nozzle can be swiveled up to 8° from the central axis of the booster to change the direction of thrust and help steer the Shuttle during ascent. Each booster has its own auxiliary power units that operate the hydraulic pumps to steer the nozzle. Guidance system sensors inform the control computer of the speed and direction of motion of the vehicle. The computer signals the booster when a change in direction of the nozzle is required. Electric power is supplied to the booster through a cable from the orbiter's electric system.

External Tank

The external tank (ET) serves two purposes: it carries the propellants for the orbiter's three main rocket engines and it is the support structure that connects the orbiter and SRBs together during ascent to orbit. Figures 8.4 and 8.5 show its structure; Table 8.4 tells its statistics. It is 154 feet long and 27 feet in diameter.

Because it carries both liquid hydrogen fuel and liquid oxygen oxidizer, the ET is really two inner tanks in one outer shell. The upper tank carries the liquid oxygen, 1.36 million pounds of it at −297 °F at liftoff. Its 19,500 cubic foot (143,000 gallon) volume is greater than that of a 2,000 square foot house. The lower tank is about 2.5 times larger (383,000 gallons) and carries about a quarter of a million pounds of liquid hydrogen at −423 °F. If these proportions do not seem to make sense, see MATHBOX 8.1.

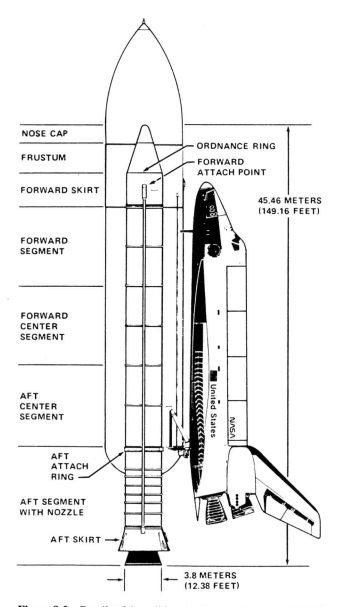

Figure 8.2   Details of the solid rocket booster. *Courtesy of NASA.*

Both inner tanks are constructed of aluminum and titanium alloys up to 2 inches thick. The oxygen tank contains baffles to keep the liquid oxygen from sloshing around during flight and throwing the Shuttle off course. The density of liquid hydrogen is so low that sloshing is not a problem. An intertank collar connects the two propellant tanks together and provides space for most of the electrical components. Since the first tank was built, the manufacturer has made some design changes, substituted titanium for aluminum in some components, and eliminated the white exterior paint. As a result, the weight of the newer tanks has been reduced by more than 6,000 pounds, allowing the Shuttle to carry a heavier payload. A still lighter ET fabricated of an aluminum-lithium alloy is currently under development which will save another 8,000 pounds.

To reduce atmospheric drag, the oxygen tank curves to a

point at the upper end of the structure. The entire outer surface of the external tank is insulated with a half inch thick cork/epoxy layer covered with 1 to 2 inches of spray-on foam. Insulation is necessary for two reasons. The propellants are very cold: liquid oxygen boils at $-297\,°F$ and liquid hydrogen at $-423\,°F$. An uninsulated tank would absorb heat from the surroundings causing uncontrolled boiling of the propellants. This poses two problems: excessive loss of hydrogen and oxygen through vent valves and buildup of excessive pressure in the tanks. Controlled boiling is necessary on the launch platform to keep the tanks pressurized for structural strength and also to assist the pumps in moving the propellants out to the engines. During flight, the tanks are pressurized by gases from the engines. In addition, because of the cold temperatures, if the tank were not insulated, water vapor in the air would readily condense as ice on the sides. At liftoff, the ice would break loose and damage the Shuttle.

Propellants flow to the engines through 17 inch pipes: oxygen at a rate of about 14,000 gallons per minute and hydrogen at nearly 50,000 gallons per minute.

The Orbiter

The *orbiter* is the airplane-rocket ship shown in Figure 8.6 and its dimensions are given in Table 8.5. It is about the size of a DC-9 jet airliner as seen in Figure 8.7. The main structure of the orbiter is constructed of aluminum, similar to the way an airliner is built. There are three main sections: the forward fuselage with the crew compartment; the midsection which includes the payload bay and its clamshell doors; and the aft fuselage which includes the engines, pods, and vertical tail. Empty it weighs about 170,000 pounds.

Five orbiters have been built. *Columbia* was the first, delivered to NASA in March 1979. It was used for the first four orbital test flights and has been modified and modernized several times as new technology developed. *Discovery* and *Atlantis* were delivered in 1983 and 1985, respectively. The newest orbiter is *Endeavour*, built as a replacement for *Challenger* which was destroyed in an accident in January 1986. Each orbiter is thoroughly inspected, refurbished, and modified about every 3 years. Current modifications include a docking port and airlock changes to prepare for docking with the Russian Mir space station and the new International Space Station. •

Payloads are carried in the cavernous cargo bay. The maximum payload weight depends on the direction of launch. Eastward launches take place from Kennedy Space Center in Florida. Polar launches were planned from Vandenberg Air Force Base, California, but no launches ever took place from there and after the *Challenger* accident in 1986, the facility was closed. Polar launches do not take place from Florida because the spacecraft would overfly land areas while under power. Because of the rotation of the

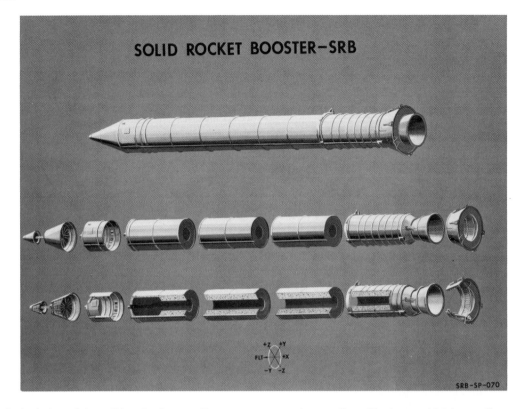

Figure 8.3 Grain design of the solid rocket booster. Four segments contain propellant. The bottom (right) propellant grain is slightly tapered and flared out at the end with the nozzle attached; exhaust gases from the entire engine must escape through this section. The two center segments have cylindrical grains which produce increasing thrust with time, as the burning surface enlarges. The upper (left) half of the topmost section contains less propellant and has slots like a star-shaped grain. This segment burns very rapidly, contributing to the high thrust needed at liftoff but burns out quickly so the maximum desired thrust level is not exceeded. The thrust-time graph is also shown. *Courtesy of NASA.*

Earth, Cape Canaveral is moving eastward with respect to space at a speed of 914 miles per hour. Given that initial speed while still on the launch platform, the Shuttle can carry a payload of 65,000 pounds into low Earth orbit on an eastward launch from Kennedy Space Center. A maximum of 39,700 pounds could be carried into polar orbit because the initial eastward velocity of Earth's rotation does not help in a polar launch. Up to 32,000 pounds of cargo can be returned from orbit to Earth.

The cargo bay is not pressurized. Its doors open like a clamshell to expose the orbiter's cooling radiators (Figure 8.8). The front radiators tilt so that heat may escape from both sides. It is absolutely essential that the doors be opened immediately on achieving orbit and that the radiators be exposed to space. Otherwise the orbiter would overheat within a few orbits and the mission would have to be aborted. On the other hand the doors must be closed for reentry or air friction would quickly destroy them, and probably the entire cargo bay. If the automatic door opener fails, a manual method can be used.

The orbiter is powered by 49 engines used in various combinations to launch to orbit, maneuver while in orbit, and return to Earth. Its three main engines are the most advanced liquid-fueled rockets ever built. Propellants are liquid hy-

drogen and liquid oxygen carried in the external tank. Each develops 375,000 pounds of thrust at 100 percent rated power. The engines can be throttled from 65 percent to 109 percent of rated thrust by varying the flow of the fuel from the tanks, just as you operate the accelerator in your car to change the engine speed.

Rockwell International built the main engines. The fuel pumps are not much larger than an automobile engine, yet they generate as much horsepower as 28 locomotives, 100

TABLE 8.3 Solid Rocket Booster Statistics

Overall dimensions		
Height	149.2	feet
Diameter	12.2	feet
Propellant composition		
Aluminum powder (fuel)	16.0	percent
Ammonium perchlorate (oxydizer)	69.8	percent
Iron oxide (catalyst)	0.2	percent
Polymers (binder)	12.0	percent
Epoxy (curing agent)	2.0	percent
Weight each booster		
Empty	192,000	pounds
Propellant	1,106,000	pounds
Total	1,300,000	pounds
Thrust at liftoff		
Each booster	3,300,000	pounds

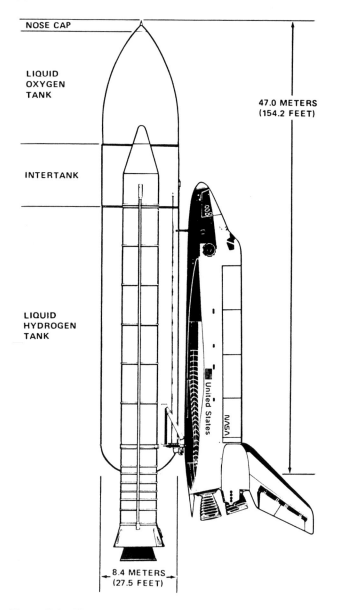

Figure 8.4   The external tank. *Courtesy of NASA.*

TABLE 8.4   External Tank Statistics

Dimensions	
Diameter	27.5 feet
Overall length	154.2 feet
Liquid oxygen tank	53.3 feet
Liquid hydrogen tank	97.0 feet
Intertank	22.5 feet
(Intertank overlaps propellant tanks)	
Total weight	
Empty	78,100 pounds
Loaded	1,668,000 pounds
Propellant weight	
Liquid oxygen	1,359,000 pounds
Liquid hydrogen	226,000 pounds
Total	1,585,000 pounds
Propellant volume	
Liquid oxygen tank	143,000 gallons
Liquid hydrogen tank	383,000 gallons
Total	526,000 gallons
Propellant densities (per cubic foot)	
Liquid oxygen	71.1 pounds
Liquid hydrogen	4.2 pounds

Orbiter Thermal Protection

When returning to Earth, the orbiter must reduce its speed from 18,000 miles per hour in orbit to a 200 mile per hour landing speed in about half an hour. Its kinetic energy of motion must be dissipated in some way. When an automobile is brought to a stop by applying the brakes, its kinetic energy is converted to heat in the brake lining. Similarly with the Shuttle; it reenters with the large, flat bottom surface leading the way so that the atmospheric frictional drag acts like a brake. During reentry into the atmosphere, the frictional heat produces temperatures on the orbiter varying from 600 °F to 2,750 °F. Without a protective covering of insulation, the aluminum would melt. Ablative heat shields were used on the Mercury, Gemini, and Apollo spacecraft, but none of those vehicles was intended for reuse. The heat shield simply charred, flaked off, and ablated away carrying the heat with it. But the Space Shuttle orbiter was designed as a 100 mission vehicle. It would be too expensive to replace the heat shield after each flight, so a reusable one had to be developed.

Figure 8.9 shows how the various areas of the orbiter are covered by different types of insulation. The nose tip and leading edges of the wings are subject to the greatest heat stress at a temperature of 2,750 °F and are protected by a carbon composite (RCC) consisting of layers of graphite cloth in a carbon matrix. The outer layers are chemically converted to silicon carbide, the same material used to make grindstones. Not shown in Figure 8.9, reinforced carbon-carbon is being added between the nose tip and the nose wheel door to provide better insulation of that area.

Perhaps the most remarkable parts of the thermal protection system are the silica fiber tiles identified as LRSI (white tiles) and HRSI (black tiles) in the diagram. This material absorbs great quantities of heat but transfers it very slowly through its interior. Therefore, the atmospheric fric-

horsepower for each pound of weight. By comparison, an automobile engine generates about 1/2 horsepower for each pound of its weight. Output pressure from the pumps could send a column of liquid hydrogen 180,000 feet in the air. The fuel pump runs at 37,000 revolutions per minute; a typical automobile engine runs at 2,500 revolutions per minute when going 60 miles an hour. Liquid hydrogen at −423 °F is the second coldest liquid on Earth (liquid helium is colder). When it burns in the combustion chamber, the temperature reaches 6,000 °F, higher than the boiling point of iron. The three engines empty the half-million gallon external tank in 8.5 minutes.

The orbiter engines are mounted in such a way that they can be gimbaled, moved to change the direction of the thrust. In conjunction with the solid rocket boosters, this provides the means for steering the Shuttle during powered flight.

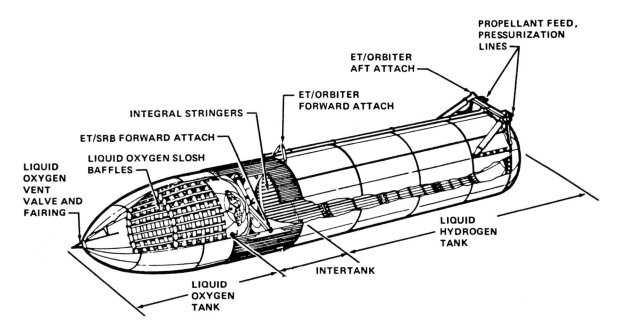

Figure 8.5 Internal structure of the external tank. *Courtesy of NASA.*

MATHBOX 8.1

Hydrogen-Oxygen Combustion

An atom of oxygen has an atomic weight of sixteen; an atom of hydrogen has an atomic weight of one. When the two react chemically (burn) they produce water.

$$2H + O \rightarrow H_2O$$

For complete combustion, then, it is necessary to have twice as many hydrogen atoms as oxygen atoms. Therefore, the hydrogen tank has more than twice the volume of the oxygen tank.

A water molecule has a molecular weight of 18 (2 for the hydrogens + 16 for the oxygen). Therefore the weight of the oxygen atoms is eight times the weight of the hydrogen atoms. If the external tank starts with 1,359,000 pounds of oxygen, then, for complete combustion it will need 1,359,000/8 = 170,000 pounds of hydrogen.

Actually, it carries more hydrogen than that (226,000 pounds) because it burns a hydrogen-rich mixture, only six parts of oxygen to one part of hydrogen, instead of the 8 to 1 ratio for perfect, complete combustion.

Thrust depends on the mass of the exhaust and its velocity out of the nozzle, so the extra hydrogen in the exhaust contributes to the thrust even if it is unburned.

---

tion during reentry heats the tiles to as much as 2,300 °F, but the aluminum skin of the spacecraft never exceeds 350 °F. A spectacular demonstration of the thermal properties of the material is shown in Figure 8.10, where a tile was heated to 2,300 °F in an oven, then picked up with bare hands while the inside was still glowing white hot. Because of the very slow rate of heat flow through the tile, the outer surface had cooled in a few seconds to near room temperature while the interior was still 2,300 °F.

About 70 percent of the orbiter is covered with tiles. Tiles on the bottom, front part of the fuselage, and leading edge of the tail are subject to heating in the range of 1,200 °F to 2,300 °F (650–1,260 °C). They are given a shiny black boron-silicate glass coating which allows 90 percent of the reentry heat to be radiated back into the atmosphere. These tiles measure about 6 inches square and vary in thickness from 1 to 5 inches. On sections of the fuselage sides, tail, and upper wing surfaces subject to temperatures in the range

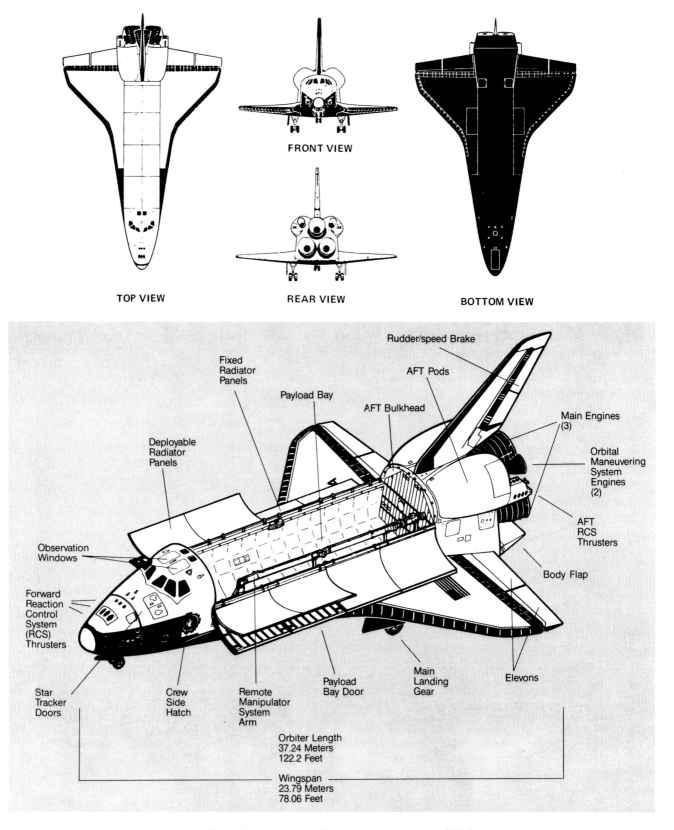

TOP VIEW

FRONT VIEW

REAR VIEW

BOTTOM VIEW

Rudder/speed Brake

AFT Pods

Fixed
Radiator
Panels

Payload Bay

AFT Bulkhead

Main Engines
(3)

Orbital
Maneuvering
System
Engines
(2)

AFT
RCS
Thrusters

Deployable
Radiator
Panels

Body Flap

Observation
Windows

Forward
Reaction
Control
System
(RCS)
Thrusters

Star
Tracker
Doors

Crew
Side
Hatch

Remote
Manipulator
System
Arm

Payload
Bay Door

Main
Landing
Gear

Elevons

Orbiter Length
37.24 Meters
122.2 Feet

Wingspan
23.79 Meters
78.06 Feet

Figure 8.6 The Space Shuttle orbiter. *Courtesy of NASA.*

TABLE 8.5 Orbiter Dimensions

Total length	122.2 feet
Height	56.6 feet
Rudder height	26.3 feet
Wing	
Span	78.1 feet
Maximum thickness	5.0 feet
Body Flap	
Width	20.0 feet
Area	135.8 square feet
Aft fuselage	
Length	18.0 feet
Width	22.0 feet
Mid fuselage	
Length	60.0 feet
Width	17.0 feet
Crew cabin	2525 cubic feet
Payload bay	
Length	60.0 feet
Diameter	15.0 feet

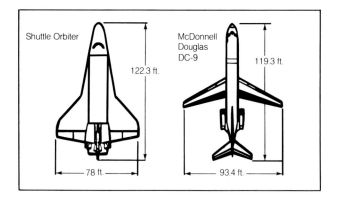

Figure 8.7 Size comparison of the orbiter. *Courtesy of NASA.*

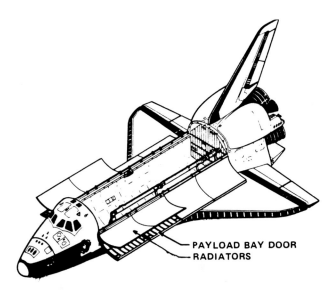

Figure 8.8 The orbiter radiators. *Courtesy of NASA.*

of 750 °F to 1,200 °F (370–650 °C), the tiles are coated with a shiny white aluminum oxide to reflect solar radiation and help keep the spacecraft cool while in orbit. These are about 8 inches square and vary from 0.5 to 2.5 inches thick.

Each of the 30,000 tiles had to be cut and shaped to conform to the curved surfaces of the orbiter. Some of them are shown in Figure 8.11; no two are exactly alike. Groupings of about 20 tiles are set into a frame after the outer surfaces are cut and coated. Then the inner surfaces are milled to shape and bonded to a felt pad which in turn is bonded to the aluminum skin of the orbiter using a silicone resin glue. The felt pad isolates the tiles from strain due to the orbiter's vibrations as it flies into and out of orbit. Excess weight due to the absorption of rain water or condensation could be a problem, so the tiles are also given a waterproofing coat. Although the tiles are soft and easily damaged, they are also easily repaired by spraying scratches and plugging larger holes.

Each tile is given a unique bar code such as on packages and cans at the grocery store. The white tiles are painted with the black lines of the bar code and the black tiles are painted with the white spaces between the bar code lines. The tiles are so important to the successful operation of the Shuttle that any work done on a tile or group of tiles on an orbiter is recorded in a computer data base. The bar codes on those tiles are scanned into the computer to maintain a complete history of each one.

The parts of the orbiter not subject to excessive heating are covered with Nomex felt thermal blankets, a nylon material coated with silicon, which affords sufficient protection where the temperature does not exceed 700 °F. These areas include the cargo bay doors, most of the upper wing surface, lower rear fuselage sides, and pods. Some of the tile-covered areas on the orbiter do not heat up as much as was originally expected, so the tiles presently in place will soon be replaced by the Nomex thermal blankets. This

will make those areas lighter, stronger, and less subject to damage.

Electric Power

Electricity for the Shuttle and its payload is produced by three fuel cells which use hydrogen and oxygen, combining them into water and generating electricity in the process. In a car battery, electricity is produced by the chemical reaction of lead and lead dioxide plates in a solution of sulfuric acid contained within the battery. In a fuel cell, hydrogen and oxygen are brought in from outside tanks.

Fuel cells were invented in 1959 and were used in the Apollo program. Shuttle fuel cells are about the same size as those used in Apollo but generate six times as much electric power. Under the floor of the cargo bay are located four liquid hydrogen tanks with a capacity of 92 pounds apiece (total 368 pounds) and four liquid oxygen tanks holding 781 pounds each (total 3,124 pounds). That is a sufficient supply to generate 1,530 kilowatt-hours of electrical energy, generally enough for an 8 day mission. As the hydrogen and

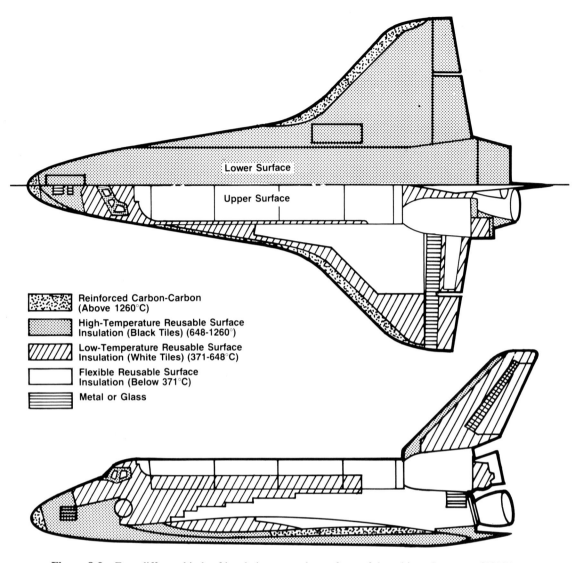

Lower Surface

Upper Surface

Reinforced Carbon-Carbon
(Above 1260°C)

High-Temperature Reusable Surface
Insulation (Black Tiles) (648-1260°)

Low-Temperature Reusable Surface
Insulation (White Tiles) (371-648°C)

Flexible Reusable Surface
Insulation (Below 371°C)

Metal or Glass

Figure 8.9   Four different kinds of insulation cover the surfaces of the orbiter. *Courtesy of NASA.*

oxygen evaporate, tank pressure is built up to force the fluids out of the tanks and into the fuel cells. As the tanks are emptied, the pressure decreases so evaporation is speeded up with electric heaters immersed in the cryogenic fluids. The tanks are actually thermos bottles consisting of an inner tank and an outer shell with a vacuum between, which keeps the cold hydrogen and oxygen in a liquid state. Pressure relief valves assure that the tanks do not explode if excessive pressure builds up.

The 3,492 pounds of pure water produced as a by-product in the fuel cells supplies all human needs with plenty left over. Any excess is dumped overboard into space at intervals.

A pallet with four additional oxygen tanks and four additional hydrogen tanks can be mounted in the cargo bay, essentially doubling the water and power available for missions lasting up to 16 days. The loaded pallet weights about 7,000

pounds and is therefore carried only when a long duration flight is planned.

Assembly

At Kennedy Space Center the various parts of the Space Shuttle are brought together and assembled inside one of the largest buildings in the world, the vehicle assembly building (VAB). It covers 8 acres, stands 525 feet tall, 716 feet long, and 518 feet wide. The large doors operate in two sections. The upper section, 342 feet high and 76 feet wide, consists of seven leaves which move vertically to open. The lower section is 114 feet high and 152 feet wide with four leaves which move horizontally. More than 70 lifting devices are in the building, including two 250-ton cranes.

The entire Space Shuttle is put together on a mobile launch platform. First the solid rocket boosters are assembled and attached to the platform and the external tank is

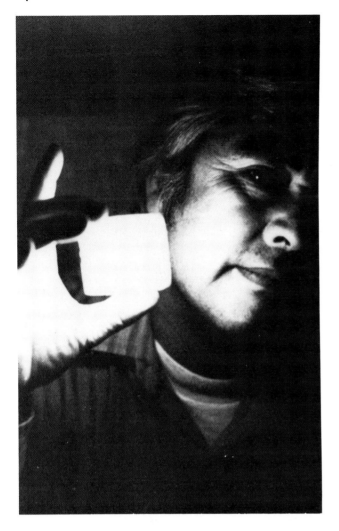

Figure 8.11 Heat protection tiles for the orbiter. Each tile is individually milled to match the contour of a specific spot; no two are alike. *Courtesy of Lockheed.*

Figure 8.10 White hot tile held in bare hands. Photograph was taken by the light of the glowing tile less than 10 seconds after it was removed from the 2,300 °F oven. *Courtesy of Lockheed.*

attached to the boosters. Horizontal payloads are mounted into the cargo bay before the orbiter is hoisted into position and connected to the external tank. Each of these components is tested individually. Then, when mated together, the entire vehicle is thoroughly checked out. A crawler-transporter (Figure 8.12) is then attached to the mobile launch platform with the Shuttle resting on top. This moves out through the doors along a 130 foot wide roadway to one of two launch pads 3.4 or 4.2 miles away, at a maximum speed of 1 mile per hour (Figure 8.13). The launch platform with the Shuttle is deposited on the pad (Figure 8.14) and the crawler-transporter moves back a safe distance. While awaiting the launch, the cold hydrogen and oxygen boil and vaporize. Excess gas must be vented, allowed to escape, so the pressure in the tanks does not build up to an unsafe level.

Typical Mission Profile

A typical sequence of events during a Shuttle launch is found in Table 8.6 and a diagram of a typical mission is shown in Figure 8.15. (Check the table and diagram as we describe the action.) By 1 hour prior to liftoff the propellant tanks are full, the crew is aboard, and computers and crew have checked out all systems to be sure they are ready. In the final minutes of the countdown the engines and nozzles are checked for proper operation and set into their launch position. The orbiter is switched to internal electric power supplied by the fuel cells.

There are several opportunities for holding the countdown if something does not appear to function properly. If an emergency occurs while the crew is in the orbiter on the launch pad, the crew can leave the vehicle to a safe location by means of the escape system shown in Figure 8.16.

Ignition and Liftoff

The three main rocket engines in the orbiter ignite one at a time at intervals of 0.12 second, the first at T − 3.46 seconds. By time T the engines have reached 90 percent thrust. The entire vehicle lurches forward about 40 inches in the direction of the tank, straining against the eight holddown bolts which attach the solid rocket boosters and the entire vehicle to the launch platform. It takes 2.64 seconds for the vehicle to rock back to a vertical position. The booster engines then ignite and come to full power by T + 3 seconds. The holddown bolts are severed by small explosive charges and the Shuttle lifts off (Figure 8.17).

Clouds of smoke and water vapor engulf the area. Much of it comes from water which is poured onto the launch pad area to suppress the intense sound waves produced by the rocket engines. Without sound suppression the orbiter could be damaged by the acoustic energy. The solid rocket boosters also contribute a great amount of smoke to the scene, but the three main rockets on the orbiter produce very little visible exhaust. Recall that they are burning hydrogen and

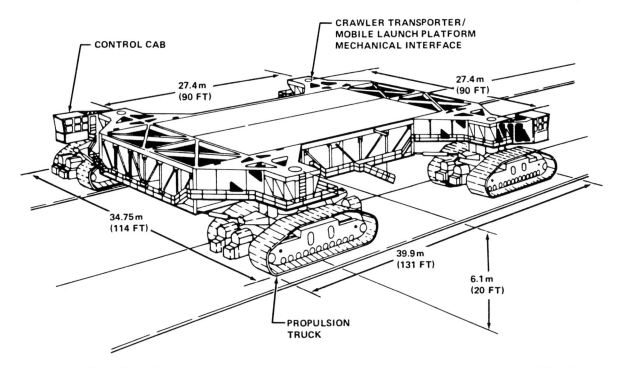

Figure 8.12   Space Shuttle crawler-transporter. Note the dimensions; the tracked drive trucks stand 10 feet tall. The entire vehicle weighs six million pounds unloads. *Courtesy of NASA.*

Figure 8.13   Assembled Space Shuttle "speeds" to the launch site at 1 mile per hour. It rests on a mobile launch platform atop a crawler-transporter. *Courtesy of NASA.*

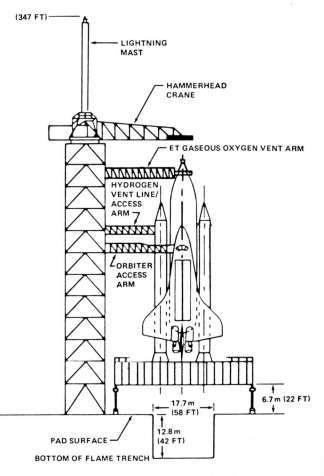

Figure 8.14   Shuttle launch pad. *Courtesy of NASA.*

TABLE 8.6 Space Shuttle Launch Events

Time	Event
T − 4 hr 30 min	Begin filling liquid oxygen tank
T − 2 hr 50 min	Begin filling liquid hydrogen tank
T − 1 hr 5 min	Crew aboard, hatch closes
T − 9 min	Automatic launch sequence starts
T − 4 min 30 sec	Orbiter on internal power
T − 3 min	Main engines move to start position
T − 2 min 55 sec	Oxygen tank at flight pressure
T − 1 min 57 sec	Hydrogen tank at flight pressure
T − 3.46 sec	First main engine starts
T − 3.34 sec	Second main engine starts
T − 3.22 sec	Third main engine starts
T − 0	Main engines reach 90% power
	Delay timer set for 2.46 sec
T + 2.46 sec	Command to start solid rocket boosters
T + 3 sec	Boosters reach 100%
	LIFTOFF

Time	Event	Altitude	Speed*	Range
T + 7 sec	Begin pitchover	545 ft	914 mph	0
T + 1 min 9 sec	Maximum dynamic pressure	8.3 mi	1,654 mph	4 mi
T + 2 min 4 sec	Booster separation	29.4 mi	3,438 mph	23.7 mi
T + 8 min 38 sec	Main engine cutoff (MECO)	73 mi	17,500 mph	829 mi
T + 8 min 50 sec	External tank separation	73.5 mi	17,498 mph	886.6 mi
T + 10 min 39 sec	OMS 1 ignition	78.3 mi	17,479 mph	1,380 mi
T + 12 min 24 sec	OMS 1 cutoff	83.2 mi	17,591 mph	1,860 mi
T + 43 min 58 sec	OMS 2 ignition	173.6 mi	17,201 mph	9,775 mi
T + 45 min 34 sec	OMS 2 cutoff	174.2 mi	17,321 mph	10,269 mi

*Speed is with respect to space. Earth at Kennedy Space Center is rotating eastward at 914 miles per hour.

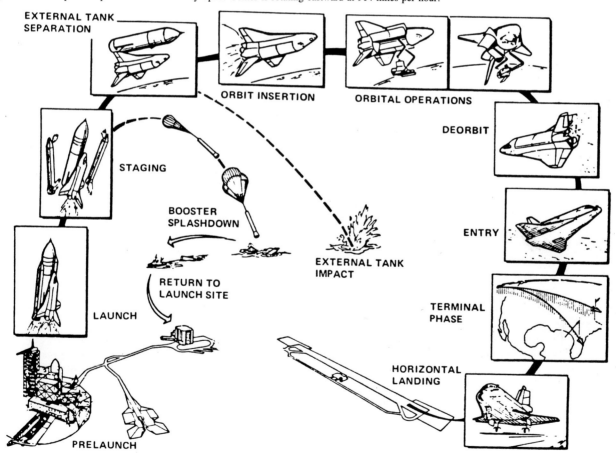

Figure 8.15 Typical launch-to-landing mission profile. *Courtesy of NASA.*

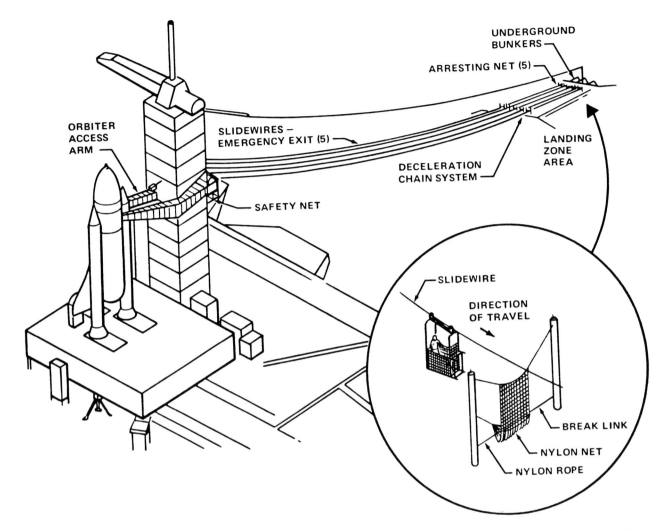

Figure 8.16 Emergency escape while on the launch pad. Crew leaves orbiter, enters baskets on slide wires, two people per basket. Near the bottom, the baskets catch and drag chains to decrease their speed before they reach the nets which stop them completely. Crew then takes cover in bunkers. *Courtesy of NASA.*

oxygen and the product of this combustion is plain water, so hot that the vapor is nearly invisible.

This sequence obviously must be computer controlled; human reaction time is not fast enough to carry it out. Sensors measure temperatures and pressures at strategic points on the main engines, continuously sending the information to the computer. As long as all readings are satisfactory, the countdown continues. If any sensor gives a reading which is out of acceptable limits, the countdown stops automatically. The liquid propellant main engines can be shut down by simply cutting off the flow of propellant, but once the solid booster engines ignite there is no way to stop. The Shuttle must lift off the launch pad.

Up to the point when the vehicle lifts off the launch pad, the entire operation is under the control of the Kennedy Space Center. Immediately upon liftoff, control of the mission is transferred to the Johnson Space Center in Houston.

Almost immediately after liftoff the vehicle rolls over so

the orbiter is beneath the external tank (Figure 8.18). At about 1 minute after launch the vehicle reaches Mach 1, the speed of sound, and experiences maximum strain as it accelerates through the dense lower atmosphere. The main engines are throttled back and the regressive grain of the solid boosters decreases their thrust to keep the acceleration below the 3 g limit. MATHBOX 8.2 shows how to calculate the average acceleration during this period.

Solid Booster Rocket Operation

The boosters burn out about 2 minutes into the flight at an altitude of about 28 miles and a speed of over Mach 4. Then the boosters are jettisoned (Figure 8.19). To prevent a collision, they are moved away from the orbiter and external tank by a 1 second burn of eight small separation rocket motors, four in the nose frustum and four in the aft skirt of each booster. Even though the boosters have separated, they are still travelling at the same forward speed as the rest of the vehicle. But, because they are no longer under power,

Figure 8.18 During ascent the orbiter rides beneath the external tank. *Courtesy of NASA.*

Figure 8.17 Space Shuttle *Columbia* lifts off from the mobile launch platform. *Courtesy of NASA.*

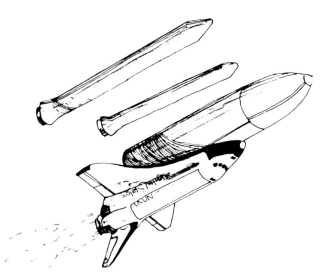

Figure 8.19 Solid rocket booster separation. *Courtesy of NASA.*

they follow a ballistic path arcing upward to an altitude of about 41 miles, then falling back downward into the more dense atmosphere. A barometer-altimeter switch ejects the nosecap at 15,400 feet altitude and the parachutes deploy, first a pilot chute which pulls away the nose cap, then a drogue chute which carries the upper conical shaped frustum. Finally three main parachutes open to lower the booster into the ocean. The main chutes, each onc 115 feet in diameter, become fully inflated at 2,200 feet altitude. Initially they are falling at over 230 mph, but by the time they reach the water they have slowed to 60 mph.

The booster rockets hanging from their parachutes strike the water bottom end first, forcing water into the empty interior and trapping air in the upper portion. Thus, they float

upright in the ocean. The parachutes disconnect automatically, beacon lights are turned on, and a radio transmitter begins sending a signal to assist aircraft and ships in finding and recovering the spent motor. A ship tows the rocket to shore where it is disassembled and sent back to Utah to be refurbished and refueled for another flight.

Main Engine and External Tank Operation

Meanwhile, the orbiter and external tank (ET) continue on their way, the three main engines accelerating the vehicle upward and downrange. During the first part of the flight, the emphasis is on gaining altitude, getting through the dense part of the atmosphere. Now it is necessary to increase horizontal velocity to reach orbital speed. When the Shuttle reaches an altitude of about 80 miles, it begins a long shallow dive to 73 miles during which its speed approaches the 17,500 mph needed for orbit.

When the hydrogen and oxygen in the ET are nearly con-

MATHBOX 8.2

Space Shuttle Acceleration

Recall **MATHBOX 2.2**. We made a rough calculation of the acceleration of the Space Shuttle at takeoff based on Newton's second law of motion. Now let us find the actual acceleration based on the data in Table 8.5.

In the first 1 minute and 6 seconds (66 seconds) after liftoff the Shuttle accelerates from rest on the launch pad at Kennedy Space Center, an initial speed of 914 miles per hour with respect to space, to a final speed of 1,654 miles per hour.

Average acceleration is the change in speed divided by the time interval required to make that change. If we use S_f to represent the final speed and S_i for the initial speed, then

$$a = \frac{S_f - S_i}{t}$$

We want our results to be in feet per second so we multiply by 5,280 feet in a mile and divide by 3,600 seconds in an hour. That is,

$$S_f = \frac{1654 \text{ miles/hr} \times 5280 \text{ ft/mile}}{3600 \text{ sec/hr}} = 2426 \text{ ft/sec}; \; S_i = \frac{914 \text{ ft/sec} \times 5280 \text{ ft/mile}}{3600 \text{ sec/hr}} = 1341 \text{ ft/sec}.$$

The acceleration is, then

$$a = \frac{2426 - 1341}{66} = 16.4 \text{ ft/sec}^2.$$

Why is this different from the acceleration we calculated in **MATHBOX 2.2**?

sumed and the vehicle is just short of orbital velocity, the orbiter's main engines cut off (MECO). A vent valve at the top of the external tank opens and oxygen escapes through the nose cap. A few seconds later the tank is disconnected (Figure 8.20), the venting oxygen causing it to pitch away from the orbiter and start to tumble. Tumbling assures that atmospheric drag will cause it to break up as it falls back to Earth and lands in the Indian Ocean. There is some uncertainty as to where it will land because of the way it tumbles. The designated impact area is an oval 2,100 miles long by 62 miles wide.

Incidentally, many people feel this is a waste of a valuable resource. External tanks are large, airtight, and structurally capable of being used as living quarters, workshops, or storage tanks. NASA has been urged to stockpile them in a parking orbit for possible future use. They already reach 99 percent of orbital velocity and only a little extra boost would inject them into orbit. There are several objections. One, the payload of the Shuttle flight would be reduced by several thousand pounds. Also, if left in very low Earth orbit, the tanks would have to be periodically boosted to higher altitudes to keep them from becoming a hazard to traffic and

from eventually burning in. A costly alternative is to strap on rockets and boost them to a higher stable parking orbit. Some planners envision them clustered together as a space station, fitted with rockets and launched to the Moon for a lunar colony, or refilled a little at a time and used as orbiting "gas stations" for vehicles heading to the outer reaches of the Solar System.

Orbit Insertion

A couple of minutes after the external tank separates, the orbital maneuvering system (OMS) engines are ignited to inject the vehicle into orbit. The two OMS engines, located in the pods just above the main engines on either side of the tail (Figure 8.21), produce 6,000 pounds of thrust each. They are propelled by monomethyl hydrazine fuel and nitrogen tetroxide oxidizer stored in the pods. These propellants are hypergolic; that is, they ignite and turn to hot gas on contact. Because no ignition system is required, the system is highly reliable. The propellants are forced out of their storage tanks and into the engines under pressure supplied by helium stored in liquid form in nearby containers.

The length of this OMS burn determines the height of

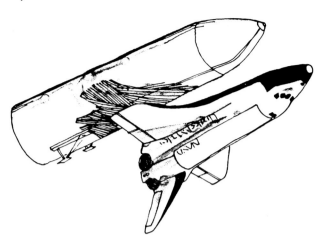

Figure 8.20 Orbiter separates from external tank. *Courtesy of NASA.*

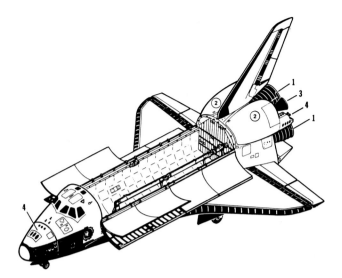

Figure 8.21 Orbiter's propulsion systems: three main engines (1), two orbital maneuvering system (OMS) engines (3), and the reaction control system (RCS) clusters of thrusters in the nose and tail sections (4). Two pods in the rear of the orbiter (2) hold the OMS engines and their fuel tanks. *Courtesy of NASA.*

apogee; a longer burn means a higher apogee. Typically the engines are fired for about 2 minutes which increases the speed by 100 miles an hour. With engines off, the orbiter coasts to apogee, near 174 miles in the example of Table 8.6. The orbit is now completely determined. After passing apogee the orbiter will return to the point where the OMS engines cut off, about 83 miles. This will be perigee.

Atmospheric drag on the vehicle at such a low perigee is sufficient to rapidly decay the orbit. Energy (speed) is lost to friction each time the orbiter passes perigee. With less speed at perigee, the vehicle will not be able to reach as high at apogee. Thus, apogee lowers with each orbit, increasing the drag even more, until finally the vehicle no longer has sufficient speed to remain in orbit. To prevent this from happening, the OMS engines are ignited again at apogee, a procedure called "apogee kick." This increase in energy raises the height of perigee above the altitude of excessive atmospheric drag. If the apogee kick is of just the right duration, perigee is raised to the same height as apogee and the orbit becomes a circle. Table 8.6 shows an OMS 2 burn of 96 seconds to circularize the orbit at about 174 miles.

On some flights, if the payload is less than maximum, the Shuttle main engines can provide enough energy to insert the orbiter directly to the desired apogee. Then the OMS 1 burn is unnecessary and only an OMS 2 burn at apogee is neeeded in order to circularize the orbit.

Attitude Control Thrusters

While in orbit the attitude (not altitude) of the orbiter, that is, its orientation with respect to Earth, is controlled by the 44 thrusters of the reaction control system (RCS). As shown in Figure 8.21, 14 primary thrusters and 2 vernier thrusters are located in the nose; 24 primary and 4 vernier thrusters are located in the rear, on the back of the two pods.

Primary thrusters produce 870 pounds of thrust; vernier thrusters, used for fine adjustment, produce just 24 pounds.

Firing the proper combination of thrusters causes the orbiter to pitch (nose up or nose down), yaw (swing nose left or right), or roll (Figure 8.22). Recall Newton's law of action and reaction. For example, firing a thruster pointed upward from the nose causes the nose to pitch down. Firing a rear end thruster toward the right causes the tail to move left and the nose to yaw to the right. The orbiter can automatically keep itself in any position, such as tail pointed straight down toward Earth and nose straight up at the sky. Different jobs require different orientations. For tasks which must be carried out without vibration, the orbiter is allowed to drift freely without firing any of the thrusters. If no particular attitude is necessary for the job at hand, the orbiter is usually oriented with the white top side toward the Sun to reduce solar heating and reduce the energy consumption of the air conditioners. The attitude which can be maintained with the least amount of fuel is the nose up, tail down attitude. Because the engines at the tail end are more massive than the cabin at the nose end, gravity tends to pull the tail end toward Earth. This is called a gravity-stabilized attitude and few or no thruster burns are needed.

Propellants for these thrusters are monomethyl hydrazine and nitrogen tetroxide, the same propellants as the OMS. In fact, propellants from the OMS tanks can be fed to the reaction control system if needed. See Figure 8.23 for a diagram of one of the pods. Hydrazine is also the propellant for the auxiliary power units in the orbiter and for the boosters' hydraulic power units.

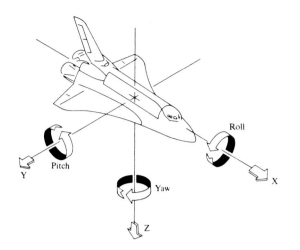

Figure 8.22 Pitch, yaw, and roll. *Courtesy of NASA.*

Because their thrust is so small, operating the reaction control system has little effect on the size or shape of the orbit. Small orbital adjustments in velocity are possible with the reaction control system, but if a large orbital change or rendezvous with another spacecraft is needed, the OMS engines (6,000 pounds thrust each) are used.

Reentry and Landing

To leave orbit and return to Earth, the orbiter must again fire the OMS engines. Over the Indian Ocean, about half an orbit (45 minutes) before touchdown, the orbiter is rotated by the reaction control system so the engines are pointed in the direction of motion. Engines are fired in that position, against the motion, in order to slow the vehicle down and transfer it to a new orbit in which perigee lies inside Earth, that is, the orbit intersects the ground. As the orbiter begins to "fall" toward Earth, its speed once again increases by two or three hundred miles per hour. The thrusters turn the vehicle again so it is falling underside first with a nose-high attitude.

Thirty-five minutes later, at an altitude of 50 miles, the orbiter encounters sufficient atmospheric drag to act as a brake. During the next 20 minutes, it drops to 15 mile altitude and slows to about 1,700 mph, as its kinetic energy of motion is transformed to heat energy. Maximum heating occurs at around 42 miles while moving at 15,000 mph. The heat ionizes the surrounding air by tearing electrons from the atoms so that a sheath of oxygen ions, nitrogen ions, and electrons, called a *plasma*, encloses the descending orbiter. It is not possible to transmit radio signals through the plasma sheath, so for 13 minutes the orbiter has no communications with the ground. It can easily be tracked by radar during that time, however, because radar pulses readily reflect off the plasma sheath.

While in the vacuum of space the wings, elevons, body flap, vertical stabilizer, and rudder (Figure 8.6) are "excess baggage"; without air they serve no useful purpose. During reentry the orbiter's altitude and speed decrease as it enters the denser atmosphere and these aerodynamic control surfaces begin to operate. Unlike the earlier manned spacecraft of the Apollo era which followed a ballistic trajectory like

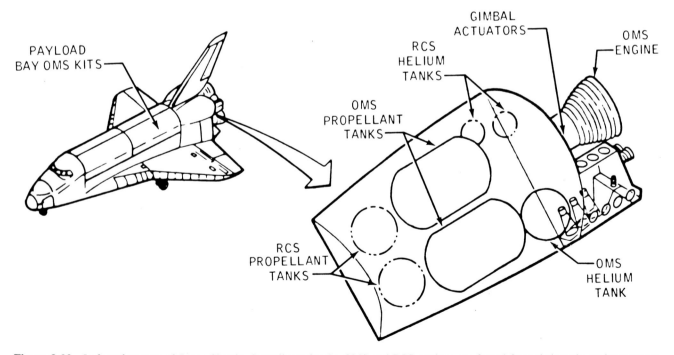

Figure 8.23 Left pod at rear of Space Shuttle. Propellants for the OMS and RCS engines are forced from their tanks under pressure supplied by the helium tanks. OMS and RCS are interconnected so the propellants from either system can feed the other. *Courtesy of NASA.*

a cannonball, the Space Shuttle orbiter can maneuver to the left or right of its entry path by over 1,200 miles. The rudder can split in half, one side moving left and the other side moving right, to act as a speed brake. The body flap is covered with tiles and doubles as a heat shield for the engines during reentry.

Once back in the atmosphere, the orbiter flies like an airplane with one important difference: it has no engines running. At this point it is better called a glider. The approach to the runway (Figure 8.24) and the touchdown must be done precisely the first time. Without power, there is no second chance. As it approaches the runway, the orbiter's glide angle is about six times steeper than a commercial jet liner. At touchdown its speed is a little more than 200 miles per hour.

A crew of technicians meets the orbiter on the runway and checks the exterior of the vehicle for safety before the astronauts emerge. In particular, they use instruments to "sniff" around the OMS engines and thrusters to be sure there is no leakage of toxic propellants. Also, it takes time for the orbiter to cool off so it can be touched. The temperature of the top surface of the orbiter is still about 200 °F and the hatch cover is 300–350 °F after landing.

The preferred landing site is at Kennedy Space Center in Florida. If bad weather prevents a Florida landing, an alternate location is at Edwards Air Force Base in the California desert. From there the orbiter is flown back to Florida, piggyback on a specially adapted Boeing 747. See Figure 8.25. At Kennedy Space Center the orbiter is refurbished, mated to a pair of reconditioned boosters and a new external tank, checked out, and made ready for another flight.

Safety

The Presidential Commission that investigated the *Challenger* accident was concerned that there were no means for the crew to escape during a launch failure. On the initial test flights, the orbiter *Columbia* was equipped with ejection seats for the two pilots. When testing was completed, they were replaced with the operational seats to save space and reduce weight. It is not possible to supply an entire seven person crew with individual ejection seats and exit hatches. In case of failure of one or more engines, the plan is to ride the orbiter to a landing back at the launch site, in southern Europe, in north Africa, or, as a last resort, to ditch in the ocean, depending on the speed of the vehicle at the time of the emergency. Even with an ejection system, it is doubtful

Figure 8.24 Space Shuttle *Columbia* approaching Edwards AFB, California, for a landing. *Courtesy of NASA.*

Figure 8.25 A modified Boeing 747 transports the orbiter piggy-back.

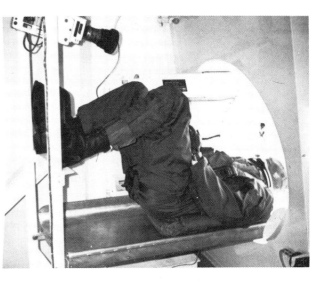

Figure 8.26 Crew escape system under test by NASA. A dummy is in place in a mockup of the orbiter's hatch area. The television camera records the ejection. *Courtesy of NASA.*

that the crew could survive a catastrophic explosion. There was no warning of trouble either at mission control or in the *Challenger* until the external tank ruptured. Then it was too late.

For crew escape, an experimental extraction system, usable during a controlled glide at altitudes between 5,000 and 20,000 feet, at a speed of about 200 miles per hour, is shown in Figures 8.26 and 8.27. A small rocket is expelled from its storage canister by gas pressure. When it reaches the end of a 10 foot lanyard, the 2,000 pound thrust rocket motor fires for a tenth of a second to pull the crew member, parachute, and survival pack free of the orbiter, clearing the wingtip by 30 feet. A different type of escape system (not shown here) makes use of a 21 foot telescoping rod extending from the side of the hatch opening. The crewmember clips a ring onto the rod and slides down and out away from the vehicle before opening the parachute. These and other actions by NASA should result in a safer, more reliable Shuttle.

Other Winged Vehicles

Buran

The former Soviet Union built a winged vehicle named Buran (translated as Snowstorm) which looks very similar to the Space Shuttle orbiter. It made an unmanned two orbit flight in 1988, lifted into space from the Baiknour cosmodrome in Kazakhstan in central Asia by the Energia booster rocket. Buran has no main engines like the Shuttle orbiter, but depends completely on the Energia to reach orbit. Onboard engines are used to maneuver, change orbit, and return to Earth.

Meanwhile, the Russians continue to use the Soyuz capsule-type spacecraft rather than winged vehicles to send cosmonauts to and from their Mir space station. In doing so, they have logged many more total hours in space than the United States. The Soyuz lands on the ground, coming to a gentle landing by parachute with retrorockets for the final touchdown.

National Aerospace Plane

A new very different sort of bus to orbit has also been under development, the National Aerospace Plane (NASP) seen in Figure 8.28. It is a single-stage-to-orbit passenger and light cargo craft, which takes off horizontally from a runway, flies into orbit, and returns to land like a standard airplane. The propulsion system is radically different—a supersonic ram jet (scramjet). No booster rockets are needed. The engine operates as an airbreathing turbojet to Mach 3 or 4. It then switches to ram jet operation to Mach 5 or 6, since a ram jet will not work from a dead stop but must be moving at high speed to ram air into the intake. The engine operates as a scramjet to speeds above Mach 12 to perhaps Mach 25. Propellants are liquid hydrogen and atmospheric oxygen. Such an engine has never been built. In fact, present jet engines are designed to avoid supersonic flow in the combustion chamber. It is difficult to keep a flame burning in supersonic flow, and the engines cannot withstand the heat and pressure associated with a supersonic shock wave. New structural materials with greater strength and lower weight are being developed for the NASP. Engineers feel confident that they can solve the problems and build a working engine, perhaps a complete plane, in the late 1990s. This same plane could be used as an airliner. Flights between any two points on Earth at Mach 12 would take less than 2 hours.

ESA's Plans

The European Space Agency (ESA) is working on a design for a shuttle-like vehicle called Hermes. It is not likely to fly before the year 2000. By that time, scramjet technology may replace shuttle-type spacecraft design. Germany's Sänger and the British/Russian Hotol are two winged vehicles with airbreathing engines and lifting wing surfaces under development by ESA. Two designs are shown in Figure 8.29.

Figure 8.27 Test of crew escape system. A tractor rocket is fired to pull the astronaut out of the orbiter. In this test, the dummy is pulled out of the mockup of the orbiter hatch and over the side of a cliff on Hurricane Mesa, Utah. *Courtesy of NASA.*

Figure 8.28 National Aerospace Plane. This vehicle under development will take off horizontally like an airliner, fly into orbit, and return to Earth to land on a runway. It will not require booster rockets. *Courtesy of NASA.*

Figure 8.29 Two winged spacecraft under study by the European Space Agency. *Courtesy of ESA.*

DISCUSSION QUESTIONS

1. Why are the main engines started first? Why not fire the solid rocket boosters first?

2. How is the orbiter steered when it is on its way to orbit? when it is in orbit? when it returns to Earth?

3. Why, do you suppose, does the orbiter fly beneath the external tank on its way to orbit?

4. In assembling the booster rockets, why not just bolt the segments together?

5. At what points in a flight would it be feasible to bail out of a disabled orbiter? When would it not be feasible?

6. What limits the altitude the orbiter can reach?

7. At liftoff, what percentage of the total weight of the Space Shuttle is propellant weight?

ADDITIONAL READING

Allaway, Howard. *The Space Shuttle at Work, NASA SP-432*, Government Printing Office, 1979. Easy reading, written before first flight.

Joels, Kerry M., and Gregory P. Kennedy. *The Space Shuttle Operator's Manual*, Ballantine Books, 1982. Very detailed and complete, popular reading level.

Kerrod, Robin. *The Illustrated History of Man in Space*. Mallard Press, 1989. Profusely illustrated, easy reading.

NASA. *Space Shuttle News Reference*, Government Printing Office. Everything you always wanted to know about the Shuttle.

NOTES

Chapter 9

Living in Space

For millions of years the human body has adapted to living on Earth with the normal force of gravity, where up and down are clearly defined, in an atmosphere of 21 percent oxygen and 78 percent nitrogen at 14.7 pounds per square inch pressure, with food and water in abundance. All these things are absent in space. If people are to survive in space, they must take a suitable environment with them, particularly air at a proper temperature and pressure. If they are to remain for an extended period of time, they must have food and water. In weightlessness the body begins to deteriorate in a number of ways and some arrangement must be made to keep physically fit by exercising or perhaps by creating artificial gravity. Protection from cosmic and solar radiation is necessary. In addition, being cooped up in a small space with a few other people for a long time can lead to sociological and psychological difficulties. All these things must be considered in designing spacecraft and in planning missions.

Cabin Atmosphere

The Space Shuttle orbiter cabin is shown in Figure 9.1. Living space is in the middeck and life support equipment is mostly in the lower deck, accessible through hatches in the floor of the middeck.

The orbiter has a "shirt-sleeve environment." That is, the atmosphere is a duplicate of the atmosphere on Earth at sea level: 21 percent oxygen, 78 percent nitrogen, 14.7 pounds per square inch (psi) pressure. An automatic system maintains the proper oxygen-nitrogen ratio. It senses the amount of oxygen in the air and adds more oxygen, more nitrogen, or both as necessary, from storage tanks.

By contrast, the Apollo spacecraft used a pure oxygen atmosphere at 5 psi pressure. Pure oxygen is very reactive and the launch pad fire, discussed in Chapter 1, showed the danger of using it unless the spacecraft is made absolutely fireproof and the pressure is kept lower than normal atmospheric pressure. Diluting the oxygen with an inert gas makes it less hazardous. The Skylab space station had an atmosphere of 75 percent nitrogen and 25 percent oxygen at a pressure of 5 psi. Other inert gases such as helium have

been considered as a substitute for nitrogen in large space vehicles, primarily because helium is a very light gas and less mass would have to be transported to orbit. For example, if the volume of a space station is 16,000 cubic feet, 264 pounds could be saved by using helium. See **MATHBOX 9.1**. On the negative side, helium gives the voice a Donald Duck-like quality which could be very disturbing after a time.

Besides providing a proper mix of oxygen and nitrogen, the life support system must remove exhaled carbon dioxide and other gases, and maintain a comfortable temperature. Poisons can accumulate in the spacecraft atmosphere. Excess methane and ammonia gas from human and food wastes could be a problem if the air purification system failed. More serious would be an increased level of carbon dioxide. At first this would cause a headache and the astronaut's breathing rate would increase noticeably. Then hearing ability would be reduced, followed by dizziness, nausea, mental confusion, convulsions, and unconsciousness.

As part of the orbiter's life support system, blowers force the air through filters that catch debris such as bits of paper, hair, and crumbs. Then about 10 percent of the air is diverted through purifying canisters containing lithium hydroxide which absorbs and removes carbon dioxide, and activated charcoal which removes other noxious gases and odors. The canisters handle about 320 cubic feet of cabin air per minute and must be changed at regular intervals as they become saturated with the gases. On Earth, air circulates because warm air is less dense than cold air. Gravity causes the cooler, denser air to sink while the warm air rises. But in orbit, when weightless, air will not circulate of its own accord Therefore the blowers are also necessary to keep air from stagnating.

Maintaining a comfortable temperature, in the range 65° to 80 °F, requires cooling, not heating. Body heat, solar heat, and heat generated by the equipment are removed by a refrigerator air conditioner. Excess heat is dumped into space via radiators located on the inside of the cargo bay doors. The doors must be opened and the radiators exposed to space for the system to work. Your refrigerator does the same job,

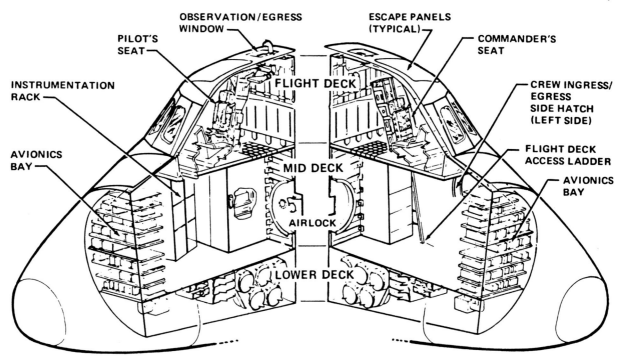

Figure 9.1 Space Shuttle orbiter cabin. *Courtesy of NASA.*

removing heat from the box and dumping it into the kitchen via a radiator on the back or bottom. Notice how hot the radiator gets when the refrigerator is running.

High humidity is not necessarily a hazard, but can be very uncomfortable. Excess moisture is removed by the air conditioning system. As the air passes over cold metal plates, the water vapor condenses out, much as water droplets collect on your bathroom mirror when you take a shower. The droplets are then blown off into a waste storage tank. Cosmonauts use the condensed water, but American astronauts found that it had a peculiar taste, probably acquired from dust and other impurities on the metal plates.

After all this recycling and processing is finished, the cabin atmosphere is cleaner and more comfortable than the air in your home. Because machinery sometimes fails, systems are designed with redundancy. Life-critical equipment generally has two backups in case the primary one goes out.

Loss of Atmospheric Pressure

Early in the space age there was concern that a meteoroid could puncture the spacecraft and let the air out. Complete loss of the atmosphere from the cabin would, of course, kill everyone. Although the concern over meteoroid impacts proved to be exaggerated, an underlying and realistic concern is that mechanical failure of some component of the spacecraft, space suit, or helmet could be disastrous. Three Russian cosmonauts died in 1971 when a valve in their Soyuz II failed.

Death comes quickly and quietly when the atmospheric pressure is suddenly reduced to almost zero as it is in the near vacuum of space. First, air in the lungs would be expelled in one quick exhalation. As blood flowed through the lungs, the oxygen in the blood would also be released to space. Symptoms of hypoxia (lack of oxygen)—dizziness and blurred vision—would be evident. Without a continuing fresh supply of oxygen, the brain would die in a few minutes.

Early in the space age it was thought that the blood would boil and a body would literally explode when exposed to near zero atmospheric pressure. Water boils at 212 °F at sea level. But at Colorado Springs, where atmospheric pressure is only about 80 percent of sea level pressure, water boils at about 190 °F. Ascending to higher altitudes, one would eventually reach the point where the pressure is so low that water would boil at 98.6 °F, body temperature. This fact led to the false conclusion that, in the near vacuum of space, blood would boil and a body would explode. Skin is a strong membrane, however, and it is more likely that the body fluids would evaporate slowly rather than explosively. The skin would blister and some blood vessels would rupture producing bruises. But the person would have been brain-dead long before.

Decompression Sickness—The Bends

In extravehicular activity (EVA), decompression sickness could be a problem. Decompression sickness, technically called dysbarism and commonly called the bends, is well known to deep sea divers. When descending in a diving suit, the pressure in the suit must be increased to keep it from collapsing under the great pressure of the surrounding water.

MATHBOX 9.1

Spacecraft Atmospheres

Air weighs a lot more than you may think. A 2,000 square foot house contains 16,000 cubic feet. Air has a density of 0.074074 pound per cubic foot at sea level pressure of about 15 pounds per square inch. The total weight of the air in the house would be 16,000 × 0.074074 = 1,185 pounds. The nitrogen, which constitutes 76 percent (by weight) of the air, would weigh 1,185 × 0.76 = 900 pounds.

Now, consider a space station the size of a house with a normal sea level atmosphere, and suppose we substituted helium for nitrogen. Helium has a molecular mass of about 4 while nitrogen has a molecular mass of about 28. Therefore, the helium would weigh only one-seventh that of the nitrogen, or 129 pounds for a saving of 900 − 129 = 771 pounds of payload.

Atmospheric pressure in a space station is likely to be held at only 5 pounds per square inch, such as was done in Skylab and Apollo. Then the total weight of the atmosphere in our house-sized space station would be 395 pounds, the nitrogen would weigh 300 pounds, and substituting helium for nitrogen would save 257 pounds.

After working under the higher air pressure for a time, the body absorbs an excessive amount of nitrogen. Oxygen is used up in metabolic processes, but nitrogen in the air is simply absorbed by the blood and body tissue. On rising to the surface the pressure reduces to normal atmospheric pressure and the nitrogen forms expanding bubbles which move along with the blood in the veins and arteries until they cannot go any further because of their size. There the bubbles form a blockage, preventing the flow of blood to the surrounding tissue. This condition is not only extremely painful, but can cause death. Divers cope with this problem by ascending very slowly, reducing the pressure slowly.

Astronauts are subject to the bends when they leave the cabin, where the pressure is 14.7 psi. Pressure in the space suit is only 4.3 psi and nitrogen in the astronaut's body will expand as it does in a rapidly ascending diver's body unless precautions are taken. First, pressure in the cabin is reduced to 10.2 psi for 24 hours prior to the planned EVA. Thus, astronauts become acclimed to a lower pressure over a period of time. Second, they breathe pure oxygen for a time before leaving the cabin to allow the body to eliminate most of the nitrogen. NASA has been developing a space suit which will hold an atmosphere at 8.2 psi. If one can be successfully built, bends will be less of a problem. It has been noted that women are at a greater risk than men.

Food

The first American to eat in space was John Glenn in 1962, squeezing applesauce through a tube into his mouth. Before the flight, some experts were concerned that, in weightlessness, food would be difficult to swallow and would collect in the throat. The concerns were unfounded, however, and Glenn had no problem except that the food was not very appetizing. When another astronaut heard about the unappetizing meal, he smuggled a ham sandwich aboard his flight.

For present day spaceflights, all food must be taken along. No running down to the corner convenience store for something you forgot! In the future, on large space stations and interplanetary spacecraft, food will undoubtedly be grown in greenhouses aboard the craft. It would be difficult to carry all provisions necessary for very long trips without resupply from Earth. The weight at liftoff would be enormous. Cosmonauts have already attempted to grow food on their space stations with mixed results. Concepts for doing this will be discussed further in Chapter 11.

Five problems and considerations arise in preparing food for Shuttle flights.

1. Achieving proper nutrition. Astronauts often experience a loss of appetite in space.
2. Preserving food from spoilage. Food must have a shelf life of 6 months at 100 °F. Five basic methods are described below.
3. Preparing and eating in zero g. Bite-sized cubes carried aboard the Mercury flights crumbled and crumbs floated about the cabin. The cubes were later coated with an edible gelatin to control the crumbling. Freeze-dried foods were difficult to rehydrate. Improved plastic packs were developed with a means for inserting the nozzle of a water gun into the package.
4. Making the food and its packaging lightweight and compact. At first, some foods were pureed and placed in aluminum tubes to be squeezed out like toothpaste. How-

ever, the tubes weighed more than the food and that concept was abandoned.

5. Developing meals that are appetizing and pleasing to eat. There were early complaints that the food tasted bland.

Water supply aboard the Space Shuttle is not a problem. It is produced by the fuel cells which generate electricity from hydrogen and oxygen. Water is a by-product, made at a rate of nearly 2 gallons per hour. This is four times the amount needed for a crew of seven, so much of it is dumped overboard. With an abundance of water, dehydrated foods are a natural choice. The food is rehydrated as needed.

Food Preparation

Normally, no refrigerator is carried in the orbiter because of the weight, although occasionally one is brought aboard for life science experiments. Therefore, food must be preserved in other ways. Five basic methods are used:

1. Rehydratables: More than 100 foods—including cereals, vegetables, beef patties, soups, spaghetti, and some fruits—can be dehydrated by freeze drying, air drying, spray drying, or other methods. This is the most commonly used method and is likely to remain so. The astronauts rehydrate some of them by simply putting them in their mouths, such as bananas and strawberries. Others, such as shrimp cocktail, scrambled eggs, and beverages are rehydrated with water. Many instant drinks are dehydrated, like tea, coffee, and chocolate. Orange juice, grapefruit juice, and milk do not rehydrate well. They get lumpy and do not mix well with the water.

2. Intermediate moisture: Some foods are partially dehydrated, such as dried apricots and dried peaches.

3. Thermostabilized: Foods are sealed in cans or aluminum-laminated pouches and cooked at temperatures that destroy bacteria. Foods processed this way can be stored while retaining their original moisture. These foods include canned fruit in heavy syrup, ground beef with relish, tuna, and stewed tomatoes.

4. Irradiated: As with thermostabilized food, bacteria are destroyed to allow food to be kept at room temperature with its natural moisture. But in this case, the bacteria are killed by exposure to ionizing radiation. Bread, corned beef, beefsteak, and smoked turkey are among the foods treated this way.

5. Natural form: Foods that are naturally low in moisture, such as nuts, hard candy, gum, peanut butter, cookies, and candy bars, can be taken on the spaceship in the same form that they are purchased at the supermarket.

Menus

The menus aboard the Space Shuttle are strictly regulated to give 3,000 calories per day, consisting of 16 to 17 percent protein, 30 to 32 percent fat, and 50 to 54 percent carbohydrate. While weightless, the body tends to lose essential vitamins and minerals, so adequate supplies of calcium, nitrogen, potassium, and other minerals are included in the diet. A typical menu may include:

Breakfast: Peaches, beef patty, scrambled eggs, bran flakes, cocoa, and orange drink.

Lunch: Hot dogs, turkey tetrazzini, bread, bananas, almond crunch bar, and apple drink.

Dinner: Shrimp cocktail, beef steak, rice pilaf, broccoli au gratin, fruit cocktail, butterscotch pudding, and grape drink.

Substitutes are allowed from the "pantry," a supply of favorite foods requested by the astronauts. There are nearly 100 foods and beverages to choose from. Tobacco and liquor are not allowed.

Food tends to taste bland in space, perhaps because of the nasal congestion often experienced by the astronauts or perhaps because the air circulation is low in microgravity. Therefore, a number of condiments are provided for seasoning such as ketchup, mustard, liquid salt and pepper, and hot sauce.

The galley aboard the orbiter is shown in Figure 9.2. Each package is labeled with a flight day and meal number and contains a meal for each crewmember. Crewmembers take turns preparing the meals. The first step is to add water to the dehydrated food by sticking a hollow needle through a small opening in the pouch and kneading it thoroughly. If it is to be heated, it is placed in the oven. A fan circulates the hot air in the oven so the food heats evenly. Trays are stuck to the wall with clamps or magnets. The food packages are set into the tray with tape or magnets and cut open with a knife or scissors. The Shuttle food tray is shown in Figure 9.3 and a meal in weightlessness aboard the orbiter *Columbia* is shown in Figure 9.4.

Food in gravy, heavy syrup, or sauce sticks together and to the container. The forces of adhesion and cohesion keep it from floating away. It also sticks to forks and spoons so eating is almost normal. There are some cautions, however. Food sticks to the side and bottom of the utensils as well as the top, so a forkful is twice as big. Sudden moves will send the food floating off into space. A straw inserted into a covered container must be used for beverages. Otherwise, when you bring an open container to your lips, the liquid keeps on going. In fact, a favorite sport in the orbiter is playing with the food. Spilled water does not fall down, but gathers into floating globules. Salt sprinkled on the food may scatter into the air. Bread crumbs scatter everywhere, so crumbly food is avoided.

Sleep

When you are weightless you do not have to lie down to sleep. Some astronauts have slept while simply floating in

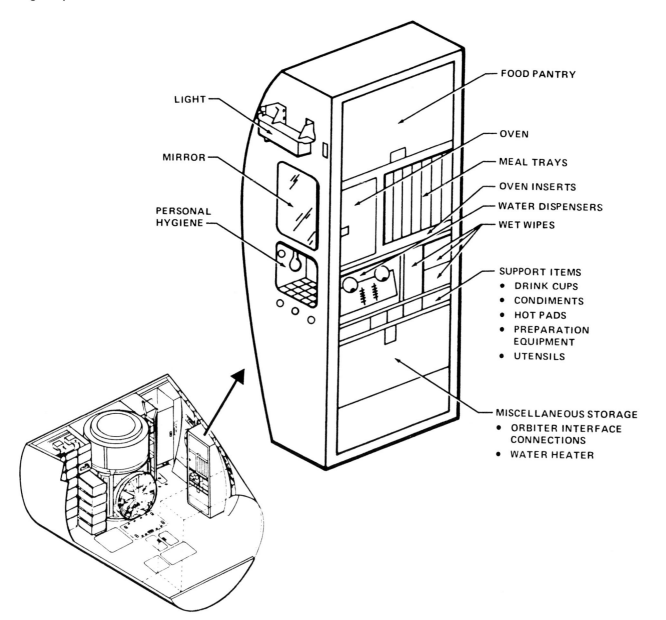

Figure 9.2   Space Shuttle pantry and personal hygiene station. *Courtesy of NASA.*

the cabin. The problem is, of course, that they drift around and get in the way of the rest of the crew.

There was concern in the early days that a person would not be able to sleep while weightless, and some astronauts have had difficulty. After sleeping for a short time, some have tried to roll over and woke themselves up flailing their arms and legs. Waking with a dizzy feeling resulted from the weightless head bobbing around. Others could not sleep well without the firm feeling of a mattress under them.

Figure 9.5 shows the arrangement of padded bunks in the orbiter. One person sleeps on the top bunk, a second on the lower bunk, a third on the bottom side of the lower bunk facing down, and the fourth vertically at the end of the other bunks. Side panels close for privacy. Fireproof sleeping

bags are attached to the bunks to keep them from floating away. Additional sleeping bags can be attached to the locker face either horizontally or vertically for additional crew-members. Of course, the words "top," "down," "horizontal," and "vertical" are only in reference to the orbiter's top and bottom. They are meaningless when you are weightless. A waist strap presses the body against the bunk to get the sensation of lying on a mattress. Slipping the hands under the waist strap keeps them from floating in space.

In a 200 mile orbit, the sun rises and sets every hour and a half, so there is no long dark night for sleeping. Eye shades and earmuffs reduce disturbing noise and light for those who want to use them. If the entire crew sleeps at the same time, two must wear communications headphones in case an emergency warning signals or the ground controllers call.

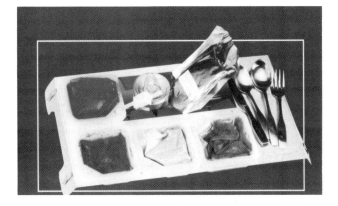

Figure 9.3   Shuttle food tray. This meal consists of (left to right, top row) fruit punch, butterscotch pudding in the can, smoked turkey in the foil bag, (bottom row) strawberries, mushroom soup, and mixed vegetables. *Courtesy of NASA.*

Hygiene

The personal hygiene station is part of the orbiter's galley, shown in Figure 9.2. A globular shaped chamber with two openings is used for washing hands without getting water all over the place. The Skylab space station was equipped with a shower. A curtain enclosed the astronaut. He used a spray to squirt water on his body and lather up, then used a vacuum cleaner device to remove it from his body, the curtain, and walls. It didn't work very well. Some water escaped from the shower stall and had to be chased around and vacuumed up.

There is no shower on the Space Shuttle, so astronauts wash up by hand. At the wash station, a handgun supplies water at a comfortable temperature between 65 and 96 °F. Water clings to the body in weightlessness and makes the job easier than you might expect. One washcloth is used to soap up and a second to rinse off. Waste water goes into a storage tank beneath the floor. For privacy, a curtain can be pulled across the front of the personal hygiene station. Towels, washcloths, and other personal hygiene items are equipped with Velcro so they can be stuck to the wall without floating away.

Dry shaving with an electric shaver in weightlessness caused problems because the whiskers floated off around the cabin causing eye and lung irritation and getting into equipment. A wind-up electric shaver with a vacuum cleaner attached to suck up the whiskers helped, but didn't give a comfortable shave according to at least one astronaut. A depilatory cream or gel can be used to melt off the whiskers. Shaving cream and safety razors seem to work best. Some astronauts give up and grow a beard.

Cleaning up immediately after a meal is more important on the Shuttle than at home. Odors and bacteria from food wastes are unwelcome in the small enclosed space of the orbiter. Garbage and trash are sealed in plastic bags. Wet wipes containing a strong disinfectant are used on utensils

Figure 9.4   Mealtime aboard the orbiter *Columbia*. A full meal tray floats at the right while an empty pudding can floats at the center. *Courtesy of NASA.*

and trays as well as around the eating area. Used clothing is also sealed in airtight plastic bags.

A question frequently asked astronauts on their public appearances is, "How do you go to the bathroom in space?" Figure 9.6 is a diagram of the Space Shuttle's commode for collecting human waste. It had to be carefully designed for use in weightless conditions by both male and female astronauts. Toeholds, handholds, and a waist restraint like an automobile seat belt hold the occupant firmly in place to assure a good seal between the user and the seat. A reading light, a nearby window, and privacy curtains are part of the commode area. Female astronauts have continually had trouble with the toilet. While it easily collects urine from a man using a tube, women have a more difficult time. It has been a constant annoyance and better designs have been sought.

Then comes the problem of what to do with the waste. The commode has separate receptacles for feces and urine. In the absence of gravity, high speed air streams carry the solid and liquid waste into their respective receptacles. Solid waste is vacuum dried, chemically treated with germicides to prevent odor and the growth of bacteria, and stored until

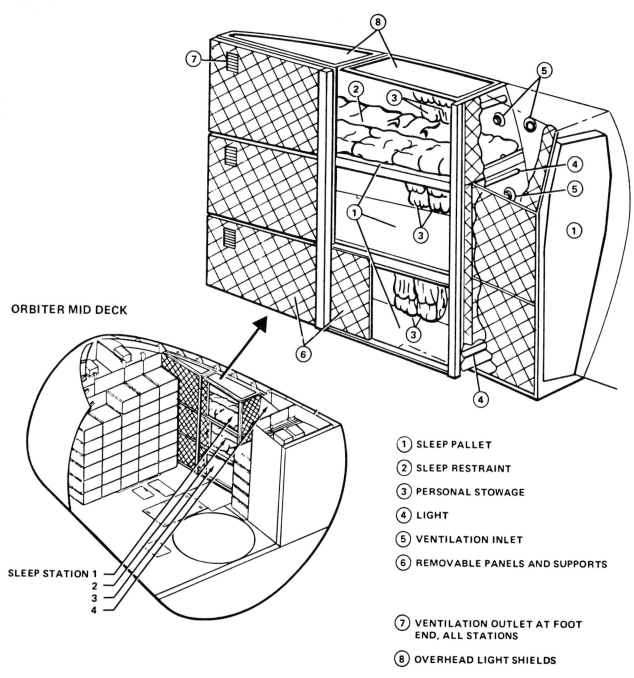

ORBITER MID DECK

SLEEP STATION 1
2
3
4

1 SLEEP PALLET

2 SLEEP RESTRAINT

3 PERSONAL STOWAGE

4 LIGHT

5 VENTILATION INLET

6 REMOVABLE PANELS AND SUPPORTS

7 VENTILATION OUTLET AT FOOT END, ALL STATIONS

8 OVERHEAD LIGHT SHIELDS

Figure 9.5 Orbiter sleep stations. *Courtesy of NASA.*

return to Earth. Liquid waste joins other waste water in the storage tank beneath the floor. When the tank fills up, it is dumped overboard.

Waste is not recycled in present-day spaceflight. Everything is saved in the orbiter and returned to Earth. On longer trips to other planets and beyond, it may be advantageous to recover whatever can be reused in order to reduce the weight of the vehicle on the launch pad. Whether or not there is an advantage to recycling depends on the weight of the recycling equipment plus the initial supplies compared to the weight of use-once-and-throw-away materials. Present re-

cycling technology gives no weight advantage for flights less than 3 years in duration. This will be discussed in more detail in Chapter 11.

Radiation

Radiation in space consists of electromagnetic waves from the Sun, high speed protons and electrons in the solar wind, cosmic rays from outer space and from the Sun, and trapped particles in the Van Allen belts. (Recall Chapter 4.) Some of these pose a serious threat to human life. On the ground, we are protected from this radiation by Earth's atmosphere

WASTE COLLECTOR

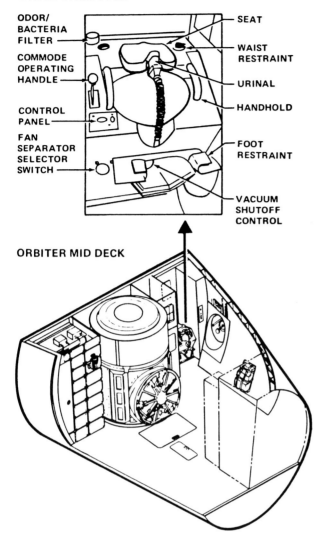

ODOR/
BACTERIA
FILTER

COMMODE
OPERATING
HANDLE

CONTROL
PANEL

FAN
SEPARATOR
SELECTOR
SWITCH

SEAT

WAIST
RESTRAINT

URINAL

HANDHOLD

FOOT
RESTRAINT

VACUUM
SHUTOFF
CONTROL

ORBITER MID DECK

Figure 9.6 Human waste collection system. *Courtesy of NASA.*

and magnetic field. The cabin of a spacecraft provides protection from electromagnetic waves and some protection from the particles. However, when a primary cosmic ray strikes the vehicle walls, it produces secondary radiation by ionizing the materials of which the walls are made. This secondary radiation penetrates into the cabin. (Secondary radiation is produced in the upper atmosphere of Earth also, when cosmic rays strike the molecules of atmospheric gases. Few primary cosmic rays reach the ground, but the secondaries do.)

When high energy protons, cosmic rays, x-rays, or gamma rays enter a living cell, they ionize the atoms and split apart the molecules of the cell, disrupting its operation and perhaps killing the cell. Recall that cosmic rays are the nuclei of atoms which have lost all their electrons. They are mostly hydrogen and helium nuclei (protons and alpha particles) with a few nuclei of heavy atoms such as carbon, oxygen, and iron. One of these heavy cosmic ray particles does more

damage than a proton because it carries more positive electrical charge and can cause more ionization in the cell. A single heavy cosmic ray particle will destroy a cell. In total, though, protons do more damage simply because there are so many more of them.

The human body is continuously manufacturing new cells and the loss of a few is no big thing, but exposure to high dosages of radiation results in the destruction of many cells in a short time. Then the body has a problem, particularly if the cells are in a vital organ.

A particular problem of a woman exposed to high radiation dosages is that one or more of her latent egg cells may be damaged. A woman is born carrying all the latent egg cells she will ever have. They develop at the rate of about one per month during her reproductive years. If one is damaged and later is impregnated, the fetus may be malformed. Male sperm cells are produced frequently and are not subject to the same lifelong concern as a woman's egg cells. However, a man can become sterile if he receives a large dose of radiation in a short time, causing the sperm-creating organs to stop functioning.

Radiation dosage is measured in rads (radiation absorbed dose), a quantity of energy absorbed per unit of body tissue. For humans the lethal dose is about 600 rads in a day. Exposure to 100 to 200 rads per day may cause nausea, vomiting, and fatigue. Biological damage from radiation is difficult to predict and the effects vary from person to person, depending on body weight and general health. To further complicate the situation, different particles have different abilities to damage biological tissue. Alpha particles do more damage than x-rays. A quality factor is applied to each type of particle to specify how much damage that particle can do. The quality factor is 1 for x-rays and electrons, 2 to 25 for neutrons and protons, and more than 15 for alpha particles and heavy nuclei. When the dose in rads is multiplied by the quality factor, the result is called the dose equivalent in rems, which is a better indicator of the damage done.

However, except in a research laboratory, there is no easy way to tell just what kind of particles are being intercepted. Under normal conditions, the radiation in a spacecraft cabin is lower than first anticipated. During the Skylab's longest mission of 84 days, total radiation received was less than 8 rads. NASA sets a limit of 75 rads per year whole body dosage and a maximum career limit of 400 rads. Astronauts carry dosimeters to monitor their exposure. Instruments in the spacecraft cabin also monitor the radiation level.

Solar Proton Storms

Large solar flares may produce cosmic ray proton storms. The radiation intensity from these events occasionally reaches levels lethal to humans outside the protection of the Earth's atmosphere. It would be particularly hazardous in a polar orbit inside the magnetosphere because the Earth's magnetic field deflects the particles toward the north and south pole. Consequently, the Space Shuttle has never been launched into a polar orbit. During a proton event an astronaut could be exposed to more radiation in half an hour than NASA standards allow for a year. For the most energetic of these events, even the cabin walls of the orbiter would not provide sufficient shielding.

A solar proton storm in October 1989 and one which lasted longer than a week in August 1972 were intense enough to have killed an astronaut on an EVA outside the spacecraft in polar orbit. On August 4, 1972, the proton flux as measured by unmanned satellites translated to over 100 rads per hour for about 8 hours. Exposed to such a high dose, the astronaut would have lived 4 to 6 days in severe pain with diarrhea and constant vomiting. Exposure to the smaller dosages inside the cabin might not have killed the astronauts, but vomiting and diarrhea, leading to dehydration and emaciation, and the destruction of blood cells would have led to death for some of them within a month.

The Space Shuttle was not flying during those two storms, but an event did occur in August 1989 while a Shuttle was in orbit. NASA considered bringing the orbiter home early, but it was at the end of the mission anyway, and increased particle radiation was observed only on the last two orbits.

Solar activity was discussed in Chapter 4. It is highly variable and not completely understood. Scientists have only a rudimentary knowledge of the causes of solar proton events, and forecasting their occurrence has only limited accuracy. Given that a flare has occurred, the forecaster can determine its intensity and can state with some reliability whether or not high energy protons have been blown into space. If they have, the forecaster can issue a warning to the astronauts. They would then have from 20 minutes to perhaps 3 hours to take cover in the best shielded part of the spacecraft. Since reentry and landing at a particular point on Earth are possible only from a certain starting point in orbit, returning home may not be a feasible emergency procedure. The inhabitants of a space station would have to ride out the storm. The only practical protection from radiation is shielding by a thick mass. Some materials are more effective shields than others, but any kind of mass will afford some protection. Lead is perhaps the best commonly available material, but is not practical on a spacecraft because of its weight. Any long duration habitat design should include shielding against normal day-to-day radiation and a heavily shielded compartment, a "storm cellar," for protection during a large solar storm.

Low Level Radiation

We have a pretty good understanding of the effects of high radiation dosage on people based on studies of the atomic bomb explosions in Japan during World War II. The effects on the body of continuous exposure to low radiation levels such as galactic cosmic rays are not well understood. We on Earth are exposed to low level radiation from cosmic rays entering the atmosphere, radioactive rocks and gases, and dental and medical x-rays. We often hear statistics such as "exposure to such radiation will lead to two deaths from cancer per 100,000 people in 50 years." The questions are who are those two and wouldn't they have died from something else in those 50 years. Certainly driving an automobile, smoking, and a dozen other human activities are more hazardous than low level radiation. But here the discussion gets emotional rather than analytical. People can limit many of their hazardous activities if they want to. Radiation is frightening to many people because it cannot be seen and it is difficult for the average person to limit radiation exposure.

The normal level of intensity of galactic cosmic rays is rather constant, but researchers differ widely in their estimates of the hazard. One source estimates that continuous exposure to galactic cosmic rays for a year would exceed the safe dosage by some 60 times. Another says there is little danger, that the total annual dose is from 6 to 20 rads. The main reason for this disparity is that researchers disagree as to the biological effect of protons. Estimates of their damage range from one to ten times the damage done by a similar dose of x-rays. While it is possible to accurately estimate the number and intensity of proton encounters, there is wide disagreement as to how to convert that to rads. We do know that for short trips the dosage is not excessive. The highest total dose received by an Apollo crew on a trip to the Moon was a little over 1 rad. The greatest hazard on a long journey, such as a 2 year trip to Mars, comes from the proton storms following large solar flares.

The trapped particles are also a serious hazard to people in space, but the Van Allen belts are well mapped and generally avoidable.

One interesting effect of cosmic rays is the occurrence of light flashes in the eyes, first reported by Buzz Aldrin on the Apollo 11 flight to the Moon. Other astronauts have also reported seeing flashes of light with their eyes closed. The most obvious explanation of this phenomenon is that cosmic rays penetrate the eye and deposit in the retina, sending a signal down the optic nerve which the brain interprets as a flash of light. However, a variety of shapes were observed such as streaks, commas, and diffuse clouds. These are difficult to explain.

In summary, people in space are exposed to higher levels of radiation than people on Earth even if the Sun is quiet. The short duration of Space Shuttle flights keeps the exposure to tolerable levels. The long term effects on any particular individual cannot be predicted. This question will become more demanding of an answer as people live for longer periods in the space station, on the Moon without an atmosphere, and on Mars with only a thin atmosphere.

Gravity and Weightlessness

A person on Earth, along with everything else, experiences a constant acceleration due to gravity of 32 feet per second per second, called 1 g. This means that any unsupported object would fall toward the center of Earth with an ever increasing speed, just as a car increases speed when the driver presses down on the accelerator. A car speeding up uniformly from a stop to 88 feet per second (60 miles per hour) in, say, 44 seconds, accelerates at the rate of 2 feet per second per second. Similarly, an unsupported object, accelerating at 1 g, would be falling at a speed of 32 feet per second after one second, 64 feet per second after two seconds, 96 feet per second after three seconds, and so on.

The 1 g acceleration of gravity is always with us on Earth. There is no way to "turn off" gravity. We are born, live, and die with it. The human body is designed and conditioned to operate in a one-g environment.

The term *g force* is somewhat misleading because g is not a force. Rather it is the acceleration due to the force of gravity. However, in common usage, "a force of 3 g" means any acceleration that gives a sensation of weighing three times more than normal. It makes no difference whether the acceleration is due to gravity or due to a rocket engine. The sensation is the same and the effect on the body is the same. We will use the term *g force* in that context.

When going into space, astronauts experience dramatic changes in g forces. First, while the engines are running and accelerating the spacecraft, they experience more than 1 g of gravity. Then when the engines shut down, the spacecraft is in free fall and they are instantly weightless. This doesn't happen gradually. The thrust suddenly stops and they immediately float! Finally, when the rocket motors are retrofired to return to Earth, they experience an acceleration again and its accompanying feeling of weight.

High g

Let us first examine the effects of a greater than normal g force on the body. Experience in high speed jet aircraft and experiments in a centrifuge and on a rocket-powered sled in the 1950s provided a good idea of the tolerance of the human body to higher than normal g forces. The record was set by Col. John Stapp when he decelerated at 35 g on a rocket sled at White Sands, New Mexico, in 1954.

While the general outer body structure is relatively rigid, the fluids within are free to move. Internal organs in the chest and abdomen are not rigidly fixed in place and can also move. Human bodies are well adapted to function in 1 g. Flexing muscles in the legs forces blood upward in the veins against normal gravity, helping the heart to keep the blood circulating. One-way valves in the veins also help prevent the blood from pooling in the lower body. At normal 1 g the internal organs hold their proper positions.

Problems occur if a person is sitting or standing erect and an acceleration greater than normal 1 g acts lengthwise, head to feet. Leg muscles cannot supply enough force to prevent blood from leaving the head and pooling in the lower part of the body. Lack of blood in the brain causes dizziness and possible blackout. You feel this sensation briefly when an elevator starts suddenly upward or when a roller coaster reaches the bottom of a curve. Col. Stapp in his rocket sled showed that very high g forces can be tolerated if the acceleration is directed front to back, perpendicular to the chest. The key factor in how much a body can withstand is the direction of the acceleration.

In the early manned flights into space, each astronaut had a contour seat specially fitted to his form. During launch into orbit he was lying back in the seat facing forward. During reentry he rode backward, again with the acceleration directed back to front, perpendicular to his chest. This allowed him to withstand over 8 g without blacking out.

The Space Shuttle is designed to operate at lower accelerations. During launch the engines are throttled so that the acceleration reaches a maximum of only 3 g twice during the rise to orbit, once just before the solid rockets burn out and separate, and a second time for about 5 minutes just before the external tank empties and drops away. This is for the safety of the spacecraft as well as the comfort of the crew. For comparison, 3 g is less than may be experienced on some carnival rides. The crew sits up in seats like riding in an airliner and the Shuttle accelerates nose first so the g forces are perpendicular to the chest. Just after liftoff, the Shuttle has greater acceleration than the Saturn boosters that carried men to the Moon. However, the Saturn attained much higher acceleration just before the first stage rocket burned out.

During reentry Shuttle crews encounter 1.5 g. Because the orbiter enters the atmosphere bottom side first, the force on the crew acts lengthwise as they sit upright in their seats. After several days or more in weightlessness the body could be especially susceptible to pooling of blood and blackout. Therefore, crewmembers wear antigravity suits during reentry. An antigravity suit looks like a pair of pants with inflatable bladders. As the g force increases, the bladders are inflated and apply pressure to the lower part of the body forcing blood toward the upper body and brain.

Weightlessness

Many of the problems a human body confronts while in space are related to what is commonly called weightlessness or microgravity or the absence of gravity or zero g (Figure 9.7). None of these terms is precisely correct. The force of gravity does decrease with distance from Earth. However, at orbital altitude an orbiter is not beyond Earth's gravity. (Recall **MATHBOX 2.1.**) At 300 miles, g is still about 9 meters per second per second, only 10 percent less than it is on the ground. A spacecraft could not remain in orbit if

Figure 9.7 Weightless astronauts working in Spacelab aboard orbiter *Columbia*. In a normal relaxed position, knees bend slightly, head raises or tilts back, shoulders hunch up, and arms float upward. It requires some effort to bend at the waist or to hold arms down to waist level. *Courtesy of NASA.*

gravity were zero. (Recall Newton's first law of motion: an object moves in a curved path only if some force deviates it from a straight line path. In the case of an orbiting spacecraft, gravity is the force.) If the astronauts were not in free fall, their weight would be nearly the same as it is on Earth. So the terms *zero g* and *microgravity* are somewhat misleading, although they do communicate the intended meaning.

As discussed in Chapter 3, an object in orbit is in free fall, falling toward Earth, but with sufficient horizontal velocity that it never strikes Earth and therefore remains in orbit. On Earth, the supporting upward force from the chair we sit on or the floor we stand on prevents us from falling downward. It is this force that gives us the sensation we call "weight." But in orbit, the floor is in free fall, too. It does not provide any support because the entire spacecraft is falling the same as the crew is. Since there is no upward supporting force as on Earth, the astronauts feel weightless.

People on Earth feel the same sensations and experiences as astronauts, but only for a fraction of a second at a time while in free fall in elevators, off diving boards, and on trampolines. Imagine that you jump off a diving board with an apple. You let go of the apple and it falls alongside you. To you it appears that the apple is floating in space.

As described earlier, the problems of high g forces on the human body are well understood and taken into account in the design of the spacecraft and in its mission profile, In contrast, problems with weightlessness are not yet solved.

Under weightlessness these are some of the effects on a human body:

a. Fluids shift from the lower part of the body to the upper. Lower body muscle structure is such that, under normal gravity, blood is forced upward to keep it from pooling in legs. Even in weightlessness these muscles continue to force blood and other body fluids upward.

b. Dehydration, loss of body fluids, and loss of lymphocytes. When the fluid shifts upward, the body attempts to eliminate some of it.

c. Cardiovascular deconditioning and red blood cell loss. The heart does not have to work so hard to move weightless blood through the arteries and veins. Total quantity of body fluids is reduced, including blood.

d. Muscular deconditioning. Muscles don't have to work at normal levels so they begin to atrophy. It takes days or weeks to recover, depending on the length of time the person was weightless and how much exercising was done during the flight.

e. Calcium loss from bones. The loss continues throughout the flight. The longer the flight, the more calcium is lost. Bones do not repair themselves as they do with normal gravity.

f. Space adaptation syndrome. Space sickness lasts a day or two, affecting about half the people who go into space.

g. Change in the number of neural connections. This increases in weightlessness and decreases in high g.

When the spacecraft reaches orbit and the engines stop, the vehicle and its contents are weightless, in free fall. At first there is a great feeling of euphoria, an absence of force on the body, freely floating about in the cabin. Without the downward force of gravity, the space between vertebrae in the backbone stretches and the average astronaut gains an inch or two in height. As a result, many astronauts have complained of back pains. The natural posture of a weightless astronaut has knees and hips bent like a crouch, with arms floating out in front. It takes considerable effort to straighten out. Of course, while in orbit there is no reason to straighten out.

After a short time, body fluids begin to shift. Muscles in the lower limbs that work against gravity on Earth, pushing blood upward in the body, continue to work. Blood vessels in the legs are soon emptied because of the "absence of gravity." Other fluids from the surrounding tissue flow into them. The upper body becomes somewhat swollen as the blood and other fluids move in that direction. The face becomes swollen in appearance as the normally loose flesh fills with fluid and "floats" on the cheekbones. Wrinkles and double chins disappear and eyes appear smaller. At the same time, the lower body, having lost the fluid, becomes thinner. Waist measurements decrease by an inch or two. Feet and legs become noticeably smaller; boots and pants may not fit as well. Astronauts called this "bird legs" and "wasp waist."

All this is accompanied by head congestion, a full stuffy head, throbbing in the neck, and perhaps a severe headache. The sensation is similar to that of hanging upside down on a gym bar.

Meanwhile, the body misinterprets what is happening. Sensors in the upper body indicate excess fluid and the kidneys begin action to eliminate it in the form of urine. One or two quarts of fluid and the dissolved salt are released this way in the first few days. The individual may not feel thirsty and does not replace the lost fluid by drinking water. Dehydration could be a problem.

Most of the symptoms described above disappear in about the first 3 days in orbit, either by the body self-adjusting its functions or by the mental attitude of the individual simply "getting used to it." The cardiovascular circulatory system settles down to a more normal operation but with a smaller volume of blood. Loss of calcium in the bones is the exception. It continues throughout the flight. A recent study suggests that this may be due to a reduced ability of the pituitary gland to secrete certain hormones which control growth. Over a period of several years, researchers sent both rat pituitary cells and live rats on flights of the Space Shuttle and the Russian biosatellites. Pituitary cells produced only half as much active hormone as similar cells on Earth. This is an interesting study, even though no connection between the hormones and calcium loss has been proven.

Prior to reentry on the last day of the flight, astronauts take salt tablets and drink about a quart of water to begin the process of renewing the blood volume to its normal Earth level. Recently a researcher concocted a drink called Astroade (after Gatorade) that requires much less volume for rehydration. In one test by an astronaut on a Shuttle mission, the drink appeared to help the dehydration problem.

On returning to Earth after a flight, about one in ten astronauts cannot stand upright to walk off the orbiter without help. Because of the low blood volume, the symptoms are similar to giving several pints at the blood bank. It may take several days for the cardiovascular system to readjust to normal gravity.

Space Adaptation Syndrome

About half the people who have ventured into space have reported becoming sick on the first or second day of the flight. Earliest onset was during the first hour. The symptoms are similar to motion sickness, being seasick or airsick. From one-fourth to one-half of the astronauts have reported vomiting, headache, malaise, fatigue, or loss of appetite, or a combination of those symptoms. The symptoms subside suddenly after 2 or 3 days and do not recur on that flight. Virtually all astronauts recover. Apparently the body adapts to whatever causes it.

This illness is not the same as motion sickness. All the early astronauts had been test pilots and were not susceptible to becoming airsick. Yet many of them suffered from space sickness. Tests on Earth do not identify those who may be susceptible. It becomes apparent only when they reach orbit. Those individuals who get sick on one flight will likely get sick if they make another flight. Conversely, those who do not get sick on one flight probably will not on another. The common motion sickness drugs such as Dramamine do not work against space sickness. Another drug has been developed to counter space sickness, but many astronauts prefer not to take it.

Space sickness is the subject of considerable research effort. Because the symptoms are similar to motion sickness, it is thought that the cause lies in the vestibular apparatus, the semicircular canals in the inner ear that give us a sense of balance and provide information to the brain on accelerations and rotations. The relationship between vomiting and this apparatus has been shown by experiment. Monkeys and people who have defective semicircular canals or have had them removed cannot be made to vomit when subjected to unusual acceleration or rotation.

A person on Earth can tell up from down and estimate direction and speed of motion from visual observation and from feeling the force of gravity While weightless in space, information sent from the eyes and from the vestibular apparatus may conflict. One may indicate you are upside down while the other says you are tumbling. After a day or two in orbit, a person adapts to the confusing signals and can function normally (almost). When they return to Earth, crewmembers have the same problem, causing difficulties with balance. It may take a few days to readapt to the gravity environment.

One NASA researcher, Dr. Patricia Cowings, equipped a laboratory with motion equipment designed to make people sick. One of the symptoms described to her by astronauts is that they have the feeling they are standing upside down. This seems related to the shift of fluids to the upper part of the body. Dr. Cowings suggested that biofeedback might help solve the space sickness problems. The subjects in her experiments learn to control their own heart rates, blood flow to their hands, facial pallor, and perspiration.

A trainer has been tested at Johnson Space Center which gives different visual cues than the vestibular system would expect. This causes confusing relationships to be fed to the brain, similar to those experienced in weightlessness. The purpose of the trainer is to teach the individual to rely only on what is seen, not on what the vestibular system is feeding to the brain. This type of training should help inexperienced astronauts to adapt more quickly to weightlessness on their first flight and may enable them to control their space sickness symptoms as well.

Space Adaption Syndrome did not show up in Mercury and Gemini astronauts, perhaps because their spacecraft were so small that their visual perception did not conflict with their vestibular system. As spacecraft became larger, it was easier for astronauts to see how they were moving and oriented in the space. Many of the Skylab astronauts suffered space sickness in their large space station. In Skylab, equipment, switches, and labels were mounted without respect to up and down. One piece of equipment was sideways to others. Even the commode was attached to a wall where one would expect a window to be. This disregard for

the normal human expectation to have a floor, ceiling, and walls in a room may play a part in the onset of space sickness. Designs for the new space station do make the living quarters and laboratories more earthlike with a familar up, down, and sideways orientation.

Exercise

A regular regimen of exercise as part of the astronaut's daily routine is essential to keep bones, heart, and muscles in condition. A stationary bicycle was used on the Skylab space station. Treadmills, stationary bicycles, and rowing machines are excellent means of exercising in weightlessness. Spring-type exercise devices also work in weightlessness, but weight lifting is useless! The treadmill shown in Figures 9.8 and 9.9 uses shoulder and waist straps to keep the astronaut in contact with the machine. Originally, the equipment was attached directly to the floor or bulkhead of the orbiter. Spaceflight revealed that an astronaut peddling a stationary bicycle can produce as much as 100 pounds of force which can cause severe vibrations of the entire orbiter. A platform for the exercise equipment was then designed with special stabilizers to absorb vibrations, reducing the force

Figure 9.8   Exercise treadmill.

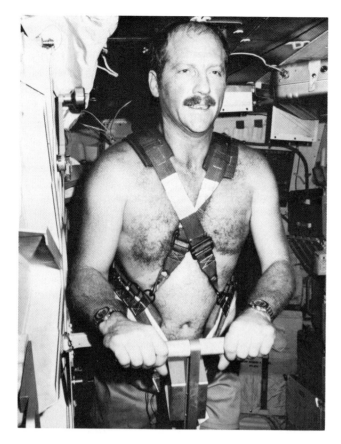

Figure 9.9 Astronaut Fred Hauck exercises on the treadmill. Straps are needed to keep the weightless astronaut in contact with the machine. *Courtesy of NASA.*

to less than 1 pound, so astronauts do not disturb sensitive microgravity experiments when they work out.

Sweat does not run off in weightlessness. It clings to the body and builds up to a thick layer. If it breaks loose, the sweat forms spherical globules floating around the cabin. If one strikes a solid surface, it spreads out like pancake batter and sticks to the surface. A strong air current from a nearby duct helps evaporate the sweat so it does not become a problem.

For Shuttle astronauts, at least 15 minutes of vigorous exercise per day is recommended for flights up to 2 weeks and 30 minutes a day for missions up to 30 days. The Russians have far more experience in spaceflight than the United States. While they are working, cosmonauts wear an exercise suit, nicknamed the penguin suit, laced up with elastic cords that create a force against the body and require exertion to do nearly any task. They eat a special diet to combat physical deterioration and keep a more rigorous exercise schedule including daily 2 mile runs on a treadmill. Even with a vigorous exercise program, calcium loss continues and muscles do not remain as strong as they would on Earth. The Russians report that they have reduced calcium loss to 5 percent at most, compared to as much as 30 percent in

early flights. They also report a wide variety of reactions to weightlessness among their cosmonauts. Following a flight of several months by two cosmonauts, one showed a 10 percent decrease in bone density in his vertebrae while his companion showed no loss at all.

Still, depending on the length of the flight, it may take days or even weeks to recover after returning to Earth. After his record 366 days in space ending in December 1988, cosmonaut Vladmir Titov reported that he was able to operate the controls of the spacecraft during reentry in a semiprone position. Back on Earth, however, he had difficulty sitting up for very long. But a few months later, Titov and Musa Manarov, the other cosmonaut on the yearlong flight, were traveling the world reporting on their experiences and appeared none the worse from their extended flight.

The human body may be the limiting factor in further exploration of space, particularly on such long missions as a 2 year flight to Mars. It may be necessary to produce an "artificial gravity" by rotating the space ship. The Russians believe that on long duration flights, psychological factors may become more significant than physical factors—more about these subjects in Chapter 11.

DISCUSSION QUESTIONS

1. A deep sea diver who experiences the bends on return to the surface is put into a pressure chamber. Why?

2. How would you package your favorite foods for eating in weightlessness?

3. Think of your daily habits of personal hygiene and cleanliness, such as combing hair, clipping fingernails, and dressing. What problems would have to be overcome while trying to do these things in weightlessness?

4. Newton's first law says that an object does not change its motion unless a force is applied. Why, then, does liquid float out of a cup and food float off a plate in an orbiting spacecraft?

5. Describe exercises that could be done in weightlessness.

6. Describe what it might be like to play basketball or handball in weightlessness.

ADDITIONAL READING

Compton, W. David, and Charles D. Benson. *Living and Working in Space, A History of Skylab*. NASA SP-4208, Government Printing Office, 1983. Very comprehensive hardback.

Kerrod, Robin. *The Illustrated History of Man in Space*. Mallard Press, 1989. Profusely illustrated, easy reading.

NASA. *Food for Space Flight*. NASA Facts, NF-133/6-82, NASA, 1982. Easy reading pamphlet.

Oberg, James and Alcestis. *Pioneering Space: Living on the Next Frontier*. McGraw-Hill, 1986. Popular book for the interested layperson.

Page, Lou and Thornton. *Apollo-Soyuz Pamphlet No. 6, Cosmic Ray Dosage*. NASA, 1977. Technical description of the experiments.

Pogue, William. *How Do You Go to the Bathroom in Space?* Doherty Associates, 1985. Delightful, fun to read; question and answer format.

Rambaut, Paul C. *Space Medicine*. Carolina Biological Supply Co., 1985. Excellent, readable pamphlet.

Rogers, Terence A. "Physiological Effects of Acceleration." *Scientific American*, February 1962. Early high-g experiments; out-of-date conclusions on weightlessness.

Steinberg, Florence S. *Aboard the Space Shuttle*. NASA EP-169, Government Printing Office, 1980. Easy reading pamphlet.

NOTES

Chapter 10

Working in Space

The Space Shuttle has made it rather easy for people to get into orbit and to do various jobs there. A list of the many impressive things done by Shuttle astronauts includes:

1. Drop off satellites in orbit.
2. Recover and repair satellites in low Earth orbit.
3. Build large structures outside the orbiter.
4. Measure the plasma environment around the orbiter.
5. Manufacture perfect spheres and crystals.
6. Prepare pharmaceutical products by separating biological cells from one another.
7. Study effects of weightlessness on the human body, plants, and animals.
8. Experimentally refuel a simulated satellite.
9. Use photography, radar, and infrared to make high resolution images of the ground.
10. Measure solar and cosmic radiation.

Some of the work has direct commercial application, some is for developing new techniques for manufacturing and processing materials in space, and some is basic research for a better understanding of the world around us.

Remote Manipulator Arm

Much of the work outside the Shuttle is done from inside the orbiter cabin using the *remote manipulator arm*. It can recover satellites from orbit and stow them in the cargo bay; lift scientific equipment from the cargo bay, drop it overboard, and later grasp it and place it back into the cargo bay; hold scientific experiments out away from the orbiter to study the space surrounding the vehicle; move an attached astronaut from place to place in the cargo bay; assemble pieces of a large structure; and make video images of inaccessible places in and around the cargo bay.

This remarkable robot device, designed and built in Canada for the orbiter, is powered by six electric motors. In many ways it resembles a human arm with a shoulder, elbow, wrist, and hand. In robotics terms, the hand is called an *end effector*. See Figures 10.1 to 10.3. When not in use, the arm

is stowed against the side of the cargo bay. From a control panel on the upper deck of the cabin, an astronaut lifts it from its cradle and moves it to any desired position and location. Movement is controlled by a joystick, similar to those used for video games. A computer reads the joystick motion and translates it to motion in the arm. If the joystick should fail, the arm can be moved by the operator with a different set of controls, one joint at a time. The operator can see what is happening through windows into the cargo bay and in the ceiling of the orbiter, and by television monitors next to the control panel. TV cameras are mounted at the wrist and the elbow of the manipulator arm and can be focused and zoomed by the operator to give the best view.

The end effector does not have fingers, but uses an ingenious rotating wire arrangement to capture and hold a payload. See Figure 10.4. A mechanical finger would be a complicated device with many joints; the wire device is simple by comparison and less likely to fail. Any payload to be manipulated by the arm must be equipped with a *grapple*, a rod-like protrusion. The end effector is slipped over the grapple and the wires tighten around it to make a firm con-

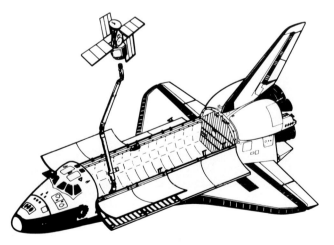

Figure 10.1 Remote manipulator arm. Fashioned after a human arm, it has a shoulder attached to the cargo bay, an elbow, a wrist, and a hand called the end effector. In this diagram it has just dropped off a satellite. *Courtesy of NASA.*

Figure 10.2 Remote manipulator arm as seen from the flight deck of the orbiter. Notice the TV camera at the elbow. *Courtesy of NASA.*

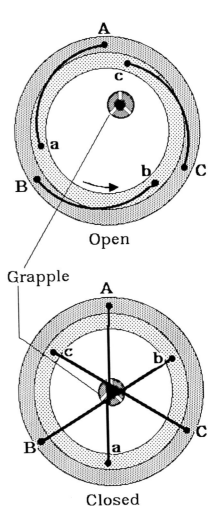

Figure 10.4 Manipulator arm's end effector. One end of each wire (A, B, and C) is fixed to the outer ring forming a triangle. The other ends (a, b, and c) are attached to a rotating inner ring; as it rotates the wires are drawn together, shrinking the triangle, to take hold of the grapple attached to the payload.

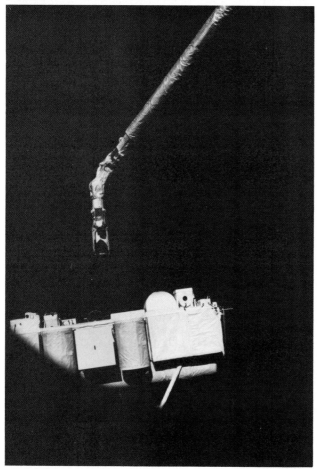

Figure 10.3 Manipulator arm moves in to grasp a small payload. The end effector is clearly seen; the grapple fixture on the payload is faintly visible. *Courtesy of NASA.*

nection. A platform with boot clamps can be attached to the end effector so an astronaut can ride the arm (Figure 10.5).

The robot arm works only in orbit. It cannot support its own weight on the ground, but in weightlessness it can manipulate any payload the orbiter can carry.

Space Suits

When work is to be done outside the cabin, in the cargo bay or away from the orbiter, the astronaut wears a space suit equipped with a self-contained life support system. It also provides some protection from both charged particles and electromagnetic radiation.

The Shuttle space suit is the outgrowth of many years of experience in constructing suits for high altitude aircraft and spacecraft. As early as the 1930s, aviation pioneer Wiley

Figure 10.5 Space suited astronaut rides the manipulator arm; foot restraints keep him from floating away. Earth is in the background. A just-released satellite is in the distance below his feet. *Courtesy of NASA.*

Post had a suit molded in the shape of a man in the sitting position. It had a helmet something like a diver's helmet and was pressurized by the engine supercharger. In it, Post set cross-country speed and altitude records with his plywood airplane.

Mercury astronauts wore space suits as backup in case the cabin lost pressure. The suits were not pressurized except in an emergency, which never happened. Gemini space suits were designed specifically for working in empty space outside the cabin, called *extravehicular activity* (EVA). They were made in several layers to give greater flexibility than the Mercury suits. An inner bladder was made of nylon coated with neoprene rubber. Over that was a net woven of nylon and Teflon strands which acted as a casing to keep the bladder from ballooning out, just as a bicycle tire acts as a casing over an inner tube. Flexibility was improved by making the outer casing slightly smaller than the inner bladder. In 1965, Ed White was the first American to leave a

vehicle while in orbit. He was connected to the spacecraft's life support system by a 25 foot long umbilical cord. Cosmonaut Alexei Leonov had been the first human to "walk in space" just 3 months before.

Apollo space suits presented other design problems because they had to work on the surface of the Moon as well as in space. See Figure 10.6. The inner structure was similar to the Gemini suit, but the outer layers had to stand up to jagged rocks, the heat of the lunar day, meteoroids falling onto the Moon, and rough handling while climbing in and out of the lunar lander. The suit also had to be flexible enough to allow stooping, bending, adjusting instruments, and sitting on the lunar rover "dune buggy." Like an inflated balloon, a space suit becomes rigid and hard to compress and bend. Bellows-like joints at the knees, elbows, and waist gave greater freedom of movement. In earlier suits, when an astronaut wanted to bend a joint, an elbow, knee, or shoulder, that entire section of the suit had to compress, reducing the volume of the enclosure and thus increasing the air pressure inside. Fighting that increased pressure quickly tired the astronaut. Bellows joints do not increase the pressure. When they bend, the outer side of the bend expands while the inner side compresses and the volume remains constant.

Like the Gemini suit, the Apollo suit was made with an inner bladder of neoprene rubber enclosed in a casing to keep it from ballooning out. Multiple layers of thin plastic alternating with fiberglass, followed by several layers of mylar and Teflon cloth provided the needed insulation and outer protection. Lunar boots fit over the space suit feet like overshoes. Backpacks carried communications and life support equipment. The suit weighed 180 pounds on Earth, but only 30 pounds on the Moon. Each one was custom made to fit its occupant at a cost of $400,000 each.

The Space Shuttle space suit is a direct descendent of the Apollo suit. See Figure 10.7. Some of the joints are equipped with ball-bearing fittings which rotate easily with little force, increasing the flexibility and reducing the fatigue of a working astronaut. It is easier to get into than previous space suits. In preparing for an EVA from the orbiter, the astronaut first dons a urine-collection device designed for either men or women with a bag attached at the groin to hold up to a quart of urine. Next comes a suit of longjohns made of spandex mesh. Air can move freely around and through this undergarment. A sealed-up space suit holds in the body heat which must be removed, so plastic tubing running through the undergarment carries cooling water to keep the suit and its occupant at a comfortable temperature.

Twenty-one ounces of drinking water is stored in a bag at the torso. A microphone and headphones are built into a tight-fitting cap, the Snoopy Cap, and are connected through the electrical harness to a radio transmitter-receiver for communicating with the cabin and with other astronauts. Instruments for monitoring heart rate and respiration are built into the suit and connected to the radio.

The space suit itself consists of a hard upper torso with an aluminum shell, a lower torso, gloves, helmet, and visor. All components are multilayered like the Apollo suit. The upper torso part of the space suit is mounted on a wall in the orbiter so a person can slip easily up into it. The astronaut puts on the lower half of the suit, then slips up into the upper torso and aligns the many connections between the two parts. The two halves lock together with metal rings to make an airtight seal. Helmet and gloves are then sealed in place, also with metal rings. A visor snaps over the helmet to provide protection from solar ultraviolet and from micrometeoroids. Work lights and a TV camera can be mounted on the helmet.

The suits are not custom made for each astronaut as in previous manned space programs. Each component comes in several sizes and all are interchangeable so any astronaut,

Figure 10.6 Apollo space suit. *Courtesy of NASA.*

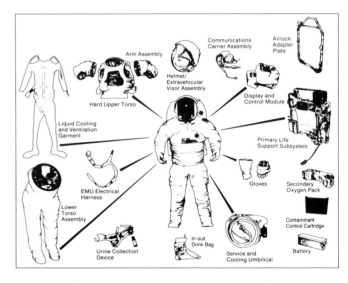

Figure 10.7 Space Shuttle space suit. *Courtesy of NASA.*

male or female, can be properly fitted. The only things that are custom made are the gloves. They pose a special engineering problem, because fingers will tire quickly if the astronaut must work against the space suit's higher pressure in ill-fitting gloves. Casts are made of the astronaut's hands in several positions and patterns for the gloves are made from the castings.

A built-in backpack contains two oxygen tanks; a fan to keep the air circulating; a contaminant control cartridge which contains filters, lithium hydroxide, and charcoal to remove carbon dioxide and other gases; cooling water and a pump to circulate it through the suit; a unit to cool the water and remove water vapor; and batteries for power. Sufficient oxygen is carried in the main tank for 7 hours; a 30 minute secondary tank provides for an emergency. The batteries power the suit for 7 hours also. Completely equipped, the suit weighs nearly 260 pounds on Earth.

A control panel is attached to the hard chest part of the suit. Seventeen sensors monitor the levels of carbon dioxide, water vapor, and other contaminants as well as the temperature, the pressure, and the supplies of oxygen, water, and power. A caution and warning system in the suit notifies the occupant of a malfunction by sounding a tone in the headphones and flashing a light on the control panel. Because a higher pressure would make the space suit less flexible and difficult to work in, pressure is kept at 4.2 psi, compared to 14.7 psi in the orbiter cabin. That gives rise to the potential problem of decompression sickness, the bends, which was discussed in Chapter 9.

Through the Airlock

Of course, astronauts cannot just open the cabin door and step out. Leaving the cabin is done through an airlock, a drum-shaped compartment 63 inches in diameter and 7 feet tall. See Figure 10.8. It may be mounted in the orbiter in one of several places shown in Figure 10.9, depending on what is being carried in the cargo bay. If a laboratory module is carried in the cargo bay, then a tunnel connects it to the middeck and, if the mission calls for EVA, the airlock would be attached to the tunnel. If the cargo bay does not hold a laboratory, the airlock may be mounted either inside or outside the cabin. The astronaut enters the compartment, dons the space suit, and attaches a tool kit and flashlight. After sealing the inner hatch, the air is pumped out. The outer hatch is then opened for exit into the payload bay, Figure 10.10.

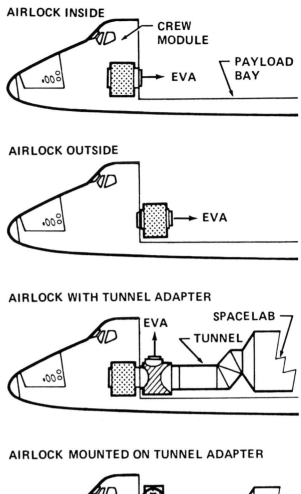

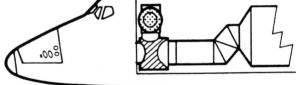

Figure 10.9 Mounting positions for the orbiter's airlock. *Courtesy of NASA.*

Handrails are conveniently located along the full length of the cargo bay near the door hinges and in several other places. See Figure 10.11. Long wires run the full length of the cargo bay just above the handrails on each side. A tether attached to the space suit and clipped to these wires keeps the astronaut from floating away, yet allows freedom of movement throughout the area. Perhaps you have tethered your dog that way, using a long wire in the backyard with the leash looped around it so the dog can run over a wide area but cannot run away.

Simple jobs become difficult when weightless. For example, turning a screwdriver is nearly impossible. When force is applied, the screw remains firmly anchored while the astronaut rotates. (Remember Newton's third law.) Foot restraints are located at strategic spots in the cabin and in

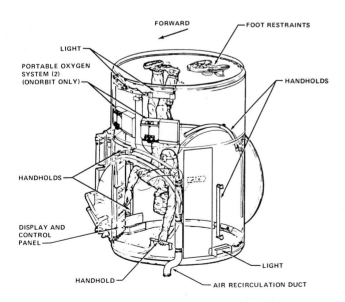

Figure 10.8 Space Shuttle orbiter airlock. *Courtesy of NASA.*

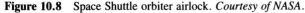

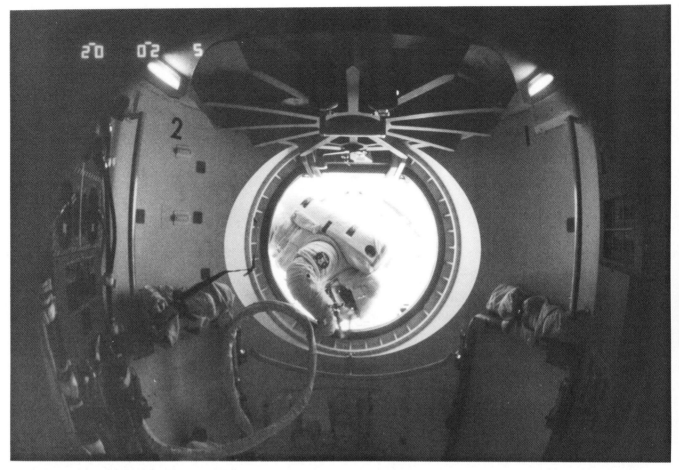

Figure 10.10   Astronaut leaves the airlock to go outside into the cargo bay. *Courtesy of NASA*.

the cargo bay, and they can be easily moved to other locations if needed. An astronaut is firmly planted in one spot by slipping both boots into the restraint. Working with hand tools is much easier when the feet stay in one place. Velcro is used abundantly to keep things from drifting off. Small objects can be tethered to straps on the wrist so they cannot float away.

Manned Maneuvering Unit

When there is work to be done beyond the reach of a person in the payload bay, the manned maneuvering unit (MMU) converts the astronaut into the world's smallest completely self-contained spacecraft. Figure 10.12 shows Bruce McCandless on the first untethered spacewalk in February 1984. He and his partner Bob Stewart flew more than 300 feet from the orbiter and returned several times. On one return McCandless flew around to the nose of the orbiter, looked in, and asked if they wanted the windows washed. Since those first flights, the MMU has been used to capture and repair several satellites.

The MMU is stowed in the payload bay against the front wall. An astronaut (in a space suit, of course) simply backs up into it, closes two latches, and flies off.

Mounted on the MMU are 24 thrusters which produce 1.6 pounds of thrust each. Nitrogen is used for propellant. Supply tanks mounted on the back carry enough propellant for about 6 hours and can be refilled from the orbiter's nitrogen supply. Flight controls are located on the ends of the armrests. At the left hand is the "throttle" and "brake" which control straight line motion: left or right, forward or backward, up or down. At the right hand is a joystick to control attitude: up and down for pitch, left and right for yaw, and rotation for roll. If the astronaut wants to stay in a particular attitude, the MMU can be put on "autopilot," freeing both hands for other tasks.

There are actually two redundant sets of controls and propulsion systems operating at all times. The thrusters work as two sets of twelve, operating independently, yet together. If thrusters or controls on one set should fail, that set can be shut down and the other can bring the astronaut back home to the orbiter.

Normal flight speed is from 0.3 to 1.0 mile per hour with respect to the Shuttle (in addition to orbital speed of 17,500 miles per hour with respect to Earth, of course).

To help the astronaut monitor what is happening, a tone is heard in the headphones anytime a thruster fires. Also, a

Figure 10.11   Astronaut moves along the side of the payload bay using handholds and a wire tether. Earth is in the background. *Courtesy of NASA.*

fiber-optic cable connects the MMU to the space suit control panel so the nitrogen supply and battery condition can be monitored. MMU silver-zinc batteries are interchangeable with those that power the space suit and can be recharged in the orbiter.

Care and Feeding of Satellites

One of the original jobs of the Space Shuttle was to act as a first stage booster taking satellites to low Earth orbit. Some satellites remain there to do their jobs; others, such as communications satellites, must be fired to higher orbits by an attached rocket engine. Certain satellites may be re-

covered and either brought back to Earth for maintenance or repaired by astronauts in orbit. The Hubble Space Telescope described in Chapter 7 is a prime example of the Shuttle's ability to launch, repair, and maintain satellites while in orbit. The HST was designed with this capability in mind.

Let us discuss this subject by way of several examples.

Satellite Launches

Figure 10.13 shows a satellite just leaving a protective cradle in the orbiter's cargo bay. The clamshell doors are kept closed to shield it from the Sun until launch time. The spacecraft rests on a turntable, and just prior to launch, a

Figure 10.12 Bruce McCandless in the manned maneuvering unit, the first man to fly untethered in space. *Courtesy of NASA.*

Figure 10.13 Communications satellite being deployed from the orbiter's payload bay. White insulated clamshell doors cover the satellite to protect it from overheating while in the Sun or getting too cold while in the dark. The egg-shaped rocket attached to the satellite will boost it to geosynchronous orbit. *Courtesy of NASA.*

motor starts the turntable and satellite rotating. Spinning the satellite serves two purposes. First, heat from the Sun is more evenly distributed over the entire surface. Intense solar heating on one side can damage the satellite. Second, a spinning object remains oriented in the same direction in space. Consider how a rotating top or gyroscope remains upright while spinning and topples over as the rotation slows.

When the satellite is rotating at about 50 revolutions per minute, springs eject it out of the cradle into space. Their slight difference in speed causes the orbiter and the satellite to slowly drift apart. In 45 minutes they will be separated by 20 to 30 miles. The Shuttle's maximum altitude is about 600 miles. If a satellite is to be sent to a higher orbit, an attached solid booster rocket is fired by a command from the ground at the proper time. This transfers the satellite from the Shuttle's orbit to an elliptical orbit with apogee at the desired altitude. When it reaches the desired altitude, an apogee motor in the satellite fires to circularize the orbit.

Retrieving Satellites

Communications satellites in geosynchronous orbit are not designed to be repaired in space because the Space Shuttle cannot reach that altitude. In February 1984, two communications satellites, Westar VI and Palapa B-2, had been dropped off in low Earth orbit, but their booster engines failed to carry them to geosynchronous altitude and they remained in low orbits, useless for communications satellites. On a Shuttle flight in November 1984, two more communications satellites were deployed and then Westar VI and

Palapa B-2 were recovered and returned to Earth for repair. See Figure 10.14.

A number of new untested procedures had to be devised. Astronaut Joe Allen flew the MMU to the spinning Palapa satellite and, using the MMU thrusters, adjusted his rotation so it matched that of Palapa. He then inserted a device called a "stinger" into the nozzle of the apogee kick motor. Attached to the satellite by the stinger, Allen fired his MMU thrusters to stop their combined rotation. He then backed away so the remote manipulator arm could take hold of the stinger and move it into the cargo bay. The next step would have been to attach a bracket to the top of the satellite for the manipulator arm to grab onto and set it into a cradle for the trip back to Earth. However, the bracket would not fit, so Allen held the satellite for a complete trip around the Earth while another astronaut, Dale Gardner, prepared it for berthing. The two together then pulled the satellite into its cradle. The satellites weigh over a thousand pounds on Earth. Although they are weightless in orbit, some force must be exerted to make them move into position (Newton's laws again).

Discovery caught up with Westar the following day. Gardner and Allen traded jobs and followed the same procedures to capture it and berth it in the cargo bay.

Insurance companies had paid out about $180 million for the lost satellites. By paying NASA about $10 million to bring them back to Earth, they expected to refurbish and

Figure 10.14 Astronauts Gardner and Allen display a "For Sale" sign for the two disabled satellites they recovered from orbit. The two satellites were brought back to Earth for repair and re-launch. *Courtesy of NASA.*

relaunch them, recouping up to $40 million of their settlement.

Solar Max Repair

Another venture undertaken by the Space Shuttle was the repair of the Solar Maximum Mission satellite. Instead of returning it to Earth, astronauts repaired it in space and released it back into orbit. Solar Max had been in orbit doing its job monitoring the Sun for about 8 months when its attitude controls and one of its instruments, a coronagraph polarimeter, failed. It remained in orbit, disabled and spinning.

In April 1984, the Shuttle *Challenger* made a rendezvous with Solar Max in orbit at about 290 miles, approached within 200 feet, and "parked." Astronaut George Nelson

donned an MMU and flew over to the rotating satellite. He approached one end where a trunion pin was protruding and, facing the pin, rolled himself at the same rate as the satellite was spinning. With both Solar Max and Nelson/MMU rotating at the same speed, there was no relative motion between them and he could reach out and grab on. But the device that Nelson carried to attach to the satellite wouldn't lock because an insulation fastener was in the way.

The plan had been to capture the satellite and fire the MMU thrusters until the rotation stopped so *Challenger's* manipulator arm could take hold. When he could not connect to the pin, Nelson tried to stop the rotation by bumping the extended solar panels. But his efforts only caused Solar Max to tumble erratically and he had to give up.

During the night, commands from the ground to Solar Max reduced the tumbling enough so that the following day *Challenger* could capture the satellite with the manipulator arm, bring it into the cargo bay, and lock it into a repair cradle. There, Nelson and James van Hoften replaced the faulty attitude control module and an electronics box for the coronagraph polarimeter. Solar Max was then released back into space and *Challenger* went home.

The original cost of Solar Max was $235 million. The cost of repair, including part of the launch costs, was about $55 million, a savings of $180 million over the price of a new satellite.

Syncom IV-3 Repair

Syncom IV-3 was another satellite that failed to reach geosynchronous orbit when the booster rocket engines would not fire on command from the ground. It, too, was stuck in a low orbit. In August 1985, Shuttle *Discovery* rendezvoused with the disabled Syncom. Astronaut James van Hoften rode the manipulator arm to a position where he could attach a handle to the satellite and pull it down to astronaut Bill Fisher. See Figure 10.15. In van Hoften's words, "This is me, . . . wondering what the hell I was doing out there. . . . The shuttle was very active, with 44 reaction control jets shooting out fireballs every which way, and I'm standing on the end of a 50-foot diving board, getting moved around, while approaching a 15,000-pound satellite, then trying to get hold of it with my hands. It was *very colorful*."

While Fisher held it steady (Figure 10.16), van Hoften attached a grapple to the satellite. The manipulator arm then took over the chore of holding it in place while the two astronauts repaired the trouble. First they shorted across some of the wiring and attached a new electronics box to bypass the defective part. Next they connected a battery power supply and sent a command to unfold the radio antenna. The satellite had been floating in space for 4 1/2 months. Therefore, a cover and temperature sensor were placed over the nozzle of the perigee-kick motor to help warm up the solid rocket propellant. A 2 month warmup period was planned before it would be safe to ignite the solid propellant and head for geosynchronous orbit. Finally, van Hoften stood on the manipulator arm and, using a handle he had attached, started the satellite rotating at three revolutions per minute and shoved it off (Figure 10.17) to drift slowly away from the orbiter.

Construction of Large Structures

Large structures, such as solar power satellites (discussed in Chapter 5) and space stations (to be covered in Chapter 11), cannot be boosted into orbit completely assembled. It will be necessary to build them in space.

Developing elements of a structure which fit together eas-

Figure 10.15   Astronaut van Hoften, riding on the remote manipulator arm, approaches the disabled Syncom IV-3 to bring it into the cargo bay for repair. *Courtesy of NASA.*

ily, Tinkertoy fashion, while weightless required several years of engineering and experimentation. Astronauts wearing space suits in a tank of water to simulate weightlessness tried working with various struts, tubes, beams, and joints until they had two designs which seemed to work well. Pieces locked together using only one simple tool, the astronaut's hands.

The designs were then tested in space during a flight of the Shuttle *Atlantis* in November and December 1985. Two structures were erected, a towerlike truss called Access and a tetrahedron called Ease.

With their boots clamped into foot restraints, astronauts Woody Spring and Jerry Ross assembled pieces of the three-sided truss into 4.5 foot sections using a rotating jig on the floor of the cargo bay. This is shown in Figure 10.18. One side of a section was assembled, then the jig was rotated and another side built, then rotated again and the third side finished. As each section was completed, it was pushed up the jig and another section assembled beneath it. Using 93 struts and 33 joints, the astronauts built ten sections, a total of 45 feet, the height of a four story building, and tore it down in an hour and 25 minutes.

Astronaut Spring, riding the manipulator arm, picked up the tower by one end and moved it in several directions to see how easily it could be maneuvered and repositioned as might be done when building a space station. Figure 10.19 shows some of the action. He found it easy to point the structure at some point on Earth and keep it in that position. He also found it easy to hold it at its center and spin it with-

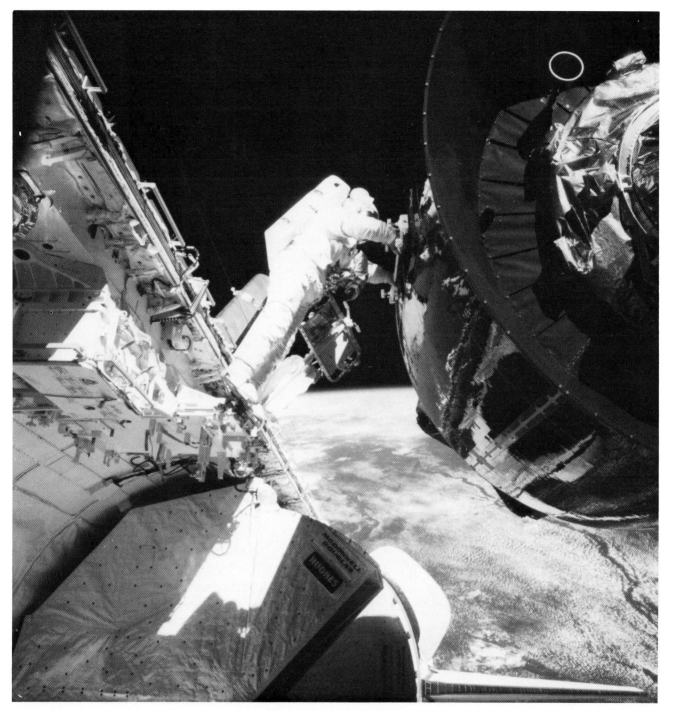

Figure 10.16 Astronaut Fisher holds Syncom in place while van Hoften, out of the picture on the other side of the satellite, attaches a grapple for the manipulator arm to grab onto. *Courtesy of NASA*.

out undue stress on the manipulator arm. Later he removed and replaced some struts in the truss to simulate repairing a damaged part. That too was not a difficult task. Finally, he ran a cable along the inside of the tower and then removed it to see how difficult a job it is in weightlessness.

Looking something like a pyramid, a tetrahedron has four triangular sides. (A pyramid has four triangular sides and a square base.) The astronauts built the Ease tetrahedron while floating freely without foot restraints, connected to the orbiter only by a tether. See Figure 10.20. The entire structure consisted of six beams, each 12 feet long, connected together with only four joints.

Trying to maintain a fixed position in space while rotating the 12 foot sections into place and snapping the corners to-

Figure 10.17 Van Hoften manually started Syncom rotating at about three revolutions per minute, then shoved it off away from the orbiter. The Moon is seen below the satellite's antenna. *Courtesy of NASA.*

gether proved to be more difficult than building the Access tower. On Earth the beams weigh 60 pounds and, even though weightless in orbit, they resist the force applied to move them (Newton, again). It was particularly tiring on the hands and wrists. The time needed to build the tetrahedron decreased with practice, of course. The first time took Spring 12 minutes; the last time required only 9 minutes. At one point he disconnected the structure from the orbiter to see how well it could be moved into various positions and found it to be more difficult to maneuver than the tower.

Scientific Research

Research in the life sciences and materials sciences, testing of prototypes of equipment for spacecraft and the space station, and development of commercially useful processes are all done in the microgravity environment of the Space Shuttle orbiter. This kind of work does not usually receive the media attention that dramatic spacewalks do because it is not as exciting or photogenic, and because much of the work is difficult to appreciate without some understanding of science or engineering.

Life Sciences

The effects of weightlessness on the human body were described in Chapter 9. Researchers are trying to find the root causes of these effects, and numerous experiments have been done on Space Shuttle missions to that end. Some measurements that are simple to do on Earth are impossible in weightlessness unless special equipment is constructed. For example, how do you weigh a weightless astronaut? A seat attached to springs is the answer. If an object is connected to a spring and the spring is stretched out, it will snap back and oscillate a number of times like a pendulum before coming to a rest. The period of the oscillations depends on the

mass of the object. That is, the more massive the object, the slower it will oscillate. The device in the orbiter has a seat which is attached to a frame by means of springs at both the front and back. An astronaut sits in the seat and begins the oscillations. A timer records how long it takes to make one oscillation; that period is directly related to the mass of the astronaut. See **MATHBOX 10.1**.

Astronauts use a bicycle ergometer for aerobic exercise. The exercising person is wired to sensors that measure physical stress, workload, and heart rate. A rotating chair is used for vestibular experiments. The subject is held tightly in place with straps and electrodes. Cameras and accelerometers record the motions while the chair is rotated in three dimensions. Also, the subject may try to keep focused on a rotating disk mounted in front to test susceptibility to dizziness, nausea, or other symptoms of space adaptation syndrome.

Internal body temperature can be monitored by a miniaturized thermometer-transmitter, about the size of a vitamin pill, small enough to be swallowed by the astronaut. An antenna and receiver are strapped on the subject and the internal body temperature is recorded for 1 to 3 days until the transmitter passes out of the body. This recently developed device offers great promise on Earth for monitoring patients in intensive care, in pregnancy and labor, and during surgery. Measurements from these and other devices are recorded for later study by physiologists who are trying to understand the response of the body to life in space.

Protein Crystal Growth

Proteins are complex compounds contained in all living cells. They provide structural support for cell walls, transport materials from place to place, promote chemical reactions, contract muscles, regulate metabolism, and do a host of other functions. Because they float immersed in body fluids, protein cells growing inside a body are nearly weightless. Similar cells growing outside the body in a laboratory are usually deformed because gravity pulls them against the bottom of the container. In the weightlessness of an orbiter or a space station, however, the cells can grow perfectly in three dimensions as they would inside a body. See Figure 10.21.

Trays for growing protein crystals aboard the orbiter have 20 experiment chambers, each of which is equipped with a syringe with two barrels; one contains the protein solution and the other contains a solution which encourages crystallization. Several hours after the Shuttle reaches orbit, an astronaut starts the process running by turning a wheel which mixes the two solutions and forces a drop out to hang on the tip of the syringe. At that point, the hanging drop is surrounded by a reservoir material saturated with a liquid. Water evaporates from the drop on the tip of the syringe and collects in the reservoir liquid which causes the protein so-

Figure 10.18   Astronaut Ross examines the tower-like truss, named Access, which he and Astronaut Spring just constructed in the cargo bay of the orbiter *Atlantis*. The white framework inside the lower part of the tower is a jig used to assist in the construction. *Courtesy of NASA.*

lution to become more concentrated and the crystal to grow. This process is called vapor diffusion. Some proteins grow to large sizes more easily if they are first seeded with a small protein crystal brought from Earth. Before landing, the chambers are photographed and the drops retracted back into the syringes. Because they are perishable, the experi-

ments must be removed from the orbiter as soon as possible after returning to Earth.

Protein crystals that been grown in space include an antibiotic for treating infections; an antibody and a drug related to AIDS; a protein used to treat diabetes; an iron-containing

Figure 10.19 Spring simulates a repair job on the truss. The Gulf of Mexico is below. *Courtesy of NASA.*

enzyme associated with liver functions; and dozens of others. Cultures of cancerous tumors, such as breast and ovarian tumors, grown in space can be used to learn how they form and grow, perhaps leading to methods for treating them growing inside a body.

Biological Science Experiments

Numerous small scientific experiments have been carried on Shuttle missions. Only a few will be mentioned here.

A colony of 3,300 honey bees in a 9 inch cube was carried on a 6 day mission to see if they would go about their usual bee-business in weightlessness. They did. They built a normal honeycomb of hexagonal cells and the queen laid 35 eggs. That experiment had been devised by a Tennessee high school student.

In another insect experiment, flies, moths, and bees were observed while weightless. Moths flew around in the chamber rather well, bees let themselves tumble without flapping their wings, and flies spent most of the time walking on the walls of the chamber.

The death of 16 chicken embryos after a 5 day ride in the Shuttle raised questions as to whether higher life forms could reproduce in weightlessness. Sixteen eggs were fertilized 9 days before the flight, and another 16 were fertilized 2 days before liftoff. The first 16 embryos survived, but 8 of the second group were dead when the eggs were opened on arrival back on Earth. The other 8 died soon after.

Sunflower seeds were carried in sealed plant containers with measured amounts of soil and water. On Earth, gravity pulls water down into the soil. Under weightlessness the water collects around the seed and drowns it. Previous experiments have failed because of this problem. The quantity of water in the soil may determine whether a seed will germinate or not.

Plants produce a substance called *lignin* which gives them structural strength to stand up against gravity. Although lignin is vital to the plants, commercially it is a relatively useless material which interferes with paper manufacturing. Seedlings have been carried on Shuttle missions to see how they grow while weightless, particularly how much lignin they produce. Better understanding of lignin production may help in more efficient utilization of wood products.

Floating algae, *phytoplankton*, in the oceans are a basic food for fish. Find the phytoplankton and you find the fish. In a Shuttle experiment, light reflected from the eastern sides of the Atlantic and Pacific Oceans was directed through a telescope to a diffraction grating which broke the light into its spectrum of colors (wavelengths). Chlorophyll in phytoplankton absorbs certain blue wavelengths and reflects certain green wavelengths. If the bright green wavelength was present while blue was absent, a concentration of phytoplankton and, therefore, fish was probably at that location. Routine methods for leading commercial fishermen to good fishing spots could come from these experiments.

Spacelab

Major scientific research is done in Spacelab, Figure 10.22 in the color section, page 191. The European Space Agency developed the cylindrical laboratory module which fits into the Shuttle cargo bay and connects to the cabin through a tunnel to the airlock hatch. This configuration can be seen in Figure 10.9. It has flown 10 times hosting American, European, and Japanese scientist-astronauts working together in a shirt-sleeve environment. Experiments can be placed either inside the Spacelab or outside on pallets toward the rear of the cargo bay and operated from inside as seen in Figure 10.23.

Hundreds of experiments have been carried on Spacelab flights in biological sciences, materials processing, Earth sensing, astronomy, solar physics, and a variety of other areas. The astronauts have voice, data, and video communications with experts and advisors on the ground so their work can be discussed as the experiments progress. If a problem comes up or if a new idea rises from the results, immediate modifications to the experiment can sometimes be made.

Many of the experiments depend on weightlessness and require a very smooth ride without bumps or jolts. Crystal growth, for example, can be disrupted by motions of the orbiter. One method of avoiding and disturbing motions is to

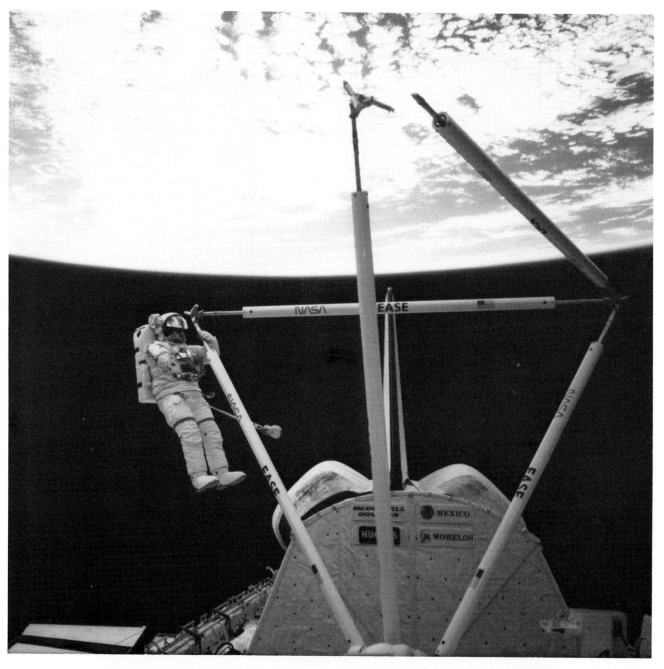

Figure 10.20 Ross floats freely while constructing the Ease tetrahedron. *Courtesy of NASA.*

let the orbiter drift free in orbit with the tail pointed down toward Earth. This attitude is called gravity gradient stabilization. Gravity acting on the engines, the most massive parts of the orbiter, keeps it oriented in that position without firing the thrusters. See Figures 10.23 and 10.24.

Industrial Research

The low gravity environment of an orbiting vehicle offers possibilities for carrying on manufacturing processes that are more difficult and less efficient than on Earth. For example, very large pure silicon crystals have been grown on Skylab and on Shuttle flights which would have been impossible to grow in normal Earth gravity. Gravity causes the crystal to sag and become malformed. In addition, impurities from the container can become part of its structure. A large percentage of the silicon crystals grown for computer chips must be thrown out because of imperfections. Gallium arsenide chips are faster and more powerful than silicon chips, but are even more difficult to manufacture. In weightlessness, crystals can be processed without a container, eliminating the impurity problem, and can be grown much larger with a near perfect molecular structure. In spite of the high cost of transportation to and from space, it may become

MATHBOX 10.1

Weighing a Weightless Astronaut

The period of oscillation of a mass hanging on a spring is given by the equation

$$T = 2\pi \sqrt{\frac{m}{k}}$$

where T is the period in seconds, m is the mass of the object, and k is the spring constant. Notice that the equation does not include the acceleration of gravity, g. If we want to find the mass of an oscillating astronaut, we solve this equation for m:

$$m = \frac{kT^2}{4\pi^2}.$$

The spring constant is a measure of the stiffness of the spring, as indicated by the force that must be applied to stretch it a certain length. It can be found experimentally on Earth by hanging the spring from a hook and then hanging masses from the spring and measuring how far the spring stretches. Using metric units, say the spring stretches 0.08 meter (8 centimeters) when a 10 kilogram mass is hung on it. On Earth, the 10 kilogram mass exerts a downward force of $10 \times 9.8 = 98$ newtons. Then the spring constant is

$$k = \frac{F}{d} = \frac{98}{0.08} = 1{,}225 \text{ newtons per meter.}$$

The spring constant is just that, a constant. It is the same in orbit as it is on the ground for any given spring.

Now an astronaut sits on the seat and oscillates with a period of 1.5 seconds. His mass is, therefore,

$$m = \frac{kT^2}{4\pi^2} = \frac{1225 \times 1.5^2}{4\pi^2} = 69.8 \text{ kilograms}$$

which is about 155 pounds Earth weight

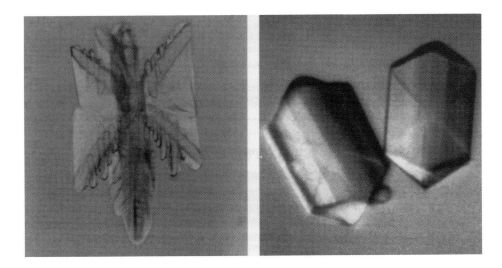

Figure 10.21 Isocitrate lyase crystal grown on Earth (left) and in space (right). *Courtesy of NASA*.

profitable in the future to manufacture such crystals in orbit.

Manufacturing high quality alloys is another possible commercial venture in space. An alloy is a mixture of two or more different metals to produce a material with different characteristics than any of its components. For example, titanium steel is a material of exceptional strength even at high temperatures. Some aluminum alloys are as strong as steel but weigh much less. When mixing an alloy in Earth gravity, the more dense metal tends to sink to the bottom before the mix cools and hardens, producing a non-uniform material. That separation would not happen in the microgravity of an orbiting spacecraft. Materials of unprecedented purity and uniformity could be prepared by allowing the mix to float in a vacuum without a container until it hardens.

Electrophoresis

One of the manufacturing processes tested for commercial use in space is electrophoresis. Electrophoresis is a method of separating materials from one another while they are immersed in a liquid or gel. Every substance has an inherent electrical charge because its molecules are made up of atoms which contain negatively charged electrons and positively charged protons. If the molecule has equal numbers of protons and electrons, then it is electrically neutral. But if one or more electrons are missing, as is often the case, the molecule bears a net positive charge, and it can be made to move through the fluid by applying an electric field. Posi-

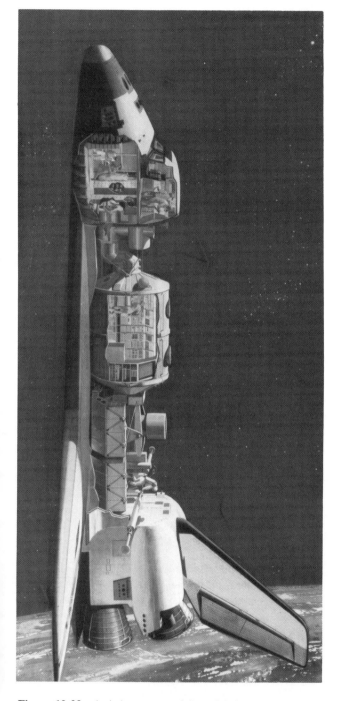

Figure 10.23 Artist's concept of Spacelab in the cargo bay of the orbiter. Scientists work inside the pressurized laboratory in a shirtsleeve environment; some of the instruments, located just to the rear, are tended by a space suited astronaut. A tunnel connects Spacelab with the orbiter cabin and the airlock for exit to space is seen attached to the top of the tunnel. (Refer back to Figure 10.9.) The orbiter is in the gravity gradient attitude. *Courtesy of NASA.*

Shuttle in gravity gradient attitude and 57° orbit plane

Figure 10.24 Shuttle in gravity gradient attitude. *Courtesy of NASA.*

tively charged particles migrate toward the negative side of the electric field. The speed at which the particles move in the fluid depends on their mass and net charge, and on the strength of the electric field.

In one type of electrophoretic machine, substances to be separated are injected into a fluid flowing continuously downward through the chamber. See Figure 10.25. An electric field is applied perpendicular to the flow. As the various substances move downward with the flow, they split into several particle streams, depending on their charge and mass, and migrate across the flow at various rates in the electric field. The substances are collected at outlets at the bottom of the chamber.

A most important application of electrophoresis is in the separation of medically useful biological substances, enzymes and hormones. For example, insulin is produced by beta cells in the pancreas. Beta cells can be a source of natural insulin if they can be separated from the pancreatic

cells. Pure natural insulin is used for research and treatment of diabetes. Similarly, kidney cells produce an enzyme, urokinase, which can help in treatment of blood clots, phlebitis, and strokes. Pituitary cells produce growth hormone. The problem comes in separating the pure hormones, enzymes, cells and other proteins from one another.

Electrophoresis works on Earth, but two problems occur, both due to gravity. First, if the biological materials are much more dense than the moving fluid, the injected stream collapses under gravity around the inlet and the purity of the separation is reduced. If they are diluted to a lower density, the purity is improved, but the quantity that can be separated in a given time is reduced.

There is a second gravity problem. If the temperature of the moving fluid is not uniform throughout, the cooler parts, being more dense, will sink while the warmer parts rise, producing convection currents. This phenomenon can be observed when looking into a pot of water on a stove. The water at the bottom of the pot is hotter because it is in contact with the burner while the top surface of the water is cooler. Convection currents can be easily observed. In an electrophoretic machine, convection cells disrupt the moving streams, reducing their purity. Using a smaller chamber slows the convection currents, but also reduces the output of the separated substances.

In a weightless environment these problems disappear. High volume, high density samples can be injected into large volume chambers and large quantities of pure substances can be produced. Two experiments were flown in the Apollo-Soyuz project in 1975, one a continuous flow system from Germany and the other a static flow system by NASA. A mixture of horse, rabbit, and human blood cells were successfully separated into the three constituents. McDonnell Douglas Corporation built electrophoretic machines which were operated by astronauts on five Shuttle missions from 1982 to 1985. The results were remarkably successful. On one of the flights the unit separated more than 700 times the material with four times the purity than was achieved in a similar machine on Earth.

Plans have been developed for an electrophoretic factory in space, either manned or unmanned. An unmanned factory would be visited several times a year by astronauts to bring up fresh materials for processing, return the separated materials back to Earth, and perform any necessary repairs and maintenance on the equipment. A manned version could be operated aboard the International Space Station about the turn of the century. Cost-effectiveness would be greatly improved because the machine would have to be launched only once and would operate continuously while in orbit.

SPACEHAB

SPACEHAB, Inc., is a privately financed company that has built laboratory modules for commercial research which fly

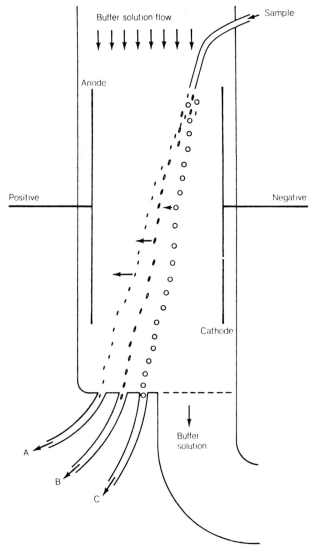

Figure 10.25 Electrophoresis. *Courtesy of NASA.*

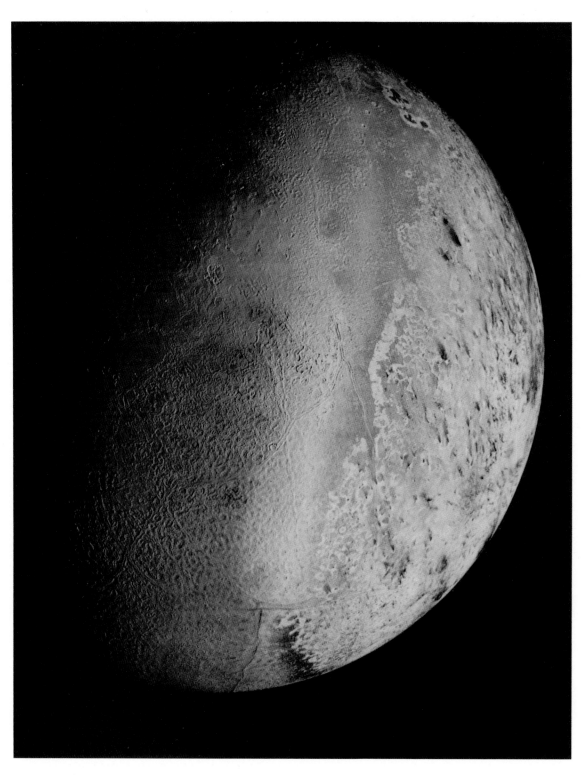

Figure 7.14 Neptune's large moon, Triton, shows an incredible variety of terrain. The bottom of the picture is the large south polar cap, probably frozen nitrogen. Farther north the darker colored area may have methane in the atmosphere, colored by ultraviolet light and radiation from the magnetosphere. The "cantaloupe terrain" in the center is unexplained. The northern polar cap is in darkness. *Courtesy of NASA.*

Figure 7.19 Earth as seen from a million miles away. South America is at the center, a bit of North America at the top, the Pacific Ocean on the left, the Atlantic on the right, and the south polar ice cap at the bottom. *Courtesy of NASA.*

Figure 7.22   Cassini ejects the European Space Agency's Huygens probe to investigate the atmosphere of Saturn's moon, Titan. *Courtesy of NASA*.

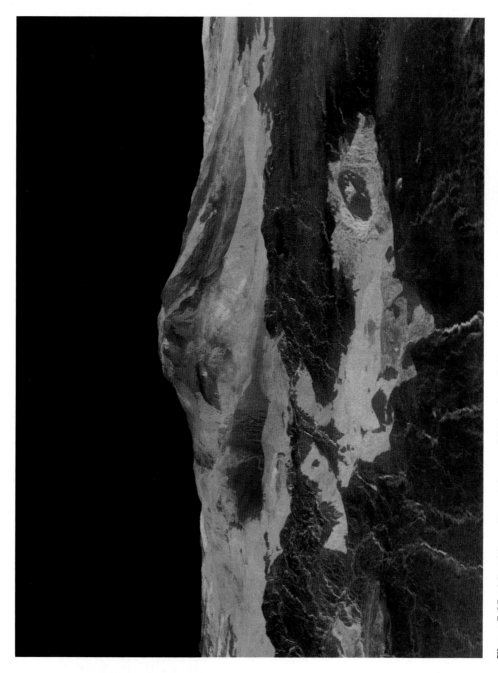

Figure 7.25 Maat Mons, Venus's largest shield volcano built up of overlapping lava flows. This three-dimensional view was produced by a computer using a combination of radar and altimeter data. Vertical height is exaggerated about 10 times. The colors are approximately true, based on the color pictures returned from the Russian Venera spacecraft that landed on Venus in the 1980s. *Courtesy of NASA/JPL.*

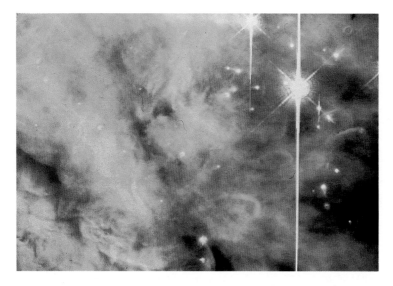

Figure 7.35 Core region of the Orion Nebula. The brightest star is θ¹C-Orionis and the next two brightest in the upper left corner are stars in the Trapezium Cluster. About a dozen low mass stars and their protoplanetary disks surround θ¹C-Orionis. Intense radiation pressure from θ¹C-Orionis sweeps the proto-planetary disks out into teardrop shapes pointing away from the star. The vertical streak is caused by overexposure of the star in the CCD. *Courtesy of Robert O'Dell and NASA.*

Figure 10.22 Spacelab in the orbiter's cargo bay. The tunnel leading from the orbiter middeck to the Spacelab can be seen in the lower foreground. *Photo ESA.*

Figure 11.3 Final configuration of the International Space Station complex in the year 2002. *Courtesy of Boeing and NASA.*

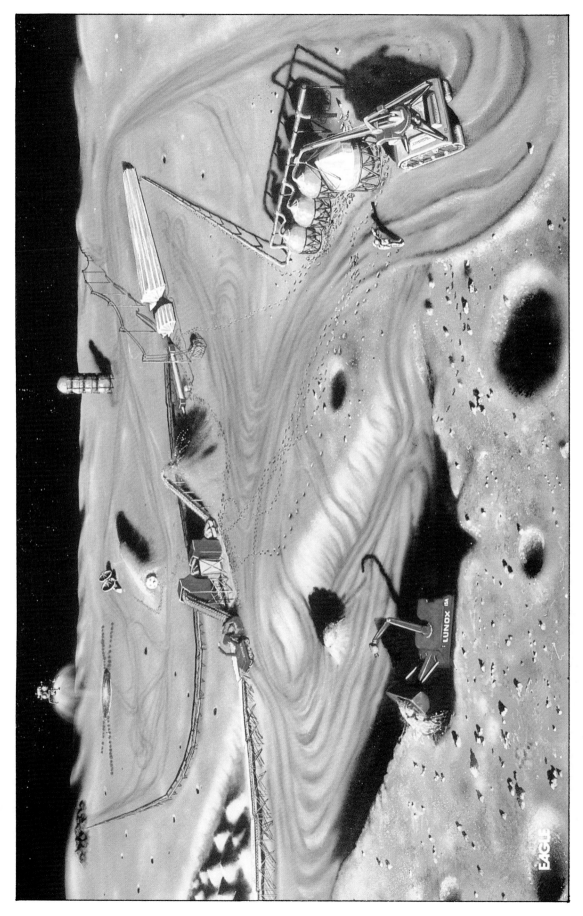

Figure 12.3 Artist's concept of a lunar mining base. Lunar soil is scooped up (red endloaders) and ground up (red structure to left); oxygen is extracted from the soil and liquified (yellow buildings), then piped to the yellow storage tanks. Waste tailings are carried away by a conveyer system. A rocket takes off in the background carrying several tanks of oxygen to a rendezvous point in lunar orbit. The habitat for a dozen technicians is just right of the communications antennas, covered by a mound of soil. Power lines come from a nuclear generator beyond the horizon. *Courtesy of Eagle Engineering for NASA; artist Pat Rawlings.*

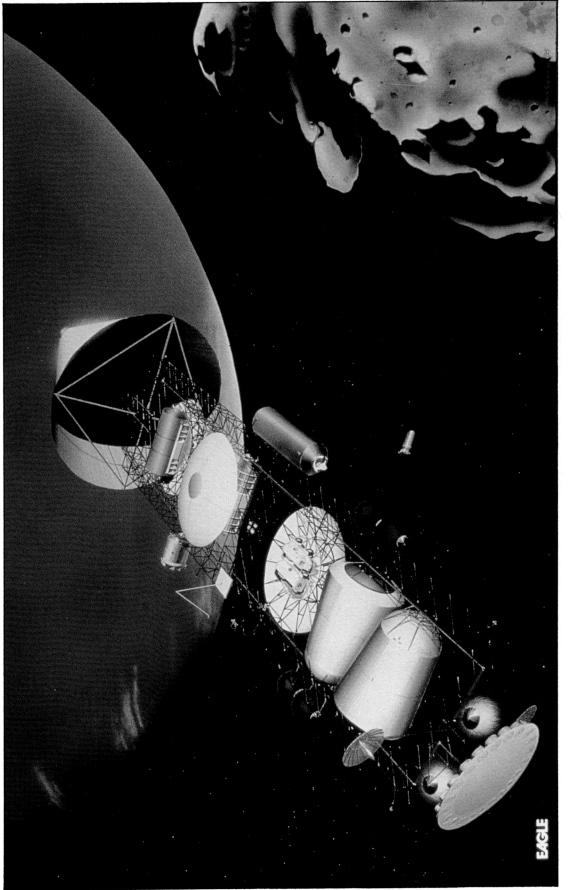

Figure 12.19 A space barge hauls a massive load of cargo from low Earth orbit to low Mars orbit. The Martian moon, Phobos, is visible at the lower right. *Courtesy of Eagle Engineering for NASA; artist Mark Dowman.*

in the Space Shuttle. The module is 10 feet long, so it occupies only part of the cargo bay, leaving space for satellites or other payloads. University and commercial researchers may lease space in the laboratory in the form of lockers or equipment racks. All the arrangements for electrical connections, data handling, temperature control, and crew-tending the experiment are provided so the researcher needs only to be concerned with the experiment itself. The first flight of SPACEHAB was in June 1993. It carried 13 experiments specifically designed to further the commercial development of space, such things as high temperature melting of metals, cell splitting, and soldering in weightlessness.

DISCUSSION QUESTIONS

1. If objects are weightless in the payload bay, what is the advantage of using the manipulator arm to move them around?

2. What kinds of jobs can robots do better than people? What jobs can people do better?

3. What limits the length of time an astronaut can work outside the cabin on an EVA?

4. Why does the pilot warn the crew when the thrusters are about to be fired to change the attitude of the orbiter in space?

5. Does it make any difference which way the orbiter is oriented when a satellite is released into orbit?

6. What commercial products manufactured in weightlessness may become available in the future for purchase in stores on Earth?

ADDITIONAL READING

Compton, W. David, and Charles D. Benson. *Living and Working in Space, A History of Skylab.* NASA SP-4280, Government Printing Office, 1983. Hardback, complete, readable.

Joels, Kerry M., and Gregory P. Kennedy. *The Space Shuttle Operator's Manual.* Ballentine Books, 1982. Easy reading, many diagrams.

NASA. *Microgravity . . . A New Tool for Basic and Applied Research in Space.* NASA EP-212, Government Printing Office, 1985. Semitechnical booklet.

Pogue, William R. *How Do You Go to the Bathroom in Space?* Tom Doherty Associates, 1985. Question and answer format, by Skylab astronaut.

Roland, Alex, "The Shuttle, Triumph or Turkey," and Overbye, Dennis, "Success Amid the Snafus." *Discover,* November 1985. The other side of the coin.

Science, July 13, 1984. Special issue on Spacelab results.

STS Mission Profiles. Comprehensive coverage of all Space Shuttle missions, including preflight guides for upcoming missions and postflight reports. Eight issues per year. Address P.O. Box 751387, Memphis, TN 38175-1387.

NOTES

Chapter 11

Space Stations

When Europeans set out westward across the Atlantic, national governments provided the necessary support. It is said that Spain's Queen Isabella even sold her jewels to finance Columbus's expeditions to the "New World." When Americans moved westward in the nineteenth century, the government provided land to settlers, incentives for building railroads, and military protection. In modern terms, the government provided or encouraged the development of an *infrastructure* to allow the westward expansion of the country. Development of air transportation also required a government-sponsored infrastructure of airports, flight control centers, radar sites, weather stations, and communications systems. Now the time has come for developing an infrastructure for expansion into the new frontier of space: a transportation system and a permanently manned space station are the first steps.

The United States had a space station in the 1970s called Skylab. See Figure 11.1 and Chapter 1. Skylab was constructed from the large fuel tank of a Saturn rocket, the type that had been used to carry men to the Moon. It weighed 100 tons, the equivalent of seven cross-country buses, the heaviest object put into orbit by the United States until the Space Shuttle orbiter which weighs 80 tons plus up to 32 tons payload. Skylab was occupied by three crews of three men, the first for a month, the second for 2 months and the third for 3 months.

The former Soviet Union has 22 years experience with space stations, far more than the United States. The Salut space stations were a continuing presence for many years; they were discussed in Chapter 1. The Mir space station was launched in 1986 and has been permanently occupied ever since (Figure 11.2).

Now an International Space Station involving the United States, Russia, Japan, Canada, and the nine member countries of the European Space Agency is being constructed in modules for launching and assembly in orbit during the late 1990s. It is a truly international endeavor. Nations will have to learn to work together for a common goal in building the space station and then to cooperate in its operation. If this international project can succeed, it certainly bodes well for the future of international affairs.

Figure 11.1 Skylab space station, photographed by the last crew to work there as they departed for Earth. The shadow of the Apollo command module can be seen on the solar array. *Courtesy of NASA.*

Why a Space Station?

Many of the scientific research tasks now being done by unmanned satellites or by astronauts on weeklong trips in the Space Shuttle can be done better aboard a permanently manned space station. Research in such areas as biomedical science, combustion physics, materials processing, Earth and atmospheric science, and fluid dynamics can be particularly fruitful when done in the microgravity conditions of low Earth orbit. The new space station will become an international laboratory in space.

Although some of this research can be carried out on unmanned satellites, they must be proven 100 percent perfect

197

Figure 11.2   Soviet Mir space station. Six docking ports enable a ferry craft, resupply vehicle, and four other modules to be attached and detached as desired. Crewmembers have individual cabins. *Courtesy of U.S. Space Command.*

before launch. If something goes wrong in orbit, unless the Space Shuttle can reach and repair it, that is the end of it. Ground testing to certify the equipment ready for space is expensive and time consuming. Such perfection is not so critical if the equipment is aboard a space station with someone tending it. When something has gone wrong on the Space Shuttle, it has often been repaired on the spot by astronauts. Also, if the results of an experiment or a measurement suggest a new or modified approach, people tending experiments on or near the space station can make changes to the experiment and test, repair, or replace the equipment. Spacelab, carried in the orbiter's cargo bay, gives a hint of this. Radio, television, and data link interactions between the on-board astronaut and the scientists on Earth have proved fruitful in many ways.

Spacehab, also carried in the orbiter's cargo bay, is a facility for doing biological and medical experiments in weightlessness while being monitored from the ground. A more thorough understanding of the long term effects of

weightlessness on people is needed before prolonged flights can be safely carried out. Since the 2 week duration of Shuttle missions is not long enough for thorough experiments, better research can be done aboard a space station. With cosmonauts in the Mir space station for more than a year at a time, the Russians have learned a great deal about how the human body reacts to extended periods of weightlessness.

Manufacturing in space has been done experimentally, electrophoresis for example (see Chapter 10), but Shuttle missions are limited to about 2 weeks. To be commercially profitable, manufacturing must be a continuous operation. Idle machinery makes no profit. Experiments and manufacturing aboard a space station could go on indefinitely.

An International Space Station

In his State of the Union Message of January 5, 1984, President Reagan directed NASA to develop a permanently manned *space station* and to do it within a decade. It didn't

happen in that time frame. International partners—Canada, the European Space Agency, and Japan—joined in the project, but the projected cost to the United States of designing, constructing, and launching the components into orbit was far beyond what Congress was willing to authorize. With the end of the cold war and the disintegration of the Soviet Union, the entire picture changed. International cooperation could now include Russia, whose space resources far exceeded other countries, even those of the United States, especially in heavy lift rocket boosters and space station experience. With the participation of Russia, an international space station will cost each participating country less than going it alone. It is hoped that the International Space Station will prove that the United States, Canada, Russia, the Europeans, Japan, and others can work together successfully on a major scientific and technical undertaking, and that peaceful cooperation can replace hostile competition. It will be a model for other large scale space ventures such as a colony on the Moon and a manned expedition to Mars. Peaceful economic and scientific gain and major leaps in research and technology are anticipated.

Design and Construction

What will the International Space Station look like? Many different designs have been proposed. Basically they all have one or more long spars with a number of things attached: living quarters, workshops, laboratories, solar collectors to produce electricity, radiators to dissipate heat, platforms (pallets) for experiments and manufacturing, communications antennas, and docking ports. The basic designs are all deliberately made flexible to allow variations and modifications, assuring station usefulness well into the 21st century.

The fully operational International Space Station is shown in Figure 11.3 in the color section, page 192. It consists of one long keel truss with six laboratories, two habitation modules, and two logistics modules, totaling more than 42,000 cubic feet of pressurized space, nearly equal to the passenger compartments in two 747 jet airliners. See Figure 11.4. An annual average of 110 kilowatts of electric power is provided by 24 solar arrays: 16 main panels and 8 panels on the Russian modules. This model space station is designed to accommodate an international crew of six engineers, scientists, and technicians. An artist's conceptions of the interiors of a laboratory module and a habitat module are shown in Figures 11.5 and 11.6. Figure 11.7 is a cutaway drawing of the ESA laboratory, Columbus, showing its attachment to the station's forward node.

The central truss and its attached modules will maintain a constant orientation with respect to the Earth for most advantageous Earth observing, while the solar arrays must be continuously aimed toward the Sun. This is achieved by means of a rotating joint at the point of connection to the central truss. While the station is in Earth's shadow, electricity will be supplied by nickel-hydrogen batteries which

are recharged when the station returns to the sunlit side of Earth. This charge-discharge cycle repeats every orbit.

The laboratory modules are located at the center of the complex where microgravity is at a minimum. Any activity that may perturb the microgravity experiments, such as docking or undocking the resupply and passenger vehicles, will be scheduled so that quiet microgravity conditions are maintained for 30 to 60 days at a stretch. Astronauts will spend about 200 hours per year of extravehicular activities (EVA) outside the station for maintenance. Such external activities will also be restricted during the microgravity quiet periods to avoid disturbing the experiments.

Much of the International Space Station will consist of prefabricated modules designed to be hauled into orbit in the cargo bay of the Space Shuttle orbiter or aboard unmanned Russian Proton and Soyuz launch vehicles. To complete the construction, 21 U.S. flights and 13 Russian flights are envisioned beginning in December 1997. The European Space Agency may provide four of the launches on its Ariane launch vehicles. Larger framework sections will be constructed in space, some automatically and some by astronauts in space suits and MMUs.

The first to go into orbit will be a Russian Salyut type of vehicle with a docking compartment for the U.S. Space Shuttle. Next come the Russian service module, docking node, and solar panels. After the first 13 flights in 8 months during 1997 and 1998, the U.S. laboratory will be attached and microgravity science work will begin. A Russian laboratory and the Canadian robotic arms will be attached next, followed by the Japanese modules, the European laboratory, and two more Russian modules. Most of these Russian components were originally intended for a second generation Mir. The station will be complete with permanent occupancy by a six person crew in 2002.

Canada's two advanced remote manipulator arms will be the primary machines used in the assembly of the space station. They will be attached to a platform which can move along a rail attached to the central keel truss, as shown in Figure 11.4. After construction is complete, the arms will be used in many space station activities such as docking the Space Shuttle orbiter, loading and unloading materials to and from the cargo bay, supporting space-walking astronauts, servicing outside payloads, and moving equipment and supplies around the station.

The International Space Station will be in a 93 minute circular orbit at about 275 miles. The area of the station is about equal to the size of two football fields. Because of this large size and the low altitude, atmospheric drag will cause the orbit to lower steadily. A boost back up to 275 miles will be needed every 3 months or so. A "space tug" will provide propulsion, guidance, and attitude control. A Russian Salyut vehicle may perform this function.

For two important reasons the inclination of the space

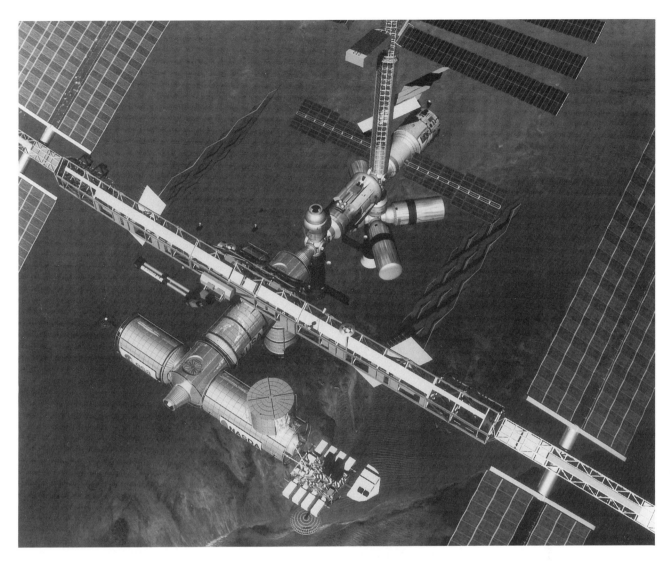

1. Main truss structure
2. Solar arrays
3. Radiators
4. Russian components
5. Russian Mir-derived service module
6. Node for attaching modules and Soyuz return vehicles
7. U.S. laboratory module
8. Habitation module
9. European Space Agency laboratory module
10. Canadian-made robotic manipulator arm
11. Italian-made logistics module
12. Japanese laboratory module
13. Japanese exposed experiment
14. Japanese logistics module

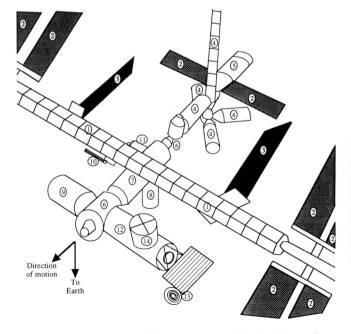

Figure 11.4   Overhead view of center section of completed International Space Station complex. The segments are identified by number in the sketch. *Courtesy of Boeing and NASA.*

Figure 11.5 Artist's concept of the interior of the U.S. laboratory module. The astronaut in the foreground is inspecting the results of a fluids experiment. Behind her another astronaut manipulates a life science experiment. Near the ceiling an astronaut pulls out a utility box. The two in the rear are working at a computer and removing a rack for servicing. *Courtesy of NASA; Harold Smelcer, artist.*

Figure 11.6 Artist's concept of the living quarters. The astronaut in the foreground is working up a sweat on a rowing machine. Another appears to hang from the ceiling as she is strapped to a treadmill for a jog. In the center, an astronaut checks a computer monitor. In the rear, one prepares to heat his dinner in a microwave oven while another looks out a window at the Earth, camera in hand. *Courtesy of NASA; Harold Smelcer, artist.*

station's orbit will be 51.6°. First, one of the major tasks aboard the station will be Earth observation. At that inclination it will overfly 85 percent of the Earth's surface and 95 percent of Earth's population. This is easily seen in Figure 11.8, the simulated ground track for 1 day plus 2 hours. Second, and equally important, all international participants will be able to launch directly into the orbit. Originally, the United States intended to use an orbit at 28.5° inclination, the latitude of the Kennedy Space Center launch site. However, the Russian launch complexes are located at higher latitudes and it is not possible to launch directly into a 28.5° orbit. Vehicles launched from their sites would need additional energy to change inclination after launch which, as a result, would reduce the payloads they could carry to orbit.

Life Support

At first, food, water, clothing, and other expendable supplies will be hauled from Earth and trash will be returned to Earth. Later it may prove more efficient to recycle at least some of the expendables. An economical means of recycling air and water in the space station is the first possibility. The Space Shuttle carries all necessities for its relatively short flights. In the Shuttle, water is manufactured by the fuel cells; hydrogen and oxygen supplies are brought along to power them. Since electricity for the space station will come from solar energy, not fuel cells, that source of fresh water will not be available. In the Shuttle, excess humidity, carbon dioxide, and other noxious gases are removed from the air, and oxygen and nitrogen are added as needed from tanks carried aboard. For a continuously operating space station, it may be less expensive in the long run to install machinery to recycle air and water rather than replenish the supplies every few weeks. A NASA study group in 1977

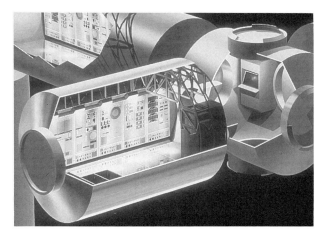

Figure 11.7 Cutaway view of Columbus, the European Space Agency's laboratory module showing its attachment to the node at the front of the space station. *PHOTO ESA.*

examined this question in some detail. The scientific principles involved are well known, but the technology to do it reliably is not perfected.

What happens to a closed and sealed atmosphere after a few years? For periods of weeks or months, we have the experience of nuclear submarines to draw on. But for longer times, research is just getting underway. Toxic chemicals tend to build up in a closed environment and must be carefully monitored. For example, nickel has been detected in fuel cell water. Excessive intake could be a problem for astronauts.

Beyond recycling air and water, the science and technology of recycling solid waste are even less well understood. It will probably not be feasible on the International

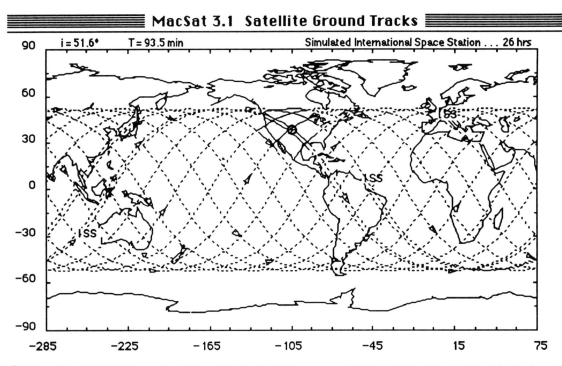

Figure 11.8   Ground track of the International Space Station for 26 hours at an altitude of 275 miles. At this altitude, the entire pattern shifts a little more than 3° eastward each day. The first 2 hours of the second day are seen over the central United States, South America, and in the far western Pacific near the Philippines. Thus, the station passes over different areas of the Earth each day and covers the entire area between 51.6° S to 51.6° N in about a week.

Space Station, but will be a necessity for very long term ventures such as large colonies in space or on other celestial bodies. We will look at this subject again later in this chapter.

Bus Service to Orbit

Once the station is completed and occupied, the U.S. Space Shuttle and the Russian Soyuz and Progress will ferry fresh supplies and crews to the station and return crews and trash to Earth about every 3 months. Crews will rotate every 6 months. The Italian Space Agency is building two pressurized logistics modules for transporting equipment and supplies to the space station and returning necessary items. One will be docked to the station while another is on the ground being refilled.

The Russian Soyuz capsule (described in Chapter 1) may be used as a "lifeboat," called an assured crew return vehicle (ACRV), to provide a fast safe way to return to Earth in an emergency. Because a Soyuz carries three passengers, two vehicles would have to be continuously docked on the International Space Station ready for instantaneous use. The Soyuz launcher has been highly reliable with a 97 percent success rate. Still, a few technical problems must be solved. The Soyuz 6 month life span must be extended to 3 years, the life of the batteries needs to be extended, and the propellants must be maintained at the proper temperature for immediate use.

In preparation for joint U.S.-Russian participation in the International Space Station, a Russian cosmonaut flew aboard Space Shuttle mission STS-60 in February 1994. Other cosmonauts will fly on future Shuttle missions and up to 10 Shuttle flights to the Mir space station are planned. American astronauts will work aboard Mir for 3 to 6 months at a time, doing joint science experiments and testing out equipment for the International Space Station. The Shuttle orbiter will be equipped with a docking ring compatible with the Mir docking port, the same kind used by Soyuz.

Future Innovations

Once the International Space Station is up and running, modifications will be relatively easy. Because of its modular design, new equipment, even entire new laboratories and living facilities can be added. Later the station can be expanded by attaching additional trusses and payloads. More power will then be required. The drag from additional solar panels can cause the station's orbit to decay quickly, so concentrating solar collectors, which produce more electricity with a smaller area, may be preferable.

Each time a Shuttle or Soyuz docks, the clunk could shake the whole station, upsetting delicate equipment and processes. One solution involves free flying platforms, nearby but not attached, to carry machinery and experiments that must be free of vibrations, protected from contamination, or require precise pointing. Machinery attached to the outside of the station or to the nearby free flying platforms could be *teleoperated*, that is, a video camera would send a

view to a person in the space station who operates the machine remotely using robot hands and arms. In advanced robotics, any movement of the operator's hands and arms is reproduced exactly by the remote machine.

We have had considerable experience using the Canadian robot arm on the Space Shuttle orbiter, primarily to pick up satellites. A large object with a large mass is not excessively disturbed when the arm comes in contact with it. On the space station, however, robot arms will also be used for smaller experiments and devices which will require considerable dexterity and precise coordination; a collision could be disastrous. An advanced robotic arm will not only allow the operator to see in three-dimensional imagery, but will also feed back the feel of the force and torque being applied. Newly developed proximity sensors attached to the robotic arm will be able to detect a nearby object and stop itself or move away to avoid a collision. Even with advanced equipment, controlling robotic devices in space will remain a challenge.

An advanced space station could also become a way-station for construction and fueling of spacecraft headed for interplanetary space.

City-Sized Space Stations

Perhaps the best known concept for a large space station is the big wheel that appeared in the movie *2001: A Space Odyssey*, based on a story by Arthur C. Clarke. Such a structure was first presented to the public by Wernher von Braun in an article in *Colliers* magazine, April 30, 1954. See Figure 11.9. The big wheel habitat of Wernher von Braun and Arthur C. Clarke may become a reality in the far future. NASA study groups in 1975 and 1977 did serious research into large space colonies and much of the material in this section comes from their work. Although NASA's greatest effort in the following decade was the Space Shuttle, some research on space colonies has continued and many questions have been answered. At least as many other questions have been raised.

Imagine 10,000 people riding in the rim of a rotating wheel a mile in diameter (Figures 11.9 and 11.10). The rim

Figure 11.9   Wernher von Braun and his conceptual space station. *Courtesy of David Christensen.*

would be a tube, like the tire on an automobile wheel, with the occupants on the inside. We live on the surface of Earth; living in the wheel would be as if Earth were turned inside out and the atmosphere and people were put on the inside. Figures 11.11 and 11.12 are artists' conceptions of what the interior might look like.

To get the funding necessary to build and support a large colony in space, there has to be some economic payoff. Certainly it would be part of the infrastructure necessary for further exploration of the Solar System. But colonists could not expect a continuous flow of supplies to come from Earth unless they offered something in return. What is in space to entice governments to build a colony and send people to live there? The International Space Station will have many useful and profitable jobs to do. That is part of the answer, but not enough to support 10,000 people.

What else is there? One major industry of a space colony could be exploitation of the Moon's natural resources to manufacture very large satellites, such as solar power satellites. In Chapter 5 we discussed satellites in geosynchronous orbit which harvest sunlight and beam it to an energy hungry Earth in the form of microwaves. The enormous cost of transporting the materials and construction crews from Earth for such a huge structure dampened the initial enthusiasm for energy satellites. To ship cargo to geosynchronous orbit requires an acceleration, called *delta-V* (ΔV), to nearly 7 miles per second from the surface of the Earth. Now suppose ore from the Moon is sent to a space colony where it is processed and fabricated into such satellites. These large structures could be transferred to geosynchronous orbit, completely built, by a ΔV of only 2.9 miles per second. As you will remember from Chapter 3, the amount of fuel consumed depends on how much you must accelerate the vehicle. The saving in fuel costs occurs because lunar gravity is only one-sixth Earth gravity and the Moon has no atmosphere.

Using lunar materials, the colony could be the site for construction of large space vehicles, "star ships," for long duration flights to the outer Solar System and beyond. The cost of an interplanetary vehicle to carry people on an expedition to Mars, for example, is greatly reduced if it originates from an orbiting space colony. The spaceship is built and stocked with supplies at the colony. Crews come aboard, check it out, and blast off. Escape velocity from Earth is about 7.5 miles per second; from the Moon it is only 1.5 miles per second, and from an orbiting colony it would be even less. Both mining the Moon and human exploration of Mars are discussed in Chapter 12.

The initial processing plant, solar power equipment, and construction crew would have to come from Earth, but the habitat itself would be built with a supply of lunar ore. When the habitat and manufacturing facility were completed, other projects, such as the two described above, could be undertaken. Building additional new colonies is another logical

Figure 11.10 One-mile diameter big wheel space colony. Ten thousand people live inside the rim of the wheel. The round object above the wheel is a mirror which reflects sunlight into the colony. The long "rod" at the lower left extends to a manufacturing facility separated from the main colony for safety. *Courtesy of NASA.*

step. Of course, the initial investment is high. An industrial plant in space and a mining camp on the Moon would not be cheap. Once in place, however, they could become profit-making ventures.

Location

Where would you locate a city in space? Low Earth orbit is not the best choice for several reasons. In the first place, some of each orbit is spent in darkness, and the colony will depend on a continuous supply of solar energy. Second, the Van Allen belts must be avoided. If the colony's industry is to operate in conjunction with a lunar mining colony, then a location near the Moon would be advantageous.

There are points in space where the gravitational attraction of Earth and Moon combine in such a way that an object placed at that point will just stay there. These places are called *libration points* or *Lagrangian points* after Joseph Lagrange, the French mathematician who discovered them theoretically. The L point locations are shown in Figure 11.13.

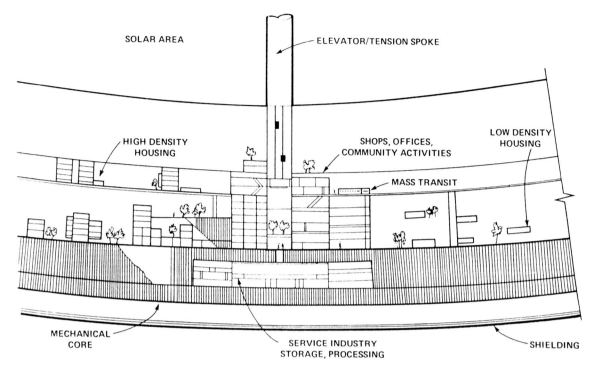

Figure 11.11 Partial cross section of the interior of the big wheel space colony. *Courtesy of NASA.*

Figure 11.12 Artist's concept of the interior of the big wheel space colony. *Courtesy of NASA.*

L1 is located between the Earth and the Moon while L2 lies beyond the Moon; L3 lies between Earth and the Sun. These three points are unstable. If displaced perpendicular to the Earth-Moon line, the space station would return to position. But if displaced away from the Earth or Moon, it would continue to drift away. Two stable points lie in the orbit of the Moon, 60 degrees ahead and 60 degrees behind the Moon. Each point is at the corner of an equilateral triangle with the Earth and the Moon at the other two corners. Placed at one of these points, the station would remain with no expenditure of energy. The point leading the Moon in its orbit is called L4 and the one trailing the Moon is L5.

To say "an object will just stay there" is misleading. It is really in the Moon's orbit, circling Earth with the same period as the Moon, about 27.5 days, but its position relative to Earth and Moon is fixed. Actually, the space station would not remain at exactly the same spot. When we include the influence of the Sun's gravity, we find that the colony would make a slow "halo" orbit. But a colony at one of these points would remain essentially fixed in the Moon's sky like a geosynchronous satellite remains in the same spot over the Earth. Let us select L5 as our colony's location.

Energy

Because the space colony is in direct sunlight, solar energy would be continuously available for smelters, mills, and fabrication factories as well as for light and agriculture inside the colony. Solar energy falling on Earth averages about 25 watts per square foot; the maximum with the Sun directly overhead is about 70 watts per square foot. In space, without

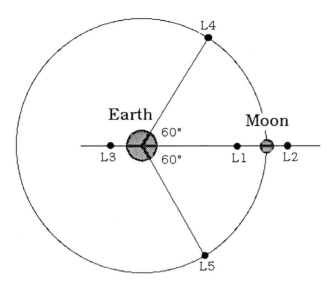

Figure 11.13 Liberation points in the Earth-Moon system.

an intervening atmosphere, a continuous 130 watts falls on a square foot surface. The only limit to the amount of energy available to the colony is the size of the array of solar cells or concentrating mirrors.

Materials

To build the L5 habitat, excluding the mandatory protective shield, will require an estimated 200,000 tons of materials (Earth weight). The proportions of the primary elements are shown in the center pie diagram of Figure 11.14. The main structure would be made chiefly of aluminum. That plus atmospheric oxygen and soil constitutes most of the

mass of the habitat. Analyses of lunar soil brought back by Apollo 16 show it to be composed mostly of oxygen, silicon, and aluminum. See the right pie diagram in Figure 11.14. This is not too surprising. Those three are the most common elements in Earth's crust, too, although the Moon has a higher percentage of aluminum, at least at the Apollo 16 site. These materials, then, would come from the Moon. Because of the Moon's lower gravity, it is much easier to lift mass to L5 from the Moon than it is from Earth. Only hydrogen to make water; nitrogen to dilute the atmospheric oxygen; carbon in the plants, animals, and humans; initial structures; and machinery would have to be brought up from Earth. As seen in the left pie diagrams of Figure 11.14, these amount to a small percentage of the total mass of the habitat.

Artificial Gravity

As pointed out in Chapter 9, weightlessness even for a few days has adverse effects on the human body. Permanent weightlessness may destroy a body, so a large colony in space must have some way of producing an artificial gravity. Rotating a big wheel space colony does just that. When you go around a corner in your car you can feel a force pushing you toward the outside of the turn. On a revolving carnival ride you get the same sensation; you need to hold on to keep from being thrown off. Objects in the tube of the rotating big wheel would feel a similar force, but would interpret it as "gravity."

What is really happening follows Newton's first law of motion; remember Chapter 2. Objects in motion will continue to move in a straight line unless acted on by an unbalanced force. Imagine you are in the tube of the big wheel,

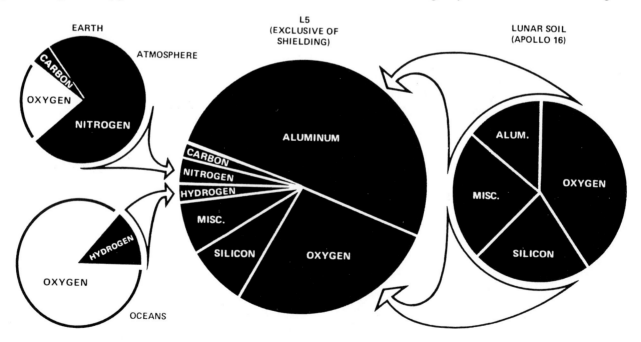

Figure 11.14 Sources of materials required for construction of the L5 habitat. Center diagram shows proportion of various elements needed; left and right diagrams show what could be supplied from the Moon and from Earth. Shielding can be the slag and waste materials from processing lunar ore. *Courtesy of NASA.*

as in Figure 11.15. Because the wheel is rotating, carrying you with it, you are in motion in the direction of A. According to Newton, you would continue to move in direction A except that the rim of the wheel curves upward, forcing you to follow the curved path B. In doing so, the rim pushes up on the bottoms of your feet, toward the center of the wheel. This force toward the center is called, appropriately, *centripetal force*. You get the same sensation that you feel on Earth when gravity pulls you down but the floor pushes up on the bottoms of your feet keeping you from falling through. Centripetal force becomes an artificial gravity. In the rim of the rotating wheel heads point in toward the central hub while feet point out. "Down" is away from the hub.

The amount of artificial gravity, that is, centripetal force, depends on how large the wheel is and how fast it is turning. A 1 mile diameter wheel rotating at one revolution per minute produces an artificial gravity at the rim approximately equal to Earth gravity. See MATHBOX 11.1 for the mathematics involved. It is possible to obtain any g force desired by changing the radius or the speed of rotation.

Living in a rotating vehicle can cause motion sickness. The problem arises when a person's head moves out of the plane of rotation of the vehicle. Physiologists have found that most people can adapt to rotation rates up to four revolutions per minute. To be conservative, most space colony designs limit the rotation rate to one revolution per minute.

The open question is: how much gravity, or rather, how little gravity can the human body tolerate permanently? Although Russian cosmonauts have spent as much as a year at a time in weightlessness, no one as yet has a conclusive answer to that question. The answer is crucial to the design of the L5 habitat. For economy and ease of construction, it would be desirable to keep gravity as low as tolerable. One g in the living area means that the buildings would weigh the same as they do on Earth and would have to be built to same specifications as if they were on the surface of Earth. The habitat shell, in turn, would have to be built to support that weight. If humans could live at one-half g, then everything would weigh only half its Earth weight, allowing the shell and buildings to be built with less structural strength. This would greatly reduce the mass of material that must be shipped out to build the colony. One of the experiments to be carried out on the International Space Station will be an artificial gravity lab which can be rotated at various speeds to find out what minimum gravity is required to keep humans in good physical condition.

Visitors to the colony would dock their spacecraft at a port on the hub shown in Figure 11.16, enter through an airlock, and take an elevator in one of the spokes out to the rim of the wheel. The hub is an interesting place because there the g force is zero, that is, weightlessness. Try to imagine that ride to the rim. You start out weightless, but as the elevator carries you "down" you feel your weight returning until, at the end of the ride, your weight is normal. The hub would make an exciting recreation center. Imagine a game of basketball in weightlessness!

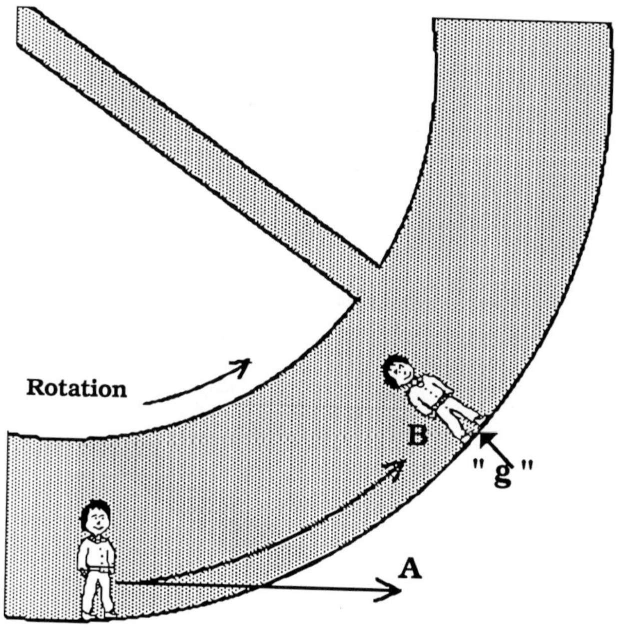

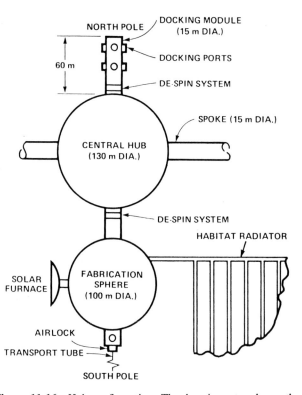

Figure 11.16 Hub configuration. The de-spin system keeps the docking ports, fabrication sphere (manufacturing plant), and radiator from rotating with the main part of the habitat. Elevators in the spokes carry passengers and materials to the living areas in the rim. *Courtesy of NASA.*

Figure 11.15 Rotation of wheel to produce artificial gravity.

Meteoroid and Radiation Protection

Possible collisions with meteoroids must be considered. A big wheel colony could lose 60 percent of its air in 1 day through a hole 1 yard in diameter. A hole that size would be discovered and patched quickly, however. Based on the density and size distribution of the meteoroid population in the Earth-Moon environment, catastrophic loss of atmosphere in a collision appears to be only a small risk. The probability of a collision creating a 1 yard hole has been calculated at one in 10 million years. Meteoroids do occur in clusters and multiple collisions in a short time are likely. Certainly, collisions with micrometeoroids would occur often and a routine program to repair small leaks would be necessary.

A more serious aspect of the meteoroid problem is the

MATHBOX 11.1

Artificial Gravity

Centripetal force is the force directed toward the center of rotation that makes an object move in a circular path. If you stand in the rim of a rotating space station you feel this force on the bottoms of your feet, pushing you into a circular path. The sensation is the same as the force of gravity produces on your feet when standing on Earth, therefore it is called artificial gravity.

The formula for centripetal force is

$$F_c = \frac{mv^2}{r}$$

where m is the mass of the object, v is its velocity, and r is the radius of the circle. Remember Newton's second law

$$F = ma$$

which relates the acceleration of an object to the applied force. Compare the two equations above and note that the acceleration due to centripetal force must be

$$a_c = v^2/r \ .$$

Solving for v we get

$$v = \sqrt{a_c r} \ .$$

Our rotating big wheel space colony is 1 mile (5,280 feet) in diameter; the radius, r, is 2,640 feet. If we desire 1 g artificial gravity, a_c must be 32 feet per second per second, the acceleration of gravity on Earth. Then the velocity of an object at the rim must be

$$v = \sqrt{2640 \times 32} = 290 \text{ ft/sec} = 198 \text{ miles/hr.}$$

This is less than half the speed of a jet airliner.

Next, let us calculate the period of rotation. First find the circumference of the wheel, that is, the distance travelled in one rotation:

$$5280 \text{ feet} \times \pi = 16,600 \text{ feet.}$$

Next, divide that distance by the speed:

$$\frac{16,600 \text{ feet}}{290 \text{ ft/sec}} = 57 \text{ seconds.}$$

So a big wheel space colony 1 mile in diameter rotating once every minute will have 1 g artificial gravity at its rim.

explosion overpressure caused by an impact. Even a small (1 gram) meteoroid would create a pressure wave of killing force for a distance of about 7 feet from the point of impact. A meteoroid shield or bumper separated from the main shell of the space station would provide protection from such pressure waves.

Protection from cosmic rays and solar protons would also be needed. Solar flares of sufficient magnitude to cause a serious radiation sickness problem occur during the high part of every 11 year solar cycle. We on Earth are protected from particle radiation by Earth's magnetic field. Similarly, a magnetic field around the space station would deflect the high energy particles. However, building a machine to produce a magnetic field of sufficient strength to protect a space station one mile in diameter is beyond present technology.

A massive shield could provide protection from both high energy particles and meteoroids, but would not hold together if rotated at one revolution per minute. It would have to be nonrotating, and therefore not attached to the rest of the space station. However, a shield placed between the Sun and the habitat to block solar particles would also block the needed sunlight. One proposed design calls for a mirror on the opposite side of the big wheel, not connected to it, reflecting sunlight into the colony. The mirror can be seen in Figure 11.10. This design would allow a massive shield to be located on the sunward side. Ten million tons of mass would be required. The slag from the industrial processing of the lunar soil would provide part, but not all of it.

Life Support

The life support question rises again as it does for any inhabited spacecraft. The design of the habitat is greatly dependent on solutions to the problems of life support: air, food, water, and protection from radiation. In Chapter 9 we discussed creating a comfortable and healthy environment in the Space Shuttle. All the supplies needed for the relatively short mission times are carried along, oxygen and nitrogen are added to the atmosphere as needed, carbon dioxide is removed by canisters of lithium hydroxide, excess water is dumped overboard, and the wastes are returned to Earth. Now we must design a city which will be completely self-sufficient, independent of Earth. Continually shipping supplies to L5 from Earth is out of the question. It would be much too expensive. The life support system must be a closed loop, everything recycled.

Atmosphere

Can a person live in an atmosphere different from normal Earth atmosphere for long time periods? Space Shuttle carries a normal Earth atmosphere, 21 percent oxygen and 79 percent nitrogen. Most previous manned spacecraft had pure oxygen atmospheres. The long term effect of a pure oxygen atmosphere on people and animals is unknown. Pure oxygen also increases the fire hazard. Until more definitive research is done, the oxygen should be diluted with some inert gas such as nitrogen. Oxygen would come from lunar ore, but the initial supply of nitrogen would have to come from Earth. Helium has been suggested as an inert diluent instead of nitrogen because it is not as dense. There would be less mass to bring from Earth, and the rotating habitat would have less weight to support. The strange quality helium gives to voices would be a consideration.

Sea level pressure is 14.7 psi, but some people, plants, and animals live their entire lives at altitudes where atmospheric pressure is only two-thirds that at sea level. Pressure in Skylab was only 5 psi, one-third normal. It would be advantageous to equip a space colony with an atmosphere at lower pressure to reduce the strain on the shell of the habitat, thus requiring less structural strength. One difficulty is that sound does not travel as far in a gas at low pressure. Skylab astronauts said that a loud speaking voice carried only about 15 feet.

For transportation, no cars would be allowed, only an electric railway or moving sidewalks, roller skates, and bicycles. The distance around the colony described here is only a little more than 3 miles, and the elevator shafts through the central hub would be shortcuts to the other side. By using nonpolluting transportation, the air would only have to be cleansed of dust, excess water vapor, carbon dioxide, and other gases. Plant photosynthesis becomes a primary factor in closing the atmosphere loop, but will not do the complete job. Cold plates and coils could condense the water vapor into droplets to remove it from the air, making it available for reuse.

Food and Water

Although much is understood about human nutritional needs, there is no certainty that all the necessary nutrients have been identified. An important question is whether loss of bone calcium, loss of fluids, muscle deterioration, and other effects of living and working in space can be compensated for by diet adjustments. Proper diet is not only necessary for good health, it is also a factor in psychological well-being.

Initial food supplies will be brought from Earth, but farming and raising livestock will begin as soon as they can be established. Plant agriculture is practical and beneficial for two reasons: food and oxygen. While animals inhale oxygen and exhale carbon dioxide, plants intake carbon dioxide and release oxygen. That is a great oversimplification, but is basically true. Photosynthesis is the chemical and metabolic process by which plants use water from the ground and carbon dioxide from the air with sunlight as an energy source to manufacture carbohydrates. Oxygen is released as a byproduct. In fact, the original atmosphere of Earth probably contained carbon dioxide and water vapor, but no oxygen. According to the geological fossil record, plants inhabited

Earth before animals did. It wasn't until plants came along that oxygen began to build up in the atmosphere and make higher forms of animal life possible.

The large variety of foods available on Earth may be impractical to produce in the colony simply for lack of space. Meats are a good example. What kind of livestock should be raised? Cattle for beef? Hogs for pork? People have different preferences. Hogs, chickens, and turkeys have been suggested because they can consume table scraps and noncellulose plant wastes. Turkeys are particularly efficient in protein production. However, these choices may not be the best because their main diet, corn, could also be eaten by humans. Energy is lost when the animal converts the food to meat and it may be more energy efficient for humans to eat the food directly.

Ruminant animals would be an excellent choice in a closed environment because they can digest cellulose plant wastes inedible for humans. Cellulose parts of plants include the structural parts, roots, husks, and hulls. Cows are ruminants and yield both dairy products and meat. A beef steer grows to about 900 pounds in 16 months. Dairy products are a particularly important source of nutrition because of the high calcium content, and dairy cows convert food to milk with greater efficiency than they convert food to meat.

Small animals such as goats, rabbits, ducks, and geese may be good choices. They reproduce frequently, mature rapidly, and would only have to be fed during their short rapid growth period, then slaughtered for their meat. Rabbits reach 7.5 pounds in about 4 months. Table scraps and by-products from vegetable production would provide the bulk of their diet. Goat milk and duck eggs would add variety to the human diet, also.

Aquatic food production is another likely possibility and has been carried on in some parts of Earth for many centuries. Animal and human waste is effective fertilizer for aquatic production. In some countries, human waste is dumped directly into lakes or ponds for cultivating water plants and fish. Fish can be harvested in a year weighing in at about 4.5 pounds.

Space Agriculture

"Weather" in a space habitat would be under complete control. In the agricultural areas, light, temperature, humidity, and carbon dioxide content would be kept at proper levels to promote most rapid growth and maturation of crops. Agricultural and animal production can be highly efficient in a small area providing diseases are eliminated or at least kept under careful control.

In one series of experiments simulating the controlled environment of a space habitat, Bruce Bugbee from Utah State University grew a common breed of wheat in hydroponic tubs with a yield five times greater than the highest productivity ever achieved in a wheat field. The most important factor is providing enough light. Plants need a light intensity 100 times greater than people need for work light. Bugbee used sodium vapor lamps and left them on 24 hours a day because wheat does not need a rest period. The range of useful wavelengths of light is 400 to 700 nanometers. (A nanometer (nm) is a billionth of a meter.) There is a sharp cutoff for photosynthesis: 695 nm works, but 710 nm does not.

Given enough light the next most important thing is an adequate supply of water and nutrients. Seeds were planted in a 3/4 inch thick layer of sterile, inert rockwool, similar to house insulation, to give support to the plant. The rockwool was fastened over the tops of 16 inch by 20 inch tubs, 4 inches deep, filled with about 5 gallons of nutrient solution. Plants in each tub would take up and transpire more than 50 gallons during the wheat's 80 day life cycle. The solution contained carefully measured amounts of the 13 essential elements for plant growth. In that time the tub full of wheat would absorb about 2 1/2 ounces of the mineral. To provide oxygen to the roots, the solution flowed through the tubs continuously.

Wheat is most productive at 63 °F; at 73 °F yield is cut in half. A relative humidity of 80 percent is optimum. A higher percentage of carbon dioxide is desirable to a point, but wheat is more sensitive than people are to excess carbon dioxide. At more than 0.2 percent, it becomes toxic to wheat. A slight breeze is needed to waft carbon dioxide to the leaves and oxygen away.

The wheat used in this experiment was not a special breed. A breed could be genetically engineered to be better adapted to a closed environment. For example, a shorter plant would save space and have less waste stalks, stems, and leaves.

While these experimental results apply to wheat, other plants would have similar needs and limitations.

About 150 acres of lunar soil would be enough to produce sufficient food for 10,000 colonists under such controlled conditions. Grown hydroponically, only half that space may be needed. As to which medium to use, the tradeoff is the high cost of delivering rockwool or other base material from the Earth versus the available space. Transporting lunar soil to L5 would certainly be cheaper but it would have greater mass and occupy more space. For a colony on the Moon, lunar soil would be the obvious medium of choice, but the cost of constructing a large lunar greenhouse must also be considered.

See the artist's conception of an agricultural area in Figure 11.17. A cross section is shown in Figure 11.18. A tiered arrangement yields more usable growing area in a smaller space. Water cascades down from the aquaculture ponds on the top tier to irrigate the lower tiers, then is pumped back up to the top to start over again. Plants must

Figure 11.17 Artist's view of agricultural area in the L5 space colony. Compare with Figure 11.18, a diagram of the area. A residential section is in the background. *Courtesy of NASA.*

also be protected from harmful radiation. They are as sensitive to this as people are.

The colony is divided into six sections, three for living and three for agriculture, alternating around the circle. See Figure 11.19. Each section can be isolated from the others. By placing the crops and livestock into three separate areas, crops can be started at various times to assure a continuous supply. Alternating residential and agricultural areas would also create a more pleasant living environment for the inhabitants. Fruit trees would grow in the residential areas for their beauty as well as their fruit.

Careful screening of everything and everybody brought to the colony would be essential to keep out infectious diseases and parasites. Infrared technology from the remote sensing satellites could be used in the agricultural areas to monitor the health of crops. If disease should occur in one area, it could be isolated and hopefully the other two would not be affected.

Common human ailments may be impossible to keep out. Immunization and quarantine are the usual tactics against the spread of infection. Using chlorine or other disinfectants to control microbes in a closed environment creates another problem, the accumulation of toxic substances. Hospitals have similar problems, but can remove toxic waste and bring in fresh air from outside.

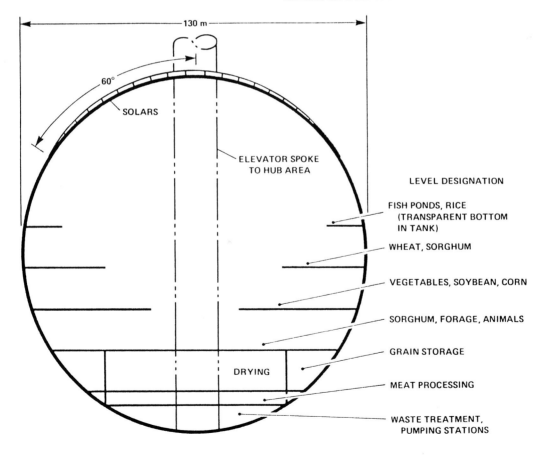

Figure 11.18 Cross section of agricultural area in the L5 habitat. *Courtesy of NASA.*

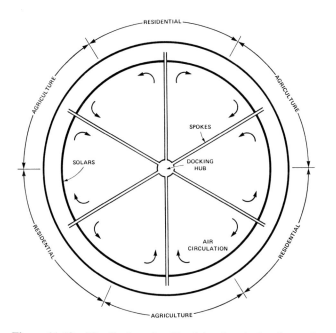

Figure 11.19 Distribution of residential and agricultural areas in the L5 habitat. *Courtesy of NASA.*

Closing the Loop

The upper part of Figure 11.20 shows on the left side what humans, plants, and animals need to live, and on the right side what they output into the environment as a result of living. The object of a closed loop life support system is to assure that all the outputs on the right side of the diagram are converted to usable inputs. If everything is not recycled, there will be a gradual buildup of waste. Such pollution would be a much more critical problem than pollution on Earth because of the small, confined volume in a space colony. Mechanical, chemical, and biological methods to convert the waste back the other way must be perfected. The lower part of Figure 11.20 shows the complexity of a closed loop life support system. Following the arrows from place to place will give an idea of what is involved. Circles represent chemical and mechanical devices which recycle the waste into usable substances.

Chemical and mechanical means for converting outputs to inputs have been devised for some but not all of the waste products. Animals and humans output water in respiration and perspiration which are recoverable mechanically from the air. Purifying urine into drinking water is easy, although there is a psychological barrier to its acceptance by many people. Wash water is also easy to reclaim and purify using mechanical and chemical processes.

Notice that the biological activity partially closes the system. For example, the respiratory output of animals and humans is partially carbon dioxide, which must be removed from the air. That is the job of the plants. The atmospheric input to plants is carbon dioxide and the respiratory output

of plants includes oxygen which is the atmospheric input to humans and animals. Plants use the carbon from carbon dioxide in building their leaves, stems, roots, and fruits. These become food for animals and humans. Animals are also food for humans. Human and animal feces make fertilizer for plant growth.

The great difficulty in completely closing the loop is that *everything* must be recycled. If it is not, there will be a gradual buildup of unusable waste and a decrease of usable materials in the system. Just imagine the accumulation of fingernail clippings of 10,000 people! Some of the waste may be toxic, causing additional problems. If the toxic substance is not removed by the recycling process, then it will accumulate, its concentration increasing with each pass through the cycle.

In the final analysis, for any particular space endeavor there is a break-even point in the means of supplying food. Whether to recycle in a closed loop system or to bring a bag lunch depends primarily on the weight at liftoff. If the recycling equipment and initial food supply weigh more than the bag lunches, then the bag lunch is the way to go. With present technology, a 3 year mission is the break-even point.

Medical and Psychological Aspects

Until the Space Shuttle era, astronauts were a homogeneous bunch of healthy, tough test pilots, who could rough it for a week or two cramped together in a small space. They were well screened in advance. Now there is a changing crew composition, less homogeneous, less screened. In the future, space travel will be for everyone. Missions are more routine. As a consequence there is a growing interest in the medical and psychological aspects of spaceflight. *Ergonomics* is the study of human factors.

Astronauts receive emergency medical training, including a limited knowledge of preventative and diagnostic procedures. In case of an emergency they could act to stabilize a situation until the person could be returned to Earth, although returning from weightless to normal gravity might cause more problems for a patient. It would be advisable on long missions to have a doctor aboard, perhaps a surgeon. At this time, no one knows whether it is possible to perform a surgical operation in weightlessness.

Data on human behavior in space is being gathered. Motivation is high at first, but during long duration flights, irritation sets in and small annoyances become big ones. Personal hygiene, use of the toilet, and lack of privacy become irritants. Astronauts also become less efficient in their work after a few months and cannot concentrate as well. Problems arise with attention span and with performing cognitive tasks. On Earth during long winters in the far northern countries, it has been noted that irritability increases and a higher suicide rate is seen.

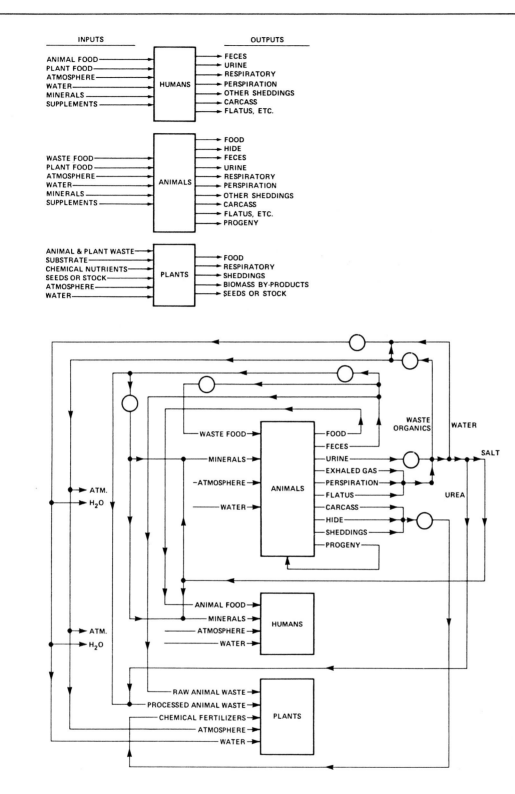

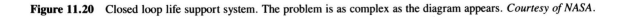

Figure 11.20   Closed loop life support system. The problem is as complex as the diagram appears. *Courtesy of NASA.*

Submarine studies also bear out this reaction to long confinement. There is a decrement in morale and mood, accompanied by higher stress, after the first 10 days. Things get better about a week before the final day of the mission when the end is in sight. People forget their petty grievances in the excitement over ending the mission. Group cohesion is important. The Russians put crews in a small car and sent them on a 30 day trip around the Soviet Union to see how well they got along! It is said that the Russians do not allow chess aboard their space station because it brings out aggressiveness.

Missile crews that work in underground ICBM launch complexes are given training as to what to look out for in their co-workers, things which indicate behavior problems. Certain changes in mood or changes in activity levels are some of the indicators. Psychological tests are available to determine profiles of candidates to predict if they might not be able to take it. Schizophrenia, delusional disorders, and hysteria are relatively easy to detect. Anxiety or emotional disorders are more difficult to diagnose. Buildup of excess carbon dioxide, toxic substances, noxious gases, or bacteria in closed environment can also lead to mental or emotional disorders or confusion.

Simple things become significant during a long stay in a relatively small volume. Even the colors of the walls and equipment have a bearing on the feelings and moods of the occupants. Some Skylab astronauts complained about the lack of color and texture in their habitat. The only colors on board were the color bars used to register the cameras. Work areas need to be well illuminated so lighter colors are logical: greys and whites with touches of blue. Also, lighter paints are more reflective and therefore reduce the amount of electric lighting needed. This reduces both power requirements and heat production, important considerations in space vehicles. Warmer, homey colors are best in living areas and darker shades in sleeping areas. Glossy paint is easy to keep clean, but may cause unacceptable glare. Semigloss and flat paints are preferable. Also, Japanese studies have shown that a large volume in a habitat is very important for psychological well-being.

During Skylab activities the astronauts sometimes complained of having too much to do and not enough time to relax and just look out the window at the Earth. One crew refused to do any work for a day. They just relaxed and did what they pleased. In longer flights the Russians have noted that comrades in a spacecraft tend to stick together, but get irritated with ground personnel in the command center. Interesting, varied tasks help relieve stress and boredom. Free time is important. Relaxation training may help relieve stress. From Soviet experience, important factors are communication with families, tapes of music of preference for each crewmember, and just being alone once in a while.

One possible problem is the disruption of diurnal circadian rhythms. A spacecraft in low Earth orbit has 16 sunrises and sunsets a day. Fortunately humans are very adaptable. Only one-tenth of the variability in most psychological factors tested is accounted for by genes. Studies of adaptation show no significant difference between men and women. Edward Teller says women are smaller and smarter than men, so we should send more women.

The bottom line is that spaceflight is expensive and performance must be maximized. Getting the job done depends on people. Psychological factors become more important as the duration of the flight increases.

Biosphere II

The most complete closed environment experiment yet devised is Biosphere II, underway near Tucson, Arizona. (Biosphere I is Earth itself.) A crew of eight people, four men and four women ranging in age from 30 to 70, lived in a completely closed ecological system for 2 years. Constructed of a framework with 8,000 panes of glass, the building covers 3 acres and has a volume of 7 million cubic feet. It is divided into seven biomes: desert, savannah, tropical rainforest, marsh, ocean, and human habitat. Over 3,800 species of plants and animals were collected worldwide to occupy Biosphere II.

Two thousand sensors connected to more than 150 computers and an array of video cameras and monitors are scattered throughout the building. Careful monitoring from both inside and outside enables the research scientists to determine how the environment is acting and reacting at any time.

Much of crew time was spent in agriculture, growing food and caring for the animals. Because of an unusually cloudy 2 years in Arizona, they were able to produce only about 80 percent of their needs. The home-grown food was supplemented by a supply previously grown in the building. Although adequate, it was not as much or as varied as some of the experimenters would have liked. Near the end they said they would give almost anything for a pizza and a beer.

In the first 2 year closure of the building, the supply of oxygen ran unexpectedly low and fresh oxygen had to be added from time to time to support the crew. Just where the oxygen went is uncertain, but it may have been absorbed by the cement used in the construction of the building.

There were some personality conflicts and quarreling among the occupants. This increased as time went by, but decreased as the end of the 2 years approached.

The Biosphere II project continues, completely funded from private sources. Although it has been criticized by some scientists as not being a properly controlled scientific experiment, the project is sure to yield some insight into the problems of building a completely closed loop environment in space or on another planet. The first 2 years were de-

signed to be a shakedown of the entire system. Final results will be many years away.

This chapter has covered only one possible configuration and location for an orbiting space colony. Many others have been considered. You can perhaps think of alternative ways to live, work, and play in space. New ideas come from the imaginations of creative people who understand enough of the basic science to know what is possible.

DISCUSSION QUESTIONS

1. Do you think the International Space Station will be worth the expense to the world? Why?

2. Discuss the advantages and disadvantages of manned versus unmanned spacecraft. What jobs are best done by each?

3. Describe some recreational activities in a nonrotating space station.

4. Describe the artificial gravity and living conditions in a spherical or a cylindrical shaped rotating space colony.

5. Would it be possible and desirable to completely all bacteria and viruses in a space colony and its population as a means of eliminating disease?

6. Compare Biosphere II to an orbiting space station. In what ways are they alike? How are they different?

ADDITIONAL READING

Allen, John, and Mark Nelson. *Space Biospheres*. Orbit Book Co., 1986.

Billingham, John, et al. *Space Resources and Space Settle-ments*, NASA SP-428, 1979. Report of NASA summer workshop, technical but readable.

Bugbee, Bruce. "Hydroponics on the Moon," *Fine Gardening Magazine*, November/December 1989. Popular article.

Grey, Jerry. *Beachheads in Space, A Blueprint for the Future*. Macmillan Publishing Co., 1983. Popular, nontechnical summary of where we are and where we are going.

Johnson, Richard D., and Charles Holbrow, eds. *Space Settlements, A Design Study*. NASA SP-413, 1977. Report of NASA summer workshop, technical but readable.

National Commission on Space. *Pioneering the Space Frontier*. Bantam Books, 1986. Recommendations to the president on future space programs.

O'Leary, Brian. *Project Space Station*. Stackpole Books, 1983. Easy reading, nontechnical.

Ride, Sally K., et al. *Leadership and America's Future in Space*. NASA, 1987. Recommendations to NASA administrator on future space programs.

Van Allen, James A. "Space Science, Space Technology and the Space Station." *Scientific American*, January 1986. Arguments against the space station.

Woodcock, Gordon R. *Space Stations and Platforms*. Orbit Book Co., 1986. Technical aspects of engineering space stations; comprehensive.

PERIODICALS

Final Frontier, The Magazine of Space Exploration. Final Frontier Publishing Co., Minneapolis, MN 55408. Bimonthly magazine, easy reading.

SSI Update, The High Frontier Newsletter. Space Studies Institute, P.O. Box 82, Princeton, NJ 08542. A bimonthly newsletter describing research sponsored by the Space Studies Institute founded by Gerard O'Neill.

NOTES

Chapter 12

Colonies on Other Worlds

Why go to the Moon or to Mars? Several often heard phrases may be appropriate: "Because it's there" or "To explore is human." But there is probably a more pragmatic reason. Perhaps "The grass is greener on the other side of the fence." Humans are always looking for greener pastures. Other species do it, too. Consider the African bees migrating northward into Texas or the Mediterranean fruit flies spreading into California. Each form of life on Earth finds an appropriate environmental niche into which it settles, perhaps moving with the seasons, perhaps suspending animation during harsh times. When times get tough, space becomes overcrowded, and food gets short, then some of the individuals leave for a better place to procreate and populate. Humans move for these social, biological, and economic reasons, too. When one leader or group makes life unbearable for other individuals or groups, the oppressed may pick up and leave.

Expanding human presence from Earth into space is sometimes compared to the great age of exploration from the 1500s to the 1800s when Europeans expanded into the Western Hemisphere. There are indeed similarities—Columbus was sponsored by the government, Queen Isabella of Spain—and there are some important differences. Though there were shortages during the long voyages as sea, Columbus knew that he would find food, water, and air at the end of his journeys. Unlike astronauts and cosmonauts, Columbus did not have continuous contact with the home port.

One of the signs of human intelligence is that, rather than search out a hospitable niche, we have learned to survive in any environment by building habitats to maintain a comfortable temperature and bringing in the necessities of life: fresh air, food, and water. All that is needed is raw materials, energy, and intelligence.

Earth has many places where people cannot survive without a habitat to protect them from an inhospitable environment. Antarctica, perpetually covered with ice and snow, provides plenty of water and air, but temperatures are above freezing only during the short summer and little or no plant or animal life is available for sustenance. The side of a

mountain may be hospitable and inviting in summer, with abundant streams of clean water, fresh air, warm sunshine, plants and animals for nourishment—all the necessities of life. But it may be impossible to survive in that same spot in winter; temperatures drop below freezing even during the day, plants wither and die, and streams freeze up and stop flowing.

Just as people living in Antarctica or on mountains, the first colonists on other bodies of the Solar System will need to build habitats to enclose environments in which they can survive comfortably. The most likely candidates for imminent human habitation are the Moon and Mars. They are something like Antarctica. Prehistoric people migrated and settled on all other continents, but Antarctica wasn't colonized until this century when the technology was available to make it liveable and governments were willing to provide support for the colonists. We are just reaching that point in space exploration. The Moon may be our next Antarctica.

Permanent Colonies on the Moon

Why live on the Moon? See Figure 12.1. It is difficult to think of a more hostile environment. Only 12 humans have ever set foot on the Moon, all of them American astronauts. In Figure 12.2 one of them takes a sample of Moon rock to bring back to Earth. The first and most obvious drawback is the lack of air to breathe. The Moon has no atmosphere. Colonists would have to live in sealed and pressurized astrodomes or semiburied habitats and would have to wear space suits while outside. Second, without an atmosphere there is no protection from ultraviolet, x-rays, gamma rays, cosmic rays, solar particles, and meteoroids. Third, all food and other supplies would have to be brought from Earth. Finally, no water has been found on the Moon.

This description sounds like space itself, and the Moon appears to have little to offer. Living conditions on the Moon would be similar to living on a space station with two important differences. First, the Moon has natural gravity, about one-sixth the force of Earth gravity, and second, there are resources to be tapped. So, why colonize the Moon?

Figure 12.1 The Moon rotates on its axis in the same time as it takes to make one complete orbit around the Earth. Therefore, the same side of the Moon always faces Earth. The features may appear in different places, however, because we view it from different angles as it moves across the sky.

Two pragmatic reasons come to the forefront: mining and science, particularly astronomy.

Moon's Natural Resources

If it were cheap and easy to go to the Moon, colonists would have done so long ago. The lunar *regolith*, the surface material, is an enormous source of valuable raw materials. Chemical analyses of lunar rocks and soil returned to Earth by Apollo astronauts show them to be rich in helium, aluminum, titanium, iron, calcium, and silicon. Pure iron and iron compounds are found everywhere as a fine powder from the bombardment of the Moon by meteorites during the past eons. It can be simply separated by a magnet as soil passes by on a conveyor belt, although large quantities of soil must be processed to produce usable quantities of iron. Aluminum and titanium are found in ores not commonly refined on Earth, so new methods of extraction would have to be developed for use on the Moon. Silicon is just what is needed for photovoltaic solar cells.

Over 40 percent of the surface soil mass is oxygen compounded with other elements. Oxygen for a colony's atmosphere and for rocket propellant would be easy to extract from surface soil simply by heating it. Hydrogen is scarce on the Moon, however. Its original supply escaped the low lunar gravity long ago, and solar wind protons are the source of the very small amount of hydrogen now found in the soil (recall that protons are nucleii of hydrogen atoms).

No water has been found, not even in chemical combination in the rocks and soils, but there is a possibility that water lies deeper in the crust or perhaps in sheltered areas in the polar regions. The Moon's axis is not tipped as Earth's axis is, so some deep polar craters and crevices may never see sunlight and temperatures may never rise higher than 40 degrees Fahrenheit above absolute zero. Not only water, but carbon dioxide, methane, and other gases may be found frozen there. Those regions are not visible from Earth and only recently have spacecraft looked into them. There is also some possibility that water lies deeper in the crust in liquid form, since the interior of the Moon is warmer than the surface. If a plentiful supply could be located, some could be electrolyzed to provide needed hydrogen for fuel. On the other hand, if water and hydrogen are not found in sufficient quantities, hydrogen could be brought from Earth and combined with lunar oxygen in fuel cells to produce electricity and a supply of pure water.

Mining the Moon's Helium

Harrison Schmitt, the astronaut shown in Figure 12.2, recently pointed to the helium found on the Moon as a potential fuel for nuclear fusion, a new energy source for Earth. Nuclear fusion in the Sun was described in Chapter 4 and **MATHBOX 4.1**. On Earth, nuclear fusion has been produced in laboratories and thermonuclear bombs. Scientists have been trying for several decades to find a practical way to control the energy so it is released slowly over a period of time instead of instantaneously in a bomb. Controlled fusion for fractions of a second has been experimentally successful in laboratories in both the United States and the former Soviet Union. It isn't easy to build a container that will hold something at 30 million degrees Fahrenheit! The trick is to use a magnetic field to keep the hot particles suspended in a vacuum away from the container walls. High temperatures can be achieved by initiating the reaction with high powered lasers focused on small pellets of fuel.

Several different fusion reactions are possible involving hydrogen and its isotopes, deuterium (H^2) and tritium (H^3). One reaction involves helium-3 (He^3) with deuterium. An isotope of helium, He^3 has two protons and one neutron in its nucleus. The most common helium is He^4 which has two protons and two neutrons. The He^3-D fusion yields a He^4 and a proton. Theoretically it produces less waste heat and radioactive by-products, two major concerns of environmentalist opponents of nuclear energy. Deuterium is found in ocean water, but He^3 is not found in any quantity on Earth, so this reaction was not considered to have any prac-

Figure 12.2   Apollo 17 astronaut-geologist Harrison Schmitt studies the rocks of the Taurus-Littrow Valley of the Moon.
Courtesy of NASA.

tical value. However, He^3, a component of the solar wind, is found on the Moon in quantities enough to have real value as a fuel. It has been estimated that the total amount of He^3 on the Moon could yield ten times more energy than all the fossil fuels on Earth.

At least half of the top 20 to 30 feet of lunar surface materials consists of fine dust. Atoms of He^3 and other gases from the solar wind striking the Moon become attached to the dust. The concentration of He^3 is low in the lunar regolith, less than 30 parts per billion in the top 10 feet, but it can be removed easily by heating to about 1,200 °F. Because of its low concentration, more than 100 million tons of regolith would have to be processed to get one ton of He^3! It would be shipped to an energy-hungry Earth for generating electricity in fusion power plants. The hydrogen and oxygen also separated in this process could be used to produce electricity in fuel cells with water as a by-product.

It has been estimated that the Moon has at least a million tons of He^3. Theoretically, a ton of He^3 can produce 10,000 megawatts of electricity for a year, enough for a city with a population of 10 million; 25 tons would produce a year's

supply of electricity for the entire United States. The United States now spends about $75 billion a year to fuel its electric generators. At that rate, He^3 is worth about $3 billion per ton! A lunar mining colony could easily become self-sufficient and profitable within half a century selling He^3 to Earth and using the by-products for its own use.

Of course, no fusion-powered electric generators have been built to use He^3. Up to now, controlled fusion in the laboratory has not turned a profit. Far more energy must be put into a laboratory machine than is produced by the machine although researchers are approaching the break-even point. Some researchers expect eventual breakthroughs and predict that nuclear fusion will be the major energy source of the 21st century. Mining the Moon's He^3 could make it so.

Mining the Moon's Metals

Consider the Moon's environment: rocks and soils, plentiful solar energy for two out of four weeks, and a near perfect vacuum. There are some advantages and some disadvantages. Many processes used for extracting metals from ores on Earth require chemical agents which are not likely to be found on the Moon. For example, gold is extracted from ore by a reaction in a sodium cyanide solution. Electrochemistry is the most likely approach for adaptation to the lunar environment.

The simplest example of an electrochemical process is the electrolysis of water. It can be demonstrated by dropping a 9-volt battery into a glass of water with a pinch of salt dissolved in it and watching the bubbles. The water, H_2O, is broken down into oxygen and hydrogen gases by the electric current. Hydrogen bubbles up at the positive terminal, the anode, and oxygen appears at the negative terminal, the cathode. (If you try this, don't leave the battery in the water for more than a few seconds or it may be destroyed by the immersion.)

Similarly, if you run an electric current into a molten mineral oxide, a compound of metals and oxygen, the metals will collect at the anode and oxygen will be given off at the cathode. For example, anorthite, a feldspar found in abundance in the lunar highlands, is a compound of calcium, aluminum, silicon, and oxygen. Its chemical formula is $CaAl_2Si_2O_8$. By weight, it is 14% calcium, 19% aluminum, 20% silicon, and 46% oxygen. (See MATHBOX 12.1.) This means that if you completely process 1,000 pounds of pure anorthite, you will get 190 pounds of aluminum. Aluminum is a most useful metal. It can be used in the con-

---

MATHBOX 12.1

Molecular Weights

The chemical formula for anorthite is $CaAl_2Si_2O_8$ which says a molecule consists of one atom of calcium, two atoms of aluminum, two atoms of silicon, and eight atoms of oxygen. Let us find the percentage by weight of these four elements. First find the atomic weights of each in a table of the elements:

calcium (Ca)	40.080	aluminum (Al)	26.981
silicon (Si)	28.086	oxygen (O)	15.999

Next find the contribution of each to the molecule of anorthite:

calcium	1×40.08 =	40.080
aluminum	2×26.981 =	53.962
silicon	2×28.086 =	56.172
oxygen	8×15.999 =	127.992

The total molecular weight of anorthite is the sum of these four: 278.206. Now we can find the percentage weight of each of the elements as follows:

$$\text{calcium:} \quad \frac{40.080}{278.206} \times 100 = 14.4\%$$

$$\text{aluminum:} \quad \frac{53.962}{278.206} \times 100 = 19.4\%$$

$$\text{silicon:} \quad \frac{56.172}{278.206} \times 100 = 20.2\%$$

$$\text{oxygen:} \quad \frac{127.992}{278.206} \times 100 = 46.0\%$$

The sum of these must be, of course, 100%

Notice that nearly half the weight is oxygen, one of the most needed elements for a colony on the Moon.

struction of buildings, satellites, mirrors, vehicles, and electrical wiring. It is easily rolled into sheets, drawn into wire, welded together, and with a low melting point, about 1,220 °F, it can be melted and cast into almost any shape.

The difficulty is that all three metals in the anorthite—the silicon, aluminum and calcium—would be collected at the cathode. Preprocessing the mineral by heating it in the vacuum of the Moon would evaporate the silicon oxide first. Electrolysis would then produce calcium and aluminum at the cathode to be later separated.

Anorthite is only one of a number of minerals which could be mined profitably. Ilmenite is a mineral compound of iron (Fe), titanium (Ti), and oxygen with a chemical formula $FeTiO_3$. It is relatively abundant on the Moon. To process the ore, it is first heated with hydrogen to 1,560 °F. The oxygen combines with the hydrogen to form water and leave a solid mixture of iron and titanium oxide. The mixture can then be ground into a fine powder and a magnet used to separate the iron from the titanium. Each of these two can be processed to yield usable construction material.

Mining the Moon's Oxygen

The electrolysis of 1,000 pounds of anorthite will also produce, as a by-product, 460 pounds of oxygen, approximately the amount of oxygen in a 1,400 square foot house. There may be an easier way to extract oxygen from other minerals on the Moon which need only to be heated and oxygen is released. For example, some oxygen would be released while processing the regolith for He^3 as described previously. If lunar mining and processing could supply the oxygen for rocket engines plying *cislunar space* (Earth-Moon space), much less mass would have to be lifted from Earth. At least three-fourths of the takeoff weight of most launch vehicles is propellant. For rockets burning hydrogen and oxygen, hydrogen is only one-seventh of the mass of the propellents; oxygen is six-sevenths.

Figure 12.3 in the color section, page 193, is an artist's concept of a lunar mining colony and Figure 12.4 shows a cargo vessel hauling liquid oxygen from the Moon to a "filling station" in low Earth orbit. The oxygen tanker is equipped with an inflatable heat shield for aerobraking in

Figure 12.4 Ferry vehicle transports liquid oxygen manufactured on the Moon to low Earth orbit. Spherical tanks are assembled at the facility in the lower right; the rocket engines and heat shield are attached. The ferry heads for Earth (upper left), the heat shield inflate to several times its illustrated size, and the vehicle aerobrakes in the atmosphere to reduce the fuel required for orbital transfer. On the return trip the ferry brings a supply of liquid hydrogen from Earth. In the upper right is a lunar space station; a lunar lander has just arrived with more oxygen from the surface. *Courtesy of Eagle Engineering; artist Pat Rawlings.*

Earth's upper atmosphere. Therefore it does not have to carry as much fuel as would be needed if the orbital change were done by rockets alone.

Automatic robotic mining machinery could be operated from inside the habitat so people would not be exposed to the harsh environment outside. In fact, it has been suggested that a completely automated mining operation could be controlled remotely from Earth and only a small maintenance crew would have to be sustained on the Moon.

Plenty of solar energy is available during the solar day for processing ores to extract the useful materials such as aluminum, silicon, and oxygen. Just spread out an array of photovoltaic cells and tap in. Since the Moon rotates on its axis once a month resulting in a two week day and a two week night, another source of energy would be needed during the long lunar night. There are several possibilities. First, use orbiting mirrors to concentrate and reflect sunlight onto the collectors like miniature suns. With four of them equally spaced in lunar orbit, one would always be in the right position throughout the two week night. Second, store some of the electricity produced during the day by spinning up massive flywheels to high speed using electric motors. After dark, connect the flywheel to a generator to generate electricity. In fact, the motor could be designed to act as the generator. Third, the brute force method, bring in a nuclear power generator from Earth. And finally, mine the ore

on the Moon and ship it for processing to a space colony which is in continuous sunlight at L5, as we discussed in Chapter 11.

The Mass Driver

Material can be lifted from the Moon to space with only 5 percent of the work needed to lift material from Earth to space. Instead of using chemical rockets to haul the ore from the Moon to the L5 space station, an electrically powered launcher on the surface can accelerate the ore to escape velocity. Such a launcher, called a *mass driver*, has been designed and tested on Earth at the Massachusetts Institute of Technology and at Princeton. The principle of a mass driver is simple. Put the ore into a bucket, then accelerate the bucket and its cargo to escape velocity, stop the bucket, and let the ore go flying off into space. We could not use such a device to launch objects off Earth. Escape velocity from the Earth's surface is about 7 miles per second. Atmospheric friction would quickly slow the object, converting its kinetic energy of motion to heat, and destroy it. However, a mass driver is entirely practical on the Moon because lunar escape velocity is only about 1.5 miles per second and there is no atmosphere. Figure 12.5 is an artist's concept of a mass driver in operation on the Moon.

One approach to accelerating the ore uses the attraction and repulsion of magnets. When an electric current flows

Figure 12.5 Advanced lunar base equipped with a mass driver which accelerates ore to escape velocity and sends it to a collection site at L2. Ore buckets enter the driver at the near end and, after releasing their load at the far end, return for another load on the track behind the mass driver. In the right foreground an astronaut sits with his son at the future site of the Apollo Museum. © *Lunar and Planetary Institute; artist Pat Rawlings, Eagle Engineering.*

through a coil, it becomes an electromagnet with a north pole and a south pole. If another magnet is brought near, with its north pole pointing toward the north pole of the coil or its south pole pointing toward the south pole of the coil, the two will repel one another. However, if a north pole and a south pole are facing one another, they will attract each other.

A container fitted with superconducting magnets is inserted into a tube and a coil is energized just ahead of the bucket to attract it and accelerate it forward. Just as the container reaches it, the coil is turned off and the next coil a little farther down the track is turned on to accelerate it further. Passing through coil after coil the loaded container is accelerated to escape velocity.

In one design, about 25 pounds (Earth weight) of lunar soil is *sintered*, (heated and compressed) to form a cylindrical slug about 18 inches long and 4.5 inches in diameter. The slug, firmly attached in its container, enters the mass driver where it accelerates at 1,000 g to escape velocity, 1.5 miles per second, in the first thousand feet. It then moves through guide rails for a distance of 6,500 feet to damp out any sideways motion. At that point the slug is released and the container is suddenly decelerated by two magnetic loops and deflected sideways away from the ore. The slug of ore, now separated from the bucket, continues on its way into orbit, while the container is directed onto a return track back to the loading point for another run.

The slug continues on its escape trajectory toward a catching device located at L2. (Refer back to Figure 11.13.) Although the processing and manufacturing plant may be located at L5, it does not seem advisable to launch the ore directly toward the colony. The best location for the mass driver would be on the front side of the Moon, near the equator, so the slug would follow a half-orbit to L2, similar to a Hohmann transfer orbit. The actual trajectory is more complicated than a simple ellipse because of the balance between Earth gravity and Moon gravity in that part of space. Recall, an elliptical orbit has only one central body, located at one focus, which supplies the gravitational force for the orbit. The velocity of the ore slugs must be very exact as they leave the mass driver; otherwise they would scatter over a wide region of space near L2. Two additional "trim stations" are located downrange from the launcher to precisely adjust the final velocity.

At L2 a device is positioned to catch the slugs as they arrive. Several designs have been suggested, from a metal mesh net to a 10 mile wide piece of styrofoam. There are two main problems to overcome. First, there is sure to be some scatter in the arriving pieces of ore, and second, the catcher would be accelerated backward from the impacts of the ore slugs. Solutions have been suggested, but both of these problems are difficult to deal with.

A space tug would haul an accumulated stock of the slugs to the L5 colony for processing. While the L5 habitat is under construction, raw ore would be sent. The waste slag from ore processing would be used for the L5 shield. Later, preliminary separation could be done on the Moon so only the useful minerals such as aluminum and titanium oxides are shipped out. Operating at full capacity, this design of a mass driver would launch 10 ore slugs per second, a total delivery of 600,000 tons a year to L2.

Once construction of the L5 space colony is completed, its citizens can go into business manufacturing and selling solar power satellites to Earth and oxygen to passing spaceships.

Astronomy on the Moon

Astronomy from a lunar base has tremendous advantages. Optical telescopes on the Moon would be unhampered by atmospheric refraction, pollution, and city lights. Low gravity and the slow rotation of the Moon would allow construction of large telescopes and simplify the construction of telescope mounts. Radio telescopes located on the far side of the Moon, pointing away from Earth, would not suffer interference from man-made radio noise. A lunar crater would be a perfect natural bowl-shaped depression in which to build a large dish antenna like the one at Arecibo, Puerto Rico (Figure 12.6). Telescopes designed for observing other parts of the electromagnetic spectrum such as infrared, ultraviolet, or x-ray, would be unimpeded by atmospheric absorption. Over a billion dollars was spent on the Hubble Space Telescope, an indication of the value astronomers place on an observing site in space. Figure 12.7 shows an array of optical telescopes forming an interferometer and a radio telescope built into a lunar crater.

Luna City

Figure 12.8 is an artist's concept of what a lunar colony may look like while under construction.

The prefabricated buildings, similar in design to space station modules, are set into trenches with several feet of lunar soil heaped over the top for shielding from solar and cosmic radiation and from meteorites. The soil also insulates the structures from extreme temperature variations during the two week lunar day and two week lunar night. Buildings are interconnected by hallway-like tunnels; all doors are airtight so if one section loses pressure it can be sealed off until the leak is repaired. A deeper "storm cellar" module is buried beneath the main habitats to provide protection during solar proton storms.

Automating most of the machinery for the mining camp and operating it remotely from Earth would require perhaps as few as 10 maintenance people at the site. With additional processing and manufacturing equipment, lunar materials could be used to build and expand a permanent city. Luna City might even become self-sustaining. Con-

Figure 12.6 The world's largest radio-radar telescope antenna hangs in a natural bowl in the mountains of Puerto Rico. The 1000 foot dish, made of nearly forty thousand 3 by 6 foot perforated aluminum panels, is precisely spherical in shape to an accuracy of less than an eighth of an inch over the 20 acre surface. The triangular-shaped 600 ton platform, suspended 435 feet above the big reflector, supports the antenna feed horns and lines. The telescope either receives natural signals from space or connects to a transmitter to act as a radar. The antenna feeds can be moved to study various parts of the sky in a wide range of wavelengths. *Courtesy of Arecibo Observatory.*

Figure 12.7 Astronomical observing site on the back side of the Moon. Two rows of optical telescopes at right angles to each other form an interferometer, used by astronomers for very high resolution imaging. At the left rear, a radio telescope antenna hangs in a crater. The covered habitat is in the foreground. Communications antennas point at a relay satellite; since the observatory is out of sight of the Earth, communications must be relayed. © *Eagle Aerospace, Inc.; artist Pat Rawlings.*

Figure 12.8 Unloading a habitat module just delivered to the lunar base. The flimsy looking crane can do the job easily because the module weighs only one-sixth its Earth weight. Other modules, delivered by other landers scattered around the scene, are already in place in trenches at the base in the background. The assembled habitat is then heaped over with lunar soil for protection and insulation. Earth hangs low in the sky. *Courtesy of Eagle Engineering; artist Pat Rawlings.*

struction would be simpler than on Earth because of the low gravity.

Once established, Luna City could engage in interplanetary commerce, selling its products to Earth and to space stations and serving as a way station for spacecraft leaving the vicinity of the Earth. As was pointed out before, it is much less costly in energy to launch supplies from the Moon with its one-sixth g than it is to launch from Earth. Only small rocket engines and a short fuel burn were needed to launch Apollo's LM from the Moon. Only small shuttle craft would be needed to deliver fresh supplies and oxygen propellant to customers parked in orbit around the Moon before they proceed to the farther reaches of the Solar System. A possible cislunar infrastructure is shown in Figure 12.9. Each part has been discussed separately in this and the previous chapter; this diagram brings it all together.

Site Selection

Of the 15 million square miles of lunar surface, how do you choose the best spot to build a colony? The decision depends primarily on the colony's function. If exploration is the primary purpose, then a scientifically interesting site would be the first choice, one with a variety of terrain, surface com-

position, and geological features. If mining is the chief objective, then a location near profitable mineral deposits is essential. If the colony is to be an astronomical research station, then a location near the equator gives a complete view of the entire sky in a month's time, although the Earth would always be in the scene at a site on the "front" side of the Moon. Radio astronomers would select a site on the back side of the Moon, free from Earthly interference.

Life support considerations are essential. The site should have loose soil and regolith to cover the habitat for protection from the normal continuous radiation as well as severe solar proton storms and small meteoroids. A nearby supply of oxygen-containing minerals would assure adequate oxygen for habitat atmospheres. Locating at a water supply, if one can be found, would be immensely valuable. It may be that an icy comet collided with the Moon and is buried somewhere. Water might be found in the polar regions, also. Locating at either the north or south pole has the major advantage of having the Sun continuously on the horizon. There, from the top of a crater rim, solar cells would produce a continuous supply of electricity.

Terrain features are important in choosing a site. Because heavy soil-moving equipment may not be immediately available, relatively flat terrain around the site would be safest, although it may not be interesting to miners and scientists. A small crater nearby would make a useful launch site. Its walls would keep much of the debris kicked up by rocket engines from spreading very far.

Access to orbit for both coming and going is an important consideration. An equatorial orbit passes over a site on the equator every time around. Similarly, a spacecraft in a 90° orbit passes over a polar site on every orbit. Sites at other latitudes are not accessible as frequently from an inclined

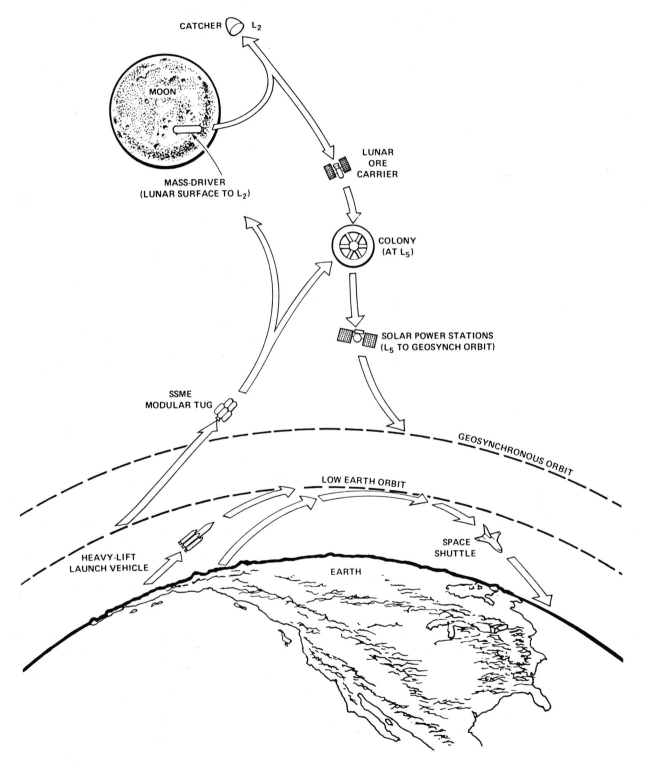

Figure 12.9   Cislunar infrastructure. (An SSME modular tug is a cargo vessel assembled from Space Shuttle main engines.)
Courtesy of NASA.

orbit. Because Earth rotates on its axis once every 24 hours, a spacecraft in an inclined orbit may overfly locations on Earth at latitudes less than the orbital inclination twice in 24 hours, once headed northward and once southward. But the Moon rotates on its axis once in about 29 days, so access to inclined orbits is not as frequent.

It is improbable that any one site satisfies all the considerations mentioned above. Much more detailed information about the Moon's surface and resources is needed before a suitable location for Luna City can be selected. The data from Clementine, described in Chapter 7, is a tremendous resource for making this decision.

Not everyone agrees that the Moon will become a permanent home to a large group of colonists. Other ideas are discussed in the additional readings listed at the end of this chapter. A lunar outpost will undoubtedly be established in the twenty-first century, but how far it will expand is the question.

Mars, the Red Planet

Mars as a place to live has been an appealing notion for centuries. Science fiction writers from Edgar Rice Burroughs to Ray Bradbury and many others have described the native Martians or human settlers who became "Martians" when they colonized the red planet.

As early as 1659, markings were seen on the surface of Mars and the observers could estimate the period of rotation of the planet by watching the markings. In 1877 Schiaparelli reported seeing linear markings which he said were like the fine threads of a spider web. He called them *canali*, an Italian word for lines or grooves.

In the 1890s, astronomer Percival Lowell built an observatory in Flagstaff, Arizona, where he devoted much time and effort to observing Mars. His observing site was better than most and he was in possession of a fine instrument. Much of the time the image of the planet was blurry and shimmering, even when looking through the clear desert sky. Sometimes the image would clear and become quite sharp and detailed.

Lowell, too, saw the markings which had been observed by Schiaparelli, but he saw much more. He saw seasonal changes in the markings and in the polar caps. The markings seemed to widen and become darker in the spring and summer. At the same time the polar caps retreated in the summer and advanced in the winter. Lowell mistakenly interpreted these seasonal variations as being due to vegetation along canals which he though had been constructed by Martians to carry water from the polar caps to their agricultural land in the warmer equatorial regions. He was convinced that intelligent beings inhabited the red planet, struggling to keep their civilization going as their planet was cooling and drying up.

Later Earth-based observations using more advanced telescopes equipped with spectrographs and other equipment told a different story. Mariner spacecraft were launched from Earth for close inspection of Mars. Mariners 4, 6, and 7 flew by for closeup looks during 1964 and 1969. Mariner 9 was injected into orbit around Mars in 1971. Then, in 1975, two spacecraft called Viking were sent. The Viking orbiters took high resolution photographs of almost the entire planet. Figure 12.10 shows a composite of orbiter pictures. The Viking landers settled softly onto the surface of Mars, sniffed the air, tested the soil, searched for signs of life, photographed the surroundings, and sent the findings back to Earth (Figure 12.11). They found no canals, no vegetation, no Martians. Mars is almost as inhospitable as the Moon. Almost. Several differences make it a more appealing place, a potential site for human communities. Two of the most important are water and an atmosphere.

Because it is 1.5 times farther from the Sun, its year is 687 Earth-days long, 1.88 Earth-years. (Recall Kepler's third law? See **MATHBOX 3.3**.) A day on Mars is 24.6

Figure 12.10 Mosaic of Viking orbiter pictures showing a section of the northern hemisphere of Mars. The horizontal gash below the center of the picture is Valles Marineris. Three large volcanoes are on the left side. (The largest volcano on Mars, Olympus Mons, is off the left edge of the picture.) The white areas on the left and top are clouds, frost, or ice. To the right side are numerous craters and small volcanoes. *Courtesy of NASA.*

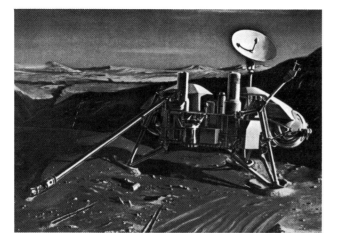

Figure 12.11 Viking lander. A long mechanical arm with a scoop for sampling Martian soil reaches out to the left. Weather instruments are on the boom extending up to the right. Imaging equipment is in the center, and the communications antenna points toward Earth. *Courtesy of NASA.*

hours long, just 37 minutes longer than Earth's day; so the Martian year is 669 Mars-days long. The tilt of its axis is 25 degrees, compared to Earth's 23.5 degrees. So Mars has seasons, but because of the long year, each season is about twice as long as on Earth; see Table 12.1. By comparison, Earth seasons last about 90 days. Also, the eccentricity of its orbit is 0.093, greater than Earth's 0.017. At perihelion it is 26 million miles closer to the Sun than it is at aphelion. Mars is at perihelion during the southern hemisphere summer so that season is shorter and warmer than northern hemisphere summer. Recall Kepler's second law?

The diameter of Mars is about 4,000 miles, half that of Earth, and its mass is only 0.1 that of Earth. Therefore, its gravity is much lower, about 0.4 g on the surface. Escape velocity from the surface of Mars is 3 miles per second compared to nearly 7 miles per second from Earth.

Martian Moons

Mars has two tiny moons. Phobos is potato shaped, about 15 miles in its longest dimension, and therefore has very low surface gravity, only 0.001 g. It orbits 5,600 miles from the planet in about 7.6 hours. Deimos is even smaller, about 3.5 by 4.5 miles, and its surface gravity is 0.0005 g. Escape speed is about 22 miles per hour; a good pitcher could throw

a baseball into orbit. It is about 14,000 miles from Mars and completes an orbit in 30.3 hours. To an observer on Mars it would appear like a bright star shooting across the sky.

In July 1988, the former Soviet Union launched two probes to Mars and its moons, both named Phobos. Because of a computer software error, Phobos 1 lost lock on the Sun, and with the solar panels pointed in the wrong direction, its batteries became depleted. Contact with the spacecraft was lost in September. Phobos 2 reached Mars in January 1989 and was injected into an orbit matching that of the moon, Phobos. It took 37 visual images of Phobos and thousands of infrared data samples of the little moon from distances of 100 to 350 miles away and of Mars itself. It also carried a number of other sensors that studied both Mars and its moon. The moon appears to be covered with a dark dust; it reflects only 7 percent of the light falling on it. A number of grooves, craters, and some rock outcrops appear in the pictures. The sensors indicated that Phobos's density is about twice that of water; by comparison, typical Earth rock is 5 to 6 times as dense as water. This would indicate that Phobos either has a porous interior or is mostly ice. The spacecraft also carried two landers which were equipped with stereo cameras and other sensors, and had rotating rods which would make them hop over the ground in 50 to 60 foot leaps to observe different spots on the moon's surface. Unfortunately, the spacecraft never got a chance to fire the probes, and contact was lost with Earth at the end of March, ending the mission after less than 2 months of operation.

It has been suggested that these two moons could be reconstructed into way stations for spaceships traveling between Earth and Mars. Rendezvous, docking, and then escaping from such a small moon would require little energy.

Martian Landscapes

Except for the lack of vegetation, the Martian landscape is strangely Earth-like. Viking lander pictures of the rocks, soil, and sand dunes have a familiar look about them, giving the impression that they were taken in a barren desert on Earth. Figure 12.12 shows the scenery surrounding Viking 1; you would almost expect to see a camel walk by. Soil analysis at the Viking lander sites detected the presence of many of the same minerals as found on Earth. Many of the iron-bearing rocks are oxidized, rusted to a red color like iron-bearing rocks on Earth. Over time, erosion has cut away sand and dust-sized grains, and wind has distributed them over the planet. For centuries Mars has been called "the red planet"; it really is red.

The southern hemisphere is covered by old, eroded impact craters much like the Moon, but the northern hemisphere shows evidence of geological activity including volcanoes, lava flows, canyons, and what appear to be dry river valleys and glacier deposits. Look back at Figure 12.10. This geological activity along with wind erosion has obliterated many of the craters. The largest of the dormant, perhaps

TABLE 12.1 Length of Martian Seasons

Northern Hemisphere	Southern Hemisphere	Earth Days	Mars Days
spring	autumn	199	194
summer	winter	183	178
autumn	spring	147	143
winter	summer	158	154
		687	669

Figure 12.12 The scene at the Viking 1 lander site looks strangely Earth-like. The structure at the right is part of the spacecraft. *Courtesy of NASA.*

extinct, volcanoes is Olympus Mons, measuring 15 miles high, 325 miles across at its base, and sporting a 40 mile wide caldera crater at the top. It is the largest volcano so far discovered in the Solar System.

A huge canyonland, Valles Marineris, stretches along the Martian equator nearly 3,000 miles, about the distance across the United States from New York to California. Parts of it have walls which appear to have been deeply eroded by running water. It is up to 150 miles wide and 4 miles deep. The Grand Canyon of Arizona would fit into one of the side tributaries.

Martian Atmosphere

Mars has an atmosphere, very thin by Earth standards, equal in density to Earth's atmosphere at about 20 miles altitude. Pressure is only 10 millibars, 0.15 pound per square inch, one-hundredth what it is at sea level on Earth. Furthermore, the Martian atmosphere is 95 percent carbon dioxide. The other 5 percent consists of nitrogen and argon, with traces of oxygen, water vapor, and other gases. Humans could not survive outside space suits or pressurized habitats. Notice, however, that the constituents are the same as Earth's atmosphere, just in vastly different proportions.

The atmosphere contains so little ozone that ultraviolet radiation reaches the surface of Mars with full intensity. Because of the distance from the Sun, the ultraviolet is much less intense than it is at Earth's distance. Nonetheless, it is sufficient to be hazardous to unprotected human beings.

Thin as it is, the atmosphere provides some protection from meteoroids. Although there are many large craters,

Viking lander pictures show no small craters. Smaller meteoroids burn up as they fall through the atmosphere.

At the landing sites, light southwest winds blew in the early morning, and light easterly winds blew in the afternoon. Maximum recorded wind speed was 15 miles per hour. Afternoon temperatures were typically −25 °F; at night temperatures regularly dropped to near −120 °F. Because of its greater distance from the Sun, Mars receives only 43 percent as much solar energy as Earth, and, because of its thin atmosphere, the little solar heat absorbed during the day is lost to space at night. In the polar regions, summer daytime temperatures reach only −95 °F. In equatorial regions, the highest temperature recorded was just above 0 °F.

Dust storms occur as winds pick up the red dust from the surface. Major storms may encompass the entire planet for several weeks. The energy contained in the wind is not great, however, because of the low density of the air. Dust storms seem to occur mostly during the southern hemisphere summer when Mars is at perihelion, closest to the Sun, and receiving maximum solar radiation. It is possible that some of the features observed by Lowell were due to the red dust settling out onto the dark bedrock at some time and then being blown off the rocks during a storm. Very fine dust remains suspended in the air for long periods of time causing the sky to appear pink rather than the blue we are accustomed to on Earth.

Water on Mars

Water is present on Mars in substantial quantities. The Mariner and Viking images show geological features that look like river beds on Earth that appear to have been made

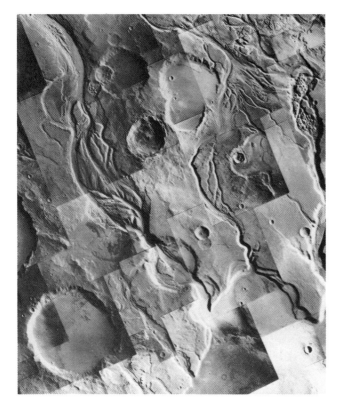

Figure 12.13 Viking orbiter picture of Martian features that have the appearance of dried up river beds. They cut across craters, indicating that water flowed on Mars after meteor bombardment formed the craters. This picture is a composite of a number of individual small photos taken by the Viking orbiter. *Courtesy of NASA.*

by running water (Figure 12.13). But there is no running water on Mars now; temperatures are too cold and the pressure is too low for water to remain in the liquid state.

Where, then, did all the water go?

A small amount is in the atmosphere. Clouds and fog appear on a number of the Viking orbiter images, especially near the equator and on the slopes of Martian volcanoes. Figure 12.14 is an example. Frost, either dry ice or water ice, appears on the rocks in some of the Viking lander pictures (Figure 12.15). It has been estimated that if all the water vapor in the Martian atmosphere could be condensed to liquid it would amount to about 0.2 cubic mile or 220 billion gallons. (For comparison, Lake Mead behind Hoover Dam holds about 10 billion gallons.)

Some water is frozen in the polar ice caps. The seasonal changes observed in the ice caps are probably the result of deposition of dry ice, frozen carbon dioxide, from the atmosphere during the severely cold polar winter season and the sublimation (evaporation) of the dry ice during the following spring. The ice caps do not disappear completely during the summer season, indicating that the residual is probably water ice mixed with Martian soil. Figure 12.16 is a Viking orbiter picture of the north polar region in midsummer when

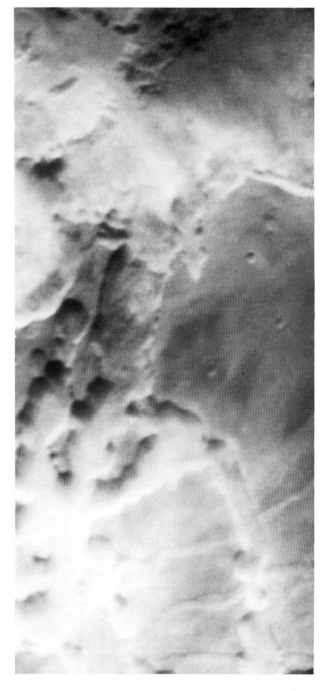

Figure 12.14 Early morning fog on Mars. The fog seems to form in the canyons and dissipate by midday. A few craters can be seen on the plateaus to the right and left of center. *Courtesy of NASA.*

the carbon dioxide dry ice has sublimated, leaving the water ice exposed.

The water vapor content of the atmosphere varies widely from season to season, indicating that water is also stored in the ground in the form of permafrost, permanently frozen soil and ground water. Permafrost is found on Earth in the northern parts of Alaska, Greenland, Canada, Scandinavia, and Siberia. In the equatorial regions of Mars it is probable that the top 3 feet or so of soil has been completely dehydrated. If permafrost water is present there, it is deep in the ground. Because the interior of the planet is warmer than the surface, liquid water may be found at greater depths, perhaps a mile down. If so, it could be tapped like groundwater is tapped on Earth, by drilling wells.

Unmanned Exploration

In spite of the detailed description on the previous pages, we know very little about Mars. Mars is a big place. For example, Valles Marinaris is 2,500 miles long! Brief human visits would be an inadequate exploratory program to lay plans for colonization. EVAs covering a few miles for a few hours will not do it. The surface area of Mars equals the land area of Earth.

We know almost nothing about the subsurface Martian rock and soil. The same is true of Martian water. We can be confident that it is there somewhere, but except for the ice caps, we do not know exactly where the water is located or in what form. No plans for human exploration and colonization of Mars can be made until further unmanned exploration reveals its composition.

In the 1998 window a Russian spacecraft will be launched to Mars carrying French-designed balloons to study the surface and atmosphere. Instruments are attached to a 300-foot rope hung below two balloons. The upper one is a closed balloon inflated with 200,000 to 300,000 cubic feet of helium. The lower one is something like a hot air balloon, open and filled with Martian atmosphere, made of a material which readily absorbs solar energy. When the Sun shines on it during the daytime, the solar balloon heats up and rises

Figure 12.15 Frost on the ground and rocks at the Viking 2 lander site. *Courtesy of NASA.*

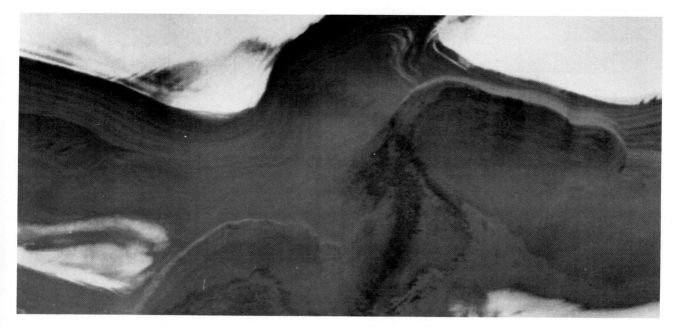

Figure 12.16 North polar region of Mars. Carbon dioxide dry ice has sublimated during the northern hemisphere summer exposing the water ice and the step-like layers in the cliffs at the edge of the ice cap. The ice must be quite thin; patches of bare ground can be seen at the top left. *Courtesy of NASA.*

into the atmosphere, drifting with the wind. After sunset, as the solar balloon cools, it slowly sinks until the instrument package reaches the ground. There it parks for the night to make measurements of soil conductivity and other things. The helium balloon remains inflated and buoyant at all times to keep the solar balloon off the ground. Next day when the Sun comes up, the solar balloon heats up again to carry the experiment to a new location. A camera package is attached for making closeup images of the ground. The balloons may be accompanied by a small robot rover and penetrating devices to be fired into the ground to make measurements. Prior to the balloon mission, the Russians plan to send a radar mapper to study the surface terrain and search for water from orbit.

NASA may send an unmanned robot roving vehicle in the 1998 window to drive around Mars for a year, picking up samples, drilling into the ground in search of permafrost,

photographing the landscape, and doing numerous other experiments. The samples would be transferred to a capsule and returned under quarantine for analysis of their chemical composition, a search for fossils, and age dating. The window for returning to Earth opens in January 2001.

Figure 12.17 shows one design of a Mars rover, a segmented vehicle consisting of three sections, each with its own pair of wheels, coupled together with swiveling pivots. It is equipped with stereo cameras in the midsection so controllers on Earth can examine the surroundings in 3-D. A rover which can traverse rocky terrain pitted with craters and cut by stream channels and steep cliffs without a driver must be a smart robot. Round trip communication time between Earth and Mars is anywhere from 8 to 20 minutes (at the speed of light) depending on their relative positions in their orbits. By the time controllers on Earth saw a problem ahead and sent a message to turn back, the rover could be a

Figure 12.17 Mars sample return mission. The lander in the background brings the rover in the foreground to the surface while the mother ship remains in orbit. The three-segmented rover wanders about the planet for a year studying the terrain and picking up samples of the soil and rock. The samples are then transferred to a small rocket attached to the lander which takes them to a rendezvous with the mother ship for return to Earth. *Courtesy of NASA/JPL.*

pile of scrap at the bottom of a canyon. The machine can climb over obstacles up to 5 feet high and climb a 35 percent grade, but it must be smart enough to recognize terrain it cannot handle. Of course, it would move very slowly, less than a mile a day, and could be instructed to move only a few feet and stop for further instructions after controllers get a new stereo picture to examine.

Some engineers think that the Russian-French balloons have a better chance of success. They can travel farther more safely in a short time, picking up ten or more samples and dropping them at one spot with a radio beacon. The rover would then land at the beacon to pick up the rock collection and return it to Earth. The complex part is steering the balloons to desired locations.

Humans on Mars

In spite of its harsh environment, Mars has the basic ingredients necessary to support a self-sufficient colony of humans. It is the most Earth-like of all the planets. Mars can be a prototype human extraterrestrial society. If we can do it on Mars, then human potential for expanding into the universe will have been proven.

Should we go? It is technically feasible and would require no major breakthroughs in technology, although breakthroughs in "exotic" propulsion systems, nuclear fusion, and bioengineering would make it easier. Indeed, the research and development concomitant with a commitment to a Mars mission would undoubtedly lead to major advances in those areas. First we must learn how to build and operate in low Earth orbit. That experience is essential for human expansion beyond Earth.

Interest and enthusiasm for establishing a colony on Mars have been steadily increasing. A particularly active group of students and teachers, dubbed the Mars Underground, held a Case for Mars conference at the University of Colorado at Boulder in 1981. Since then the group has joined with the Planetary Society in forming a Mars Institute to promote studies and research for Mars exploration. Three more Case for Mars conferences have been held in Boulder, attracting more and more researchers and engineers. Studies related to transportation systems, construction in orbit and on Mars, use of Mars resources, and human factors have been done and are continuing. The Russians are planning several unmanned research flights to Mars and a cooperative mission between the United States and the Russians is likely.

It seems probable that a human settlement on Mars could become a reality within the next 50 years. Indeed, the first Martians are probably alive on Earth today! We need to set a deadline and stick to it. The Apollo Moon landing took 8 years; that was all the time allocated. If 15 years had been given, every minute of it would have been used. It seems to be human nature to accomplish a task in the time available to do it.

Ultimately we may have self-sufficient cities on Mars where people live as normally as they do on Earth, going about their business of growing food, manufacturing products, doing scientific research, and enjoying themselves while doing it.

The First Settlement

The first problem that comes to mind is, of course, the atmosphere. It would be a major accomplishment if we could modify the Martian atmosphere so it could support our human needs. *Terraforming* describes the process of modifying a planet to make it more Earth-like so humans can live a normal existence. It will be discussed later in this chapter. The first colonists, however, will have to live in closed, sealed habitats as they would on a space station or lunar colony. Work outside the habitat will be done in space suits. Figure 12.18 is an artist's concept of a Mars base.

The first habitat modules and supplies of food, air and water will be imported along with machinery for mining ore, extracting usable minerals, and fabricating construction materials. Continuing dependence on Earth for fresh supplies would be far too expensive, so eventually colonists will have to learn to live off the land as our ancestors did when they migrated to new territory. In the case of Mars, the land is barren, but the technology to make it productive is available.

Buildings constructed on permafrost and heated on the inside could have structural problems if the permafrost melts beneath the building. On Earth, such buildings are built on piers above the ground, insulated from the ground. But on Mars it is necessary to protect the inhabitants from solar and galactic cosmic rays. As on the Moon, the solution is to dig a trench, put a habitat module in it, and cover it with several feet of soil. Storm cellars will be needed, also. After ore is dug from the mines, the tunnels can be sealed off and used as habitats.

Power for the colony could come from solar cells, concentrating solar electric generators, methane manufactured from hydrogen and carbon dioxide, or nuclear power plants. Because Mars is one-and-a-half times farther from the Sun than Earth is, solar energy is only 43 percent as intense as it is at Earth. Therefore, arrays of solar cells will have to be twice the size of arrays near Earth to produce the same amount of electricity.

Nuclear power generators could be installed in the polar ice cap to generate electricity and their waste heat could melt the ice to provide water for the colony in a warmer climate near the equator. Then Mars might finally have its canals.

Interplanetary Transportation

Shipping all the machinery, materials, and supplies directly from the surface of the Earth would be enormously expensive in energy, especially if cargo and crew are sent together

Figure 12.18 Artist's concept of a Mars base. Partially buried habitat is at the center, greenhouses to the right, and mine tunnels to the left rear. The vertical structure near the tunnel entrances is a well drilling rig. A "wagon train" vehicle made in segments climbs the hill in the foreground while a rocket lifts off at the right rear of the scene. Dish antennas on the mesa are for off-planet communications; the mast antenna is for communicating with roving exploration vehicles. Because Mars has an atmosphere, a very thin one, remotely piloted aircraft with very large wings can be used for transportation and exploration. In the lower right a geologist examines a newly discovered fossil. *Courtesy of Eagle Engineering; artist Pat Rawlings.*

in a fast transfer orbit. Several ideas have been explored. One calls for shipping cargo and fuel in smaller batches first to a space station in low Earth orbit. There the pieces would be assembled into a larger unmanned cargo vessel and sent ahead to Mars via a Hohmann transfer, the least-energy transfer, requiring nearly a year transit time. See Figure 12.19 in the color section, page 194. At Mars it would be injected into a low orbit to wait the arrival of the smaller passenger craft coming by a faster, high energy route, taking about 5 months. An important advantage to separating the cargo from the crew is that the cargo would be in place in orbit around Mars before the crew leaves Earth, giving greater assurance of a successful mission.

The interplanetary passenger ship would rendezvous with the cargo ship and materials would be taken to the surface in small batches. Aerobraking in the Martian atmosphere reduces the fuel requirements for retrorockets to inject the spacecraft into orbit around Mars.

The Hohmann transfer requires a minimum of fuel but takes a longer time. The slow route is all right for cargo, but not for people, for two reasons: (1) the weight of a year's supply of food, air, and water that must be lifted from Earth, and (2) the physiological and psychological effects of a yearlong trip on the human body. On the other hand, a fast trip requires more fuel at both ends, to accelerate the craft to higher speed at departure and to slow it on arrival. Although the mass of life support supplies is less for a quick trip, mass of fuel is greater. An optimum must be determined.

If a lunar colony is established first, it could provide the oxygen for the trip and some of the necessities for establishing the Mars base at a lower energy cost than supplying them from Earth.

Robert Zubrin of Martin Marietta Co. suggests sending robotic equipment ahead to land on Mars and manufacture methane gas and oxygen for powering the base and for propellants for the return trip home. That saves carrying hundreds of tons of fuel all the way to Mars. Zubrin's equipment would include a 14 ton crew return vehicle, a 50 kilowatt nuclear power generator, and 6 tons of hydrogen. It sets itself up and starts running automatically, intaking carbon dioxide from the Martian atmosphere and combining it with the hydrogen to produce methane and water. The chemical reaction is $4H_2 + CO_2 \rightarrow 2H_2O + CH_4$. The water is electrolyzed into hydrogen and oxygen. The oxygen and methane are stored, and the hydrogen is used to make more methane. The machinery runs for 10 months. Then if all is working well and 96 tons of methane and oxygen are in storage, the crew is sent with its habitat and supplies for the stay on Mars. A second methane factory and crew return vehicle are also sent separately to begin manufacturing methane and oxygen for the following mission. Methane produced from Martian carbon dioxide and hydrogen sent from Earth be-

comes the primary energy source for Mars surface activities and for the return trip home.

Zubrin and his colleagues have successfully built a prototype unit which produces about 1.5 pounds of methane per day, is more than 94 percent efficient, yet weighs only about 45 pounds. The engineers think they can cut the weight of the unit in half. To prove the concept, they suggest that their machine be sent to Mars on a sample return mission. In 2 years of operation on Mars it would produce enough fuel to bring back 9 pounds of Martian rock and soil. Just as important, it would demonstrate the feasibility of scaling the machinery up to a size capable of bringing a crew back from the red planet.

It is unlikely that any one nation will establish a base on either the Moon or Mars alone. Indeed, it is not likely that the first human expeditions to Mars will be carried on by a single nation. The costs are simply too great for any one country to afford. Like the International Space Station, human presence on the Moon and beyond will probably be the result of international cooperative efforts.

1999 Trajectories

Every 2 years Earth and Mars are properly aligned in their orbits for a voyage. Because Earth is traveling faster in its smaller orbit closer to the Sun, it overtakes and passes Mars every 780 days. When that happens, Mars and Earth are lined up on the same side of the Sun, a position called *opposition* by astronomers, and minimum energy voyages can be made.

Three of many possible trajectories to Mars during the 1999 opposition are shown in Figure 12.20. Figure 12.20a is a flyby without a landing, a 1 year flight. A complicated mission taking the spacecraft past Venus for a gravity assist on the way to Mars is shown in Figure 12.20b. The gravity of Venus accelerates the spacecraft so it "cracks the whip" around that planet and heads for Mars. Less energy is required for a larger payload, including provisions and vehicles for a 2 month stay on Mars, but the trip takes more time. Total mission time is 22 months. A 16 month stay on Mars is included in the mission shown in Figure 12.20c. It uses Hohmann transfers to Mars at the 1999 conjunction and back to Earth at the 2001 conjunction. Mission time is more than 2 years and 8 months.

Buzz Aldrin, the second man on the Moon, has investigated the idea of a "bus" service between Earth and Mars by placing large rotating spacecraft into an orbit around the Sun that intersects the orbits of the two planets. For example, see Figure 12.20a. At each encounter with Earth, a gravity assist boost transfers it to an orbit that reaches the Martian orbit at the point where Mars will be the next time around. The gravity boost would accelerate it so that the trip to Mars would take only 4 months. The cycling bus would continue out well beyond Mars orbit, then return to

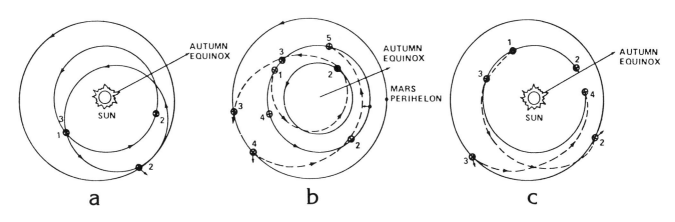

MARS FLYBY FOR 1999 OPPOSITION
ONE YEAR MISSION

1 EARTH DEPARTURE, APRIL 2, 1999
2 MARS PASSAGE, AUGUST 8, 1999
3 EARTH ARRIVAL, APRIL 2, 2000

OUTBOUND VENUS SWINGBY 1999 OPPOSITION

1 EARTH DEPARTURE, JAN. 26, 1998
2 VENUS PASSAGE, JULY 9, 1998
3 MARS ARRIVAL, JAN. 16, 1999
4 MARS DEPARTURE, MARCH 17, 1999
5 EARTH ARRIVAL, NOV. 18, 1999

CONJUNCTION CLASS MISSION 1999 OPPOSITION

1 EARTH DEPARTURE, DEC. 17, 1998
2 MARS ARRIVAL, SEP. 28, 1999
3 MARS DEPARTURE, JAN. 25, 2001
4 EARTH ARRIVAL, SEP. 2, 2001

Figure 12.20   Three possible trajectories to Mars during the 1999 opposition. Follow the numbers in sequence in each figure. (**a**) is a flyby without landing, a transfer orbit with a period of exactly 1 year; the spacecraft returns to meet Earth at the same point as at departure. The trajectory shown in (**b**) includes a flyby of Venus. The spacecraft picks up enough energy in its fall toward Venus to whip past that planet and head out for Mars; a greater payload can be carried with less takeoff energy. This mission takes 1 year and 10 months, including a 2 month stopover on Mars. (**c**) shows a 9 month outbound transfer orbit, 16 months on the planet, and about 7 months to return. *Courtesy of NASA.*

Earth for another gravity boost to meet Mars again. By continuously cycling, the need for accelerating and decelerating the large vehicle at each end of the trip is eliminated.

Passengers bound for Mars leave Earth in a passenger craft with a first stop at the space station in low Earth orbit. From there they ride a small shuttle craft to a rendezvous with one of the cycling ships already equipped with an atmosphere and provisions supplied from the Moon or Mars. The refueled shuttle craft is hangared in the cycler for use at the other end of the trip. At Mars the shuttle craft takes the passengers from the cycling bus orbit to a space station in low Mars orbit (Figure 12.21). A Mars lander carries the passengers to the surface (Figure 12.22).

Once the bus line is established, no fuel would be required to keep the cyclers running in their orbits. Small amounts of propellant would be needed to adjust the course for the gravity assist. Power would be needed only for the life support equipment. Shuttle craft fuel requirements would be comparatively small.

Terraforming

Someday human Martians may stand by the side of a flowing stream, gaze at the blue sky above and the flowers below

and think how much it reminds them of pictures they have seen of Earth. We have been discussing building comfortable human habitats in colonies on other worlds. Is it possible to change an entire world into an Earth-like human habitat? Technically, yes. But it requires time—decades or centuries.

Earth is warm and comfortable primarily because of its distance from the Sun and because of the composition of its atmosphere. The Moon is the same distance from the Sun, but, without an atmosphere, it is cold, dry, and bombarded with lethal radiation. Mars, farther from the Sun and with only a thin atmosphere, is also cold and bleak. On the other side, Venus, closer to the Sun and with a dense carbon dioxide atmosphere, is a 700 °F inferno.

The air blanket around Earth is transparent to the visible wavelengths of sunlight, the region of the spectrum which contains most of the solar energy. Visible light can penetrate almost undiminished through the air to warm the ground. On the other hand, heat from the Earth does not readily escape back to space because water vapor and carbon dioxide, both minor constituents of the atmosphere, strongly absorb the infrared wavelengths emitted by the warm ground. You are probably aware that deserts cool rapidly after sunset because the absence of water vapor in the air allows the heat

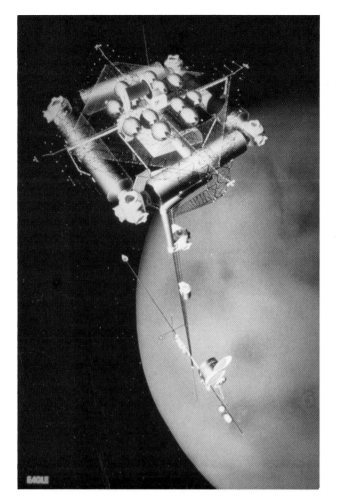

Figure 12.21 A spaceport in low Mars orbit for refueling ships on their way to and from the planet. Docking and hangar facilities and "motel" habitats are part of the port. *Courtesy of Eagle Engineering; artist Mark Dowman.*

Figure 12.22 Aerobraking Mars shuttle vehicle lands vertically using retrorockets. Because Mars has a thin atmosphere, the aerobraking system must have a large area. This one opens wide for reentry and folds up against the sides to make a smooth streamlined vehicle for launch back to orbit. *Courtesy of Eagle Engineering; artist Pat Rawlings.*

to escape. Similarly on a clear, cloudless night, the temperature drops quickly for the same reason. Thus, these trace gases in the atmosphere act as a blanket, keeping the heat in.

To make Mars more Earth-like, then, the first step is to increase the density of the atmosphere and change its composition. Oxygen is needed by humans, but that may be difficult to find and it will not help warm the planet. So let us begin with water vapor and carbon dioxide, two "greenhouse" gases which absorb infrared radiation and hold in the heat. Both are found on Mars, although the quantity of each is uncertain. We can begin by melting the ice caps. Orbiting mirrors can concentrate solar energy into the polar regions. White ice reflects most of the sunlight, so first a coating of dark material should be spread over the ice to absorb rather than reflect the energy. Where do we get the dark material? Phobos and Deimos are both made of dark carbonaceous matter and their gravity is so low that mass drivers can easily propel the stuff off the moons onto the planet. These moons

probably contain water also, which would evaporate and aid in building the Martian atmosphere.

The carbon dioxide dry ice evaporates first, at a temperature of $-108\,°F$. As the carbon dioxide content of the atmosphere increases, the heat loss to space decreases and the planet gradually warms. As the temperature goes up, still more carbon dioxide is released, further warming the planet and releasing carbon dioxide from the regolith. This positive feedback, referred to as a runaway greenhouse effect, continues until all the carbon dioxide is liberated into the atmosphere. It has been calculated that a temperature rise of only 8 degrees Fahrenheit at the south pole would start the process going and that the polar dry ice would evaporate very quickly, perhaps in only 10 years. How much the temperature rises depends on how much carbon dioxide is present on Mars, a largely unknown quantity at this time.

Introduction of other greenhouse gases, such as chlorofluorocarbons, to accelerate the process has also been suggested. Chlorofluorocarbons are the gases used in refrigerators and spray cans that are being accused of destroying Earth's ozone layer. They absorb different wavelengths of infrared than carbon dioxide does. One way to rid the Earth of them and to speed up the terraforming of Mars may be to introduce them into the Martian atmosphere at a very early stage. They would initiate the runaway greenhouse effect and, after a few hundred years or so, would have broken down and dissipated.

Because Mars is farther from the Sun, an atmosphere twice as dense as Earth's would be needed to bring the planetary temperature above freezing, a temperature rise of 100 degrees Fahrenheit. Once the temperature goes above freezing during the summer, water would again flow on Mars. Melting the water from the ice caps and permafrost is the second step in raising the planetary temperature. Water vapor in the air will contribute a great deal to the blanket. Mars could have a comfortable planetary temperature in perhaps 50 years.

People could not breathe the carbon dioxide/water vapor air, but they could go outside without pressurized space suits, carrying only oxygen bottles. As the atmospheric density increases and the temperature rises, we introduce primitive life, beginning with lichens and cyano-bacteria (blue-green algae) which photosynthesize carbon dioxide from the air and nitrogen from the soil, building their own mass, releasing oxygen, and breaking rock down into fertile soil. Later, grasses, shrubs, trees, and food crops can be introduced to speed up the process. Slowly the oxygen content increases until the atmosphere is breathable by insects, then by birds and mammals. Because carbon dioxide acts as a blanket to keep the planet warm while oxygen does not, the quantity of oxygen-producing plants will have to be carefully monitored.

This biological creation of a breathable atmosphere has happened before. The sequence of events described above is similar to the evolution of Earth's atmosphere. How long would it take on Mars? Thousands of years, probably, but in that time the Martians may tire of living in closed habitats and will find some shortcuts. For example, genetic engineering of a particular life form suited to the harsh Martian climate could begin the oxygenation process at an earlier stage.

Other planets and moons could perhaps be terraformed, but with more difficulty than Mars. For a planet with plenty of oxygen but no water, hydrogen could be imported by tanker from Saturn. Or a water-bearing asteroid could be steered on a collision course. It would vaporize on impact yielding, at least temporarily, a water vapor atmosphere.

On to the Stars

Human travel beyond the Solar System is not at all feasible with our present science and technology. The distances involved are enormous. (See Figure 12.23.) The next nearest star after the Sun, Alpha Centauri, is 4 light years away. That is, light travelling at 186,000 miles per second, takes 4 years to travel that distance. The speed of light is a cosmic speed limit; nothing can travel faster. However, strange things begin to happen as one accelerates to high speeds. In particular, time slows down for the high speed object relative to its surroundings. In his theory of relativity, Einstein postulated that this strange thing would occur and experiments have shown it to be true. It is not in the realm of science fiction; it really happens.

If some of the more exotic propulsion systems could be made to work to accelerate spaceships to speeds approach-

Figure 12.23 On to the stars—the belt and sword of Orion. The bright star in the lower right is 815 light years away. By comparison, the star nearest to Earth is about 4 light years away.

ing the speed of light, then it may be possible to send humans on interstellar flights. For example, by accelerating to 90 percent of the speed of light during the first half of the trip, then turning around and decelerating for the last half, it would take about 9 Earth-years to make the round trip to Alpha Centauri. But on the spaceship only a few months would pass. The crew would return from the trip a few months older while everyone on Earth would have aged 9 years. A round trip to the next large galaxy, in the constellation Andromeda, would require 40 or 50 years ship time, depending on how fast it accelerated and decelerated. But 5 million years would pass on Earth. Not many of their friends would be around to greet the crew when they returned! Perhaps in a century or two, the capability of travel at near light speed will be commonplace. But now, in the last part of the twentieth century, there is no propulsion system capable of accelerating to that speed.

Even if we had a spaceship traveling near the speed of light, we have no idea in which direction to head to find a habitable planet, or whether planets even exist around other stars. Even through large telescopes on Earth, the stars appear as just points of light. Looking for planets around other stars is especially difficult from Earth because the atmosphere scatters starlight, increasing the glare around a star's image. Planets only reflect starlight and are so dim that they are lost in the glaring brilliance of the star. With the high resolution of the refurbished Hubble Space Telescope, above the Earth's atmosphere, it may be possible to distinguish between a distant star and a planet in orbit around it.

Attempts have been made to infer the presence of large planets around stars by looking for nonuniform motion of those stars. If it has a large planet, both the star and the planet will orbit around their common center of mass, and the irregular motion should be observable. Several such searches have thus far been inconclusive.

Indirect evidence of other planetary systems has come from the Infrared Astronomical Satellite which has shown the presence of cool dust clouds around several stars. These observations are consistent with the theory that planetary systems form when a star is born from a collapsing cloud of gas and dust. Most of the gas goes into the star while some of the remaining dust condenses into planets. Figure 7.35 in the color section, page 191, is a visual image of these dust clouds around a star in the Orion Nebula made by the Space Telescope.

Even if we find a planet around another star, we do not now have the means of determining whether it is habitable by human beings. To be habitable, the planet should have surface gravity similar to Earth gravity. Less than Earth gravity may be satisfactory, but much more than one g would be difficult to adapt to. A breathable atmosphere and suitable temperatures would certainly be desirable although we could build comfortable habitats and go out only when necessary. Water and oxygen are the two essentials. It would be difficult if not impossible for Earth to provide continuing support to a colony of human beings light years away.

Any native inhabitants would, we hope, be friendly. If intelligent, they would have to be willing to accept our colonists. If not intelligent, they must not be hostile. If only microbes, they must not cause incurable disease.

People who are headed for the stars must be equipped with the best total education possible. Everyone will have to work. A group of colonists should include not only specialist scientists, engineers, and test pilots, but people with a broad knowledge of the humanities and the arts as well as the sciences. We are, first of all, human and we must take our humanity with us. Colonizing the Moon and Mars will provide experience as to the endurance of the human body and mind. One curmudgeon said that we will probably take the two worst possible things into space: germs and management.

DISCUSSION QUESTIONS

1. Why does Earth have an atmosphere, while Mars has only a thin one and the Moon has none?

2. Compare and contrast a colony on Mars, a settlement on the Moon, a big wheel space station at L5, and the bases in Antarctica.

3. How would you select the first group of settlers for a colony on the Moon or on Mars? What skills? What ratio of women to men? What age groups? What health standards? Any mental health screening?

4. In terms of Kepler's second law, why is the southern hemisphere summer on Mars shorter than the northern hemisphere summer? Is the southern hemisphere winter also shorter than the northern hemisphere winter?

5. In terms of Kepler's third law, why is the Mars year nearly 2 Earth years long?

6. If you were going to Mars, what one personal item would you take with you besides the basic necessities of life? Why?

7. How would you govern a permanent settlement on Mars? Should it become an independent state? At what point in time?

8. Describe some recreational activities on the Moon or Mars.

9. Design a calendar for Mars considering the length of the day, year, and seasons. Choose some logical length for something like weeks and months.

10. How could you terraform Venus?

11. What do you think is the most compelling reason for establishing a colony on the Moon? What criteria would be important for selecting a site for your colony?

ADDITIONAL READING

Boston, Penelope J., ed. *The Case for Mars*. Univelt for American Astronautical Society, 1984. Conference papers.

Haberle, Robert M. "The Climate of Mars." *Scientific American*, May 1986.

Hartmann, William K., Ron Miller, and Pamela Lee. *Out of the Cradle: Exploring the Frontiers Beyond Earth*. Workman Publishers, 1984. Beautifully illustrated, popular reading.

Joels, Kerry M. *The Mars One Crew Manual*. Ballentine Books, 1985. Imaginatively prepared, technically accurate, readable.

Kasting, James F., et al. "How Climate Evolved on the Terrestrial Planets." *Scientific American*, February 1988. Factors which produce a comfortable climate.

McKay, Christopher, ed. *The Case for Mars II*. Univelt for American Astronautical Society, 1985. Conference papers.

National Commission on Space. *Pioneering the Space Frontier*. Bantam Books, 1986. Presidential commission report on future goals in space.

Oberg, James E. *New Earths, Restructuring Earth and Other Planets*. New American Library, 1981. Popular treatise on terraforming.

Oberg, James E. *Mission to Mars*. New American Library, 1982. Popular book based on first Case for Mars conference.

Reiber, Duke B., ed. *The NASA Mars Conference*. Univelt for American Astronautical Society, 1988. Conference papers.

Ride, Sally K., et al. *Leadership and America's Future in Space*. NASA, 1987. Recommendations to NASA administrator on future national goals in space.

Schmitt, Harrison, and Gerald Kulcincki. "Helium-3: The Space Connection," Proceedings of the Ninth National Space Symposium, 1993. Summary paper with additional references.

Stafford, Thomas, and the Synthesis Group. *America at the Threshold, America's Space Exploration Initiative*. Government Printing Office, 1991. Report to the president charting a course for future exploration of space, particularly the Moon and Mars.

Zubrin, Robert M., and Christopher P. McKay. "Pioneering Mars." *Ad Astra*, September/October 1992. Original

and unique ideas about exploring, settling, and terraforming Mars.

PERIODICALS

Ad Astra, the Magazine of the National Space Society. National Space Society, 922 Pennsylvania Ave., SE, Washington, DC 20003-2140. Bimonthly magazine, easy reading.

Final Frontier, The Magazine of Space Exploration. Final Frontier Publishing Co., Minneapolis, MN 55408. Bimonthly magazine, easy reading.

Mars Underground News. The Planetary Society, 65 N. Catalina Ave., Pasadena, CA 91106. A quarterly newsletter.

The Planetary Report. The Planetary Society, 65 N. Catalina Ave., Pasadena, CA 91106. Bimonthly magazine of The Planetary Society.

SSI Update, The High Frontier Newsletter. Space Studies Institute, P.O. Box 82, Princeton, NJ 08542. A bimonthly newsletter describing research sponsored by the Space Studies Institute founded by Gerard O'Neill.

NOTES

Chapter 13

Life in the Universe

Do living beings exist elsewhere in the universe? The answer is either "yes" or "no," and either one of them would be bewildering. How should we look for other life? Is the probability of finding it so low that searching would be a waste of time?

There are several ways to search: by going and looking, or by remote sensing. We have sent astronauts and robot spacecraft to look throughout the Solar System. Astronauts have brought rocks and soil back from the Moon for examination; robot laboratories have tested the soil of Mars. So far, nothing living has been found anywhere off Earth.

Space travel is expensive and time consuming. It took almost a year for the Viking spacecraft to reach Mars. The fastest spacecraft, Voyager, launched in 1977 took 1 1/2 years to reach Jupiter, over 2 years to Saturn, 9 years to Uranus, and 12 to Neptune. At that speed, Voyager would take some 50,000 years to reach the next nearest star. Exploration of the Solar System may be practical in a lifetime, but until faster spacecraft are developed, an alternative method must be found to look beyond.

We have also looked for signs of life on the other planets of the Solar System by remote sensing. We examine the electromagnetic energy coming from them, determine their temperatures, analyze the composition of their atmospheres, and try to deduce whether living organisms could possibly endure. By listening intensely to the radio frequencies, we are searching for signals which would proclaim an intelligent source. That type of search is, of course, limited to intelligent extraterrestrials who have discovered radio communications. None has been found, as yet.

Both deliberately and unintentionally we have been signalling our presence to the universe. We have attached messages in the form of audio and video recordings onto several spacecraft. Daily we send radio, radar, and semi-intelligent television signals which could be detected by sensitive receivers many light years away. If some intelligent, hopefully friendly, aliens pick them up, they will learn about Earth and its inhabitants.

The Chemistry of Life

Before we can search for life elsewhere, we need to define what it is we are looking for and to understand how it manages to flourish on Earth.

What do all living things have in common? First, they all have the ability to reproduce, to replicate themselves cell by cell. Instructions for reproducing a cell are contained in the *DNA molecule* which is an integral part of each cell. The DNA splits lengthwise, unzips, so to speak, each half joining with other atoms and molecules until two identical DNA replicas are made. Each builds a duplicate cell. And so the growth of new cells and replacement of old dying cells continue.

Second, living organisms take in nourishment and release waste materials; they interact with their environment. On Earth, animals inhale oxygen and exhale carbon dioxide. They eat food and eliminate gas, liquid, and solid waste. Plants perform basically the same processes as animals. But, additionally, plants take in carbon dioxide from the air and nutrient-laden water from the soil to manufacture carbohydrates. During this process, which requires light energy falling on their leaves, they return more oxygen back to the air than they consume.

Earth life is carbon-based. That is, all living organisms are composed of long chain molecules made up of carbon, oxygen, hydrogen, and other elements. It is not essential that life be carbon-based, but the chemical characteristics of carbon make it the most likely building block. Carbon atoms combine with each other and with other elements in a multitude of ways to form large complex molecules which remain stable at Earth temperatures. In contrast, silicon, sometimes proposed as a element base for alien life, cannot form stable long-chain molecules. It tends to fall apart into useless fragments.

Atoms of elements bind together through the electrical force of attraction to form molecules. The nucleus of an atom carries a positive electrical charge and is surrounded by an electron cloud bearing an equal but negative charge.

Thus, an atom is electrically neutral. If some of the negative charge is missing, the atom becomes a *positive ion*. If an excess of negative charge is present, the atom becomes a *negative ion*. Because particles bearing opposite electrical charges attract one another, positive and negative ions move toward each other and bind together to form ionic compounds. For example, a positive sodium ion and a negative chlorine ion bind together to form the compound of sodium chloride, common table salt.

Ions of sodium are devoid of one electron. They therefore carry a single positive charge and are said to have a valence of +1. The *valence* indicates how many and what kind of excess charges the ion carries. Oxygen has a valence of −2 while hydrogen has a valence of +1. Thus, two hydrogen ions will bind with one oxygen ion to form a molecule of water or two hydrogen ions will bind with two oxygen ions to form a molecule of hydrogen peroxide. Salt molecules and hydrogen peroxide molecules are the "end of the line"; they are the largest molecules that can be formed from those elements. Carbon is unique in that it has a valence of −4 or +4 and can combine with other carbon ions as well as with ions of other elements to form very large molecules. Molecules built on carbon are called *organic* molecules and their study is called organic chemistry.

Molecules of water and salt may be represented by diagrams as follows:

$$H \text{ — } O \text{ — } H \qquad Na \text{ — } Cl$$

H stands for hydrogen, O for oxygen, Na for sodium, and Cl for chlorine. These diagrams, called structural formulas, show how the atoms are "hooked together." Some simple organic molecules are represented as follows:

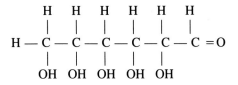

methane   chloroform

These two compounds involve only one carbon atom each, but carbon atoms can also join with other carbon atoms. A simple sugar looks like this:

OH is the hydroxyl radical; it consists of the union of an oxygen atom, valence −2, and a hydrogen atom, valence +1.

Thus the hydroxyl radical, taken as a unit, has a valence of −1. The double line connecting the O to the C indicates oxygen's valence of −2. A simple sugar, then, is a chain of six carbon atoms combined with hydrogen atoms, hydroxyl radicals, and an oxygen. More complex carbohydrates are formed by linking together simple sugar molecules.

A glycerine molecule is similar, but starts with a chain of three carbons and contains no oxygen:

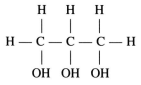

Simple fatty acids are structured like this:

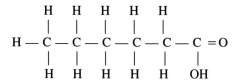

and simple amino acids are structured:

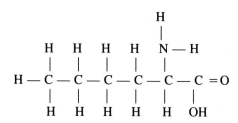

N stands for nitrogen, and the combination of N and two H is an amino group. The combination of a C, an O, and a hydroxyl is a carboxyl group. Complex fats and proteins are made of these simpler molecules linked together into long chains.

The significance of this discussion is that all life on Earth, all life that we know about, is composed of these organic molecules. Every plant, every animal that has ever been studied, from simple bacteria to complex human beings, all are made of carbohydrates, fats, and proteins. There is, of course, a limit to the length of a molecular chain that can be constructed in this way. Temperature is a factor. In the optimum environment provided by typical Earth temperatures, some organic molecules consist of hundreds of atoms. At higher temperatures, they simply fall apart. Boiling water or pasteurizing milk destroys bacteria because the heat breaks up their molecules.

Organic molecules are easily manufactured by nature. At the University of Chicago in the early 1950s, Harold Urey, a Nobel Prize winning chemist, and Stanley Miller, a graduate student, mixed together methane (natural gas), ammonia, water vapor, and hydrogen, and ran an electric spark through the chamber. After several days, a brown coating of amino acids and other organic compounds was found on the inside. The experiment is easy to perform; it is now done in undergraduate biology labs. But the conditions they created are also found commonly in natural environments. The gas mixture is thought to be similar to the original atmosphere of the Earth. Similar atmospheres are found on Jupiter, Saturn, and Titan. Electric sparks, lightning, were observed by Voyager as it passed by Jupiter. The experiment also succeeds if ultraviolet light is used instead of electric sparks. Ultraviolet is found in both sunlight and starlight.

Organic molecules were discovered in interstellar dust clouds by radio telescopes and have been found in meteorites. Although these organic molecules are not living things, they do show that organic chemistry takes place in other parts of the Galaxy as well as on Earth, and they support the idea that if life exists elsewhere, its chemistry is most likely carbon based.

The presence of organic molecules on a planet does not necessarily mean that living organisms are present. The step from molecules to a living organism is a giant one, not at all well understood, and impossible to duplicate in the laboratory. Even the simplest bacteria are marvelously complex living machines which have never been made from simpler molecules in a laboratory.

Life on Earth

Earth is certainly a beautiful place: green plants, blue oxygen-nitrogen atmosphere, colorful birds and flowers, oceans of water and ice. There is none other like it, at least not in our Solar System.

What set of circumstances allows life to flourish on Earth? Earth was not always as it is today. Its original environment was conducive to the development of primitive living things which, in turn, altered the environment of Earth. If a planet is to support life as we know it, its atmosphere must contain the elements found in organic molecules: carbon, oxygen, nitrogen, and hydrogen. Earth's original atmosphere was mostly hydrogen, methane (CH_4), sulfur oxides, ammonia (NH_3), and water vapor—gases which were expelled from the interior of the planet through volcanic vents. The essential elements were there, but they needed to be ordered and combined into the stuff of life. There was little or no free oxygen. That came much later, as a waste product from the first organisms living on Earth. In fact, free oxygen would have been lethal to early life forms.

The importance of liquid water to life as we understand it cannot be overemphasized. In a liquid, molecules are mobile and can meet and join together into cells, and cells can join together into colonies. Equally important, water absorbs ultraviolet radiation which can disintegrate organic

molecules. Earth's oceans provide protection from solar UV, allowing complex molecules to form and remain stable. The early atmosphere, without oxygen, afforded no such protection. Later when oxygen molecules were released into the atmosphere, they were split into atoms by the UV, and then recombined into three-atom-molecules of ozone. Ozone is a strong absorber of UV and provides the needed shield for living things to survive out of water on dry land.

Probably the most important circumstance conducive to nurturing life was Earth's location just the proper distance from the Sun so that its temperature was just right for liquid water to exist. Recent study indicates that the zone around a star which has a satisfactory temperature may be very narrow. A little farther from the Sun, a little colder, and Earth would have been frozen in the grips of a perpetual ice age. A little closer to the Sun, a little warmer, and the water would have remained a vapor, perhaps even escaped Earth's gravity and drifted off into space.

As discussed earlier, temperature also determines the length of a molecular chain that can be constructed from the basic sugar, fatty acid, and amino acid molecules. At Earth temperatures organic molecules may contain hundreds of atoms. At higher temperatures they disintegrate. Other planets may not have such appropriate temperatures.

The Drake Equation

Frank Drake of Cornell University has been asking questions about life in the universe since the 1950s. At one point he organized his thinking by setting the unknowns into an equation to calculate the number of advanced civilizations in our Milky Way Galaxy that would be capable of communicating with us. The equation and some estimated answers to the question are shown in MATHBOX 13.1. Let us examine the ideas Drake incorporated into his equation.

Many astronomers think that the creation of planets and moons always accompanies the birth of a star, that they all form simultaneously out of the same cloud of dust and gas. Since there are about 100 billion stars in our Galaxy, there may be that many planetary systems. It seems like living creatures should thrive in many places, but other factors interfere. About half the stars have companions, two or three or more stars in orbit around a common center of mass. Planets in such star systems would undergo extremes of temperatures as they moved around in the star group. About a quarter of the stars are very old, born when the Galaxy was just forming out of the primordial hydrogen. They could have no planets because no heavy elements existed at that time for the creation of planets. That still leaves 25 billion candidate stars to consider.

On Earth simple forms of life date back about 3 billion years. It has taken that long for life forms to evolve to their present state of intelligence. At 5.5 billion years of age, our Sun is about halfway through its lifetime. But many stars

MATHBOX 13.1

The Drake Equation

Drake's equation calculates N, the number of advanced civilizations now existing in our Galaxy which would be capable of transmitting and receiving radio signals.

$$N = R * P * E * L * I * C * T$$

R is the number of stars formed per year in the Galaxy. There are about 100 billion stars in the Galaxy. Therefore, since the Galaxy is about 10 billion years old, the average rate of production of stars during the lifetime of the Galaxy would be about 10 per year. They may not be forming at that rate now, however.

P is the fraction of stars that have planets. If all stars have planets, then P equals 1.0; if one in ten have planets, P is 0.1.

E is the number of planets per star that have suitable conditions to support life. Judging from our Solar System, the only one we know anything about, E could be about five.

L is the fraction of those planets on which life actually does develop. L equals 1.0 if life develops inevitably in any suitable environment. If life is a very fragile and unlikely occurrence, then L would be very small.

I is the fraction of those life forms that develop intelligence.

C is the fraction of intelligent species which become technically developed and discover radio communications; and

T is the time in years that a technically developed civilization manages to survive.

Values for these last three factors are even more speculative. Any numbers could be used depending on your viewpoint. There is no scientific basis or experience for making a choice.

To find how many advanced civilizations exist in the Galaxy, multiply all these factors together.

Let us, then, use the following optimistic values: $R = 10$; $P = 1$; $E = 5$; $L = 1$; $I = 0.1$ (1 in 10); $C = 1$; and $T = 5,000$ years. Multiplying these together gives $N = 25,000$. Twenty-five thousand intelligent civilizations! But if we take more conservative numbers of, say, $L = 0.1$ (1 in 10), $C = 0.2$ (1 in 5) and $T = 100$ years, then N is only 1: us.

To consider the possibility of any type of life, intelligent or not, including bacteria, algae, or other primitive forms, then we omit I and C from the equation. T becomes the time that any kind of life form manages to survive. On Earth, simple forms of life date back about 3 billion years. Using the conservative numbers from the previous example and 3 billion years for T gives an N of 15 billion places where life now exists in the Galaxy.

The value of Drake's equation lies not so much in the answer you get for N; you can get any result you want depending on your point of view. The importance of the equation is that it provides a starting point for considering the factors involved in the search for life in the universe.

---

burn at a furious rate and live only a few million years. Life may not have sufficient time to develop on their planets.

Our Solar System has nine planets and about 50 moons. Most of these objects are too cold, too hot, too small, or too something else to support living organisms. We know life exists only on one, our Earth, although other environments may be hospitable to primitive life forms: Mars, Titan, the seas of Europa, or the atmosphere of Jupiter. Some scientists believe that given a suitable environment, it is inevitable that life will develop. Others believe life to be a fragile and unlikely occurrence.

No one knows how intelligence arises. In our Solar System, only Earth has the proper environment for supporting life forms of sufficient complexity to be called intelligent. Presumably, if a species is intelligent, it will discover the natural laws of the universe and become technologically capable. Of course, a species could be intelligent and yet not be able to build radio transmitters. We consider dolphins intelligent and attempt to communicate with them, but they could not communicate across interstellar space.

How long an intelligent species can survive is the next question. Humans have survived plague, pestilence, climate changes, and meteoroid impacts. Can we survive our own technology in the form of nuclear weapons, environmental destruction, and pollution? Perhaps technical civilizations promptly destroy themselves within a hundred years or so after discovering nuclear power. But perhaps those species that develop technologically are also smart enough to figure out how to survive for millions of years.

Scientists are strongly divided on the question. Some would say that we humans are indeed the only intelligent beings in the Galaxy. Others believe that life is a natural part of the physics and chemistry of the universe and that the Galaxy must be teeming with life. If we search and find another advanced civilization, it will be perhaps the most important discovery ever made. On the other hand, if we come to the realization that we are alone, that too will be a most profound conclusion.

The search for life thus far has taken several forms: performing on-site experiments on Mars, transmitting a radio message, sending videodisks on spacecraft, and listening for intelligent radio signals.

The Mars Experiments

Influenced by Percival Lowell's observations of Mars at the beginning of this century, many people were convinced that Martians truly lived. Notions of what a Martian would look like led to many descriptions and drawings. If Mars had low gravity then its inhabitants would probably have long, spindly legs. A thin atmosphere meant large lung capacity. Sound would not propagate well in a thin atmosphere so large ears would be needed. And because the planet's dryness, a long nose would be needed to filter out the dust.

In 1976, two Viking spacecraft carried robot laboratories to the surface of Mars to search for something that could be considered alive. They landed during the northern hemisphere summer, one at 23° north latitude, the other at 48° north, on opposite sides of the planet. A mechanical arm (Figure 13.1) scooped up Martian soil (Figure 13.2) and did various tests to see if it contained living organisms.

The Mars experiments were designed on the assumption that any life forms encountered would be carbon based. Experiments done prior to the Viking missions had showed that

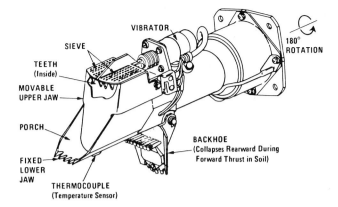

Figure 13.1 Viking scoop for sampling Martian soil. The upper jaw opens and the scoop, on the end of a mechanical arm, digs into the soil. With the upper jaw closed, the scoop is moved to the instrument, rotated upside down, and soil is sifted into the instrument's intake. A backhoe on the bottom enables the scoop to dig deeper into the soil. *Courtesy of NASA.*

Figure 13.2 Trenches dug into the Martian soil. The boom at the center holds weather-observing sensors. Trenches to the right of the boom are up to 12 inches deep. The boom's shadow can be seen just to the left. *Courtesy of NASA.*

organic chemistry could take place in Martian conditions simulated in Earth laboratories. When the experiments were carried out on the surface of Mars, reactions were observed, some of them almost violent reactions, but the results were ambiguous, inconclusive, and could be explained as simple chemical reactions.

In the *labeled release experiment* (Figure 13.3) a nutrient "soup" which contained radioactive carbon-14 atoms was mixed with Martian soil. The idea was that if living organisms were in the soil they would eat the soup and exhale or

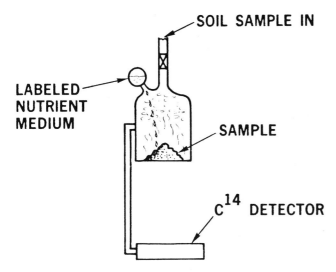

Figure 13.3   Labeled release experiment. *Courtesy of NASA.*

excrete gases made up of molecules which included the carbon-14. Sensors would then look for the radioactive carbon. A large quantity of carbon-14, more than expected, was released. The problem was that the reaction rose to a high level almost immediately, then stopped and did not start again when more soup was added. If the gases were biologically produced, they should have continued to increase as the organisms metabolized, grew, and reproduced. The observed results seemed to indicate that the reaction was chemical rather than biological. Something in the soil with a high oxygen content, perhaps a peroxide, fizzed and bubbled up when the soup was added. Whatever produced the reaction was used up at the first introduction of the moist nutrient and therefore when more was added, nothing happened. To be sure that any positive result was not simply a chemical reaction, other samples of soil were used as controls. They were first heated to temperatures up to 320 °F, then mixed with the soup, and the results were compared with the unheated soil. The heated samples showed less activity, indicating that the reaction may be biological. Heat destroys microorganisms so heated samples would show less action.

The second experiment, the *gas exchange experiment* (Figure 13.4), looked for signs of respiration. The chamber was filled with an inert gas that does not react chemically or biologically, then the soil and nutrient were introduced. If anything was exhaling carbon dioxide, nitrogen, oxygen, or other gases, it could be detected. At first, only a small amount of nutrient was injected into the chamber just to increase the humidity without saturating the soil. Carbon dioxide and oxygen were released immediately and rapidly, then ceased abruptly. Martian soil is probably full of carbon dioxide absorbed from the atmosphere and the moisture drove it out in this experiment. The production of oxygen is more difficult to understand, but was probably a chemical reaction like that seen in the labeled release experiment. Later, enough nutrient was added to saturate the soil and the mixture was allowed to incubate for nearly 7 months. Fre-

quent measurements were made, but nothing further of significance happened.

The *pyrolytic release experiment* (Figure 13.5) searched for photosynthesis, the biological activity of Earth plants. Soil was sealed in a chamber along with a simulated Martian atmosphere of carbon dioxide and carbon monoxide that contained carbon-14. Artificial light which simulated sunlight at Mars illuminated the chamber for 5 days. The gas was then removed and the soil sample baked at high temperature to break any organic molecules that formed. If the radioactive gas had been consumed, the molecule fragments released by the baking would contain carbon-14. Some was detected in the experiment, but it may have been simply absorbed by the soil rather than metabolized by living organisms.

The Viking television pictures were, of course, examined in minute detail for any sign of life. Although the desert landscape would have been a perfect backdrop for a camel or a prospector with a mule, no hint of life, large or small, was seen (unless Martians are shaped like rocks and sit very quietly).

Also, a mass spectrometer analyzed soil samples to determine what molecules were present. Absolutely no organic molecules were found at either site. The apparatus was sensitive enough to detect organic molecules in concentrations as low as a few parts per billion. It was thought that the apparatus would detect at least a few organic molecules, some from meteorites if nothing else. It is now thought that, because of the thin atmosphere, the solar ultraviolet waves reach the ground with enough intensity to destroy all exposed organic molecules.

In summary, the experiments all showed some unexpected reactions taking place, different from any observed on Earth or with lunar soil, but none could definitely be associated with biological activity. This does not say that there is nothing living on Mars; it only says that there was no firm indication in the results at these two locations. There may be more favorable environments elsewhere, perhaps in deep

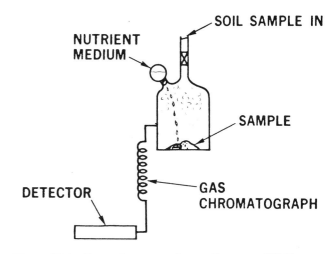

Figure 13.4   Gas exchange experiment. *Courtesy of NASA.*

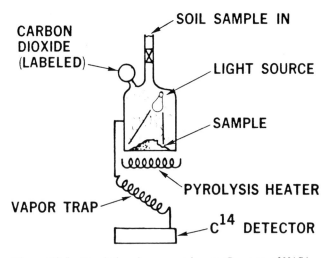

CARBON
DIOXIDE
(LABELED)

SOIL SAMPLE IN

LIGHT SOURCE

SAMPLE

PYROLYSIS HEATER

VAPOR TRAP

C^{14} DETECTOR

Figure 13.5 Pyrolytic release experiment. *Courtesy of NASA.*

canyons, at the polar caps, or deep in the soil. Scientists designed the experiments based on what they knew about life on Earth. Perhaps there are living organisms on Mars, but they just didn't like the soup we sent. Their metabolism may be quite different from Earth organisms. Possibly there was once life on Mars, but it became extinct as the atmosphere dwindled, water dried up, and climate changed. Our first crew to land on Mars should include a paleontologist to search for fossils!

We must take care in sending our spacecraft to other planets that we do not send Earth life with them, not even microorganisms. Not only could they contaminate the spacecraft's experiments or future experiments, but they could conceivably grow, spread, and compete with the planet's indigenous life. Component parts of the Viking spacecraft were cleaned and sterilized and were assembled in a clean room. The entire spacecraft then underwent an elaborate and expensive heat-sterilization process. These procedures added about 10 percent to the cost of the mission.

Search for Extraterrestrial Intelligence (SETI)

To look for living organisms in general, we must either go there and look, or we must infer their presence from their biological activity or their organic chemistry. But to search for intelligent creatures, we have alternate methods. An advanced civilization will almost certainly discover the electromagnetic spectrum and the ways to produce electromagnetic waves: light, radio, infrared, and so on. In doing so, they give themselves away. The waves, travelling at the speed of light, spread throughout the Galaxy to be intercepted by anyone with a sensitive enough detector and the patience to search carefully and thoroughly.

Earthlings have been generating radio waves since the early 1900s, radar pulses since World War II, and television since the 1940s. The earliest radio broadcasts were weak and were of such a low frequency that they could not penetrate the ionosphere into space. However, for at least 60 years broadcasts from Earth have been travelling through space at the speed of light. The first signals to penetrate the ionosphere are now 60 light years from Earth and have passed at least 3,000 stars. Has anyone been listening out there? Does anyone know we are here? Perhaps someone on a planet 60 light years from us has just received one of those old programs and is sending a response in our direction. Sixty years from now it will reach us. Two-way communications across interstellar distances is a very slow process indeed.

But perhaps there are civilizations that have been transmitting radio waves for a longer time. If we do receive a transmission from a planet, say, a thousand light years away, they certainly will not be as backward as we are. Their message would have originated while we were still in the Dark Ages. In those intervening thousand years they may have progressed to something beyond radio or television, something we have not yet discovered and cannot imagine. Maybe they are no longer listening.

It would seem easy enough to conduct a search of the heavens for intelligent signals, but it is not as simple as it first appears. Where do we point our antenna? Where do we tune the dial? What kind of signal do we look for? The electromagnetic spectrum is a gigantic place. For example, the FM radio band (Figure 13.6) covers 20 megahertz, from 88 to 108 megahertz on the dial. (A *hertz* is one oscillation per second, and *mega* means million.) There are 20 million individual frequencies in the FM band alone. Each television channel has a bandwidth of 6 million hertz and there are 82 such channels in the television bands. In addition there are the police, fire, and emergency bands, the satellite communications bands, and the microwave bands. There are a total of 100 billion frequencies in the entire radio band. So, where do we begin?

We certainly do not want to start in the FM radio band. It would be difficult to pick up an extraterrestrial through the rock music. The television bands are out, too, for a similar reason. By international agreement, a number of fre-

Figure 13.6 FM radio band from 88 to 108 megahertz. The AM radio band is between 540 and 1600 kilohertz.

quency bands have been set aside as quiet bands in which no human transmissions are made. They are used primarily by radio astronomers for scientific research. Their selection has been somewhat arbitrary, but one in particular is at the natural resonant frequency of the hydrogen atom, near 1,420 megahertz, a frequency used by radio telescopes to locate clouds of hydrogen gas in interstellar space. If intelligent extraterrestrials think like human scientists, they may think of the hydrogen resonance frequency as an interstellar hailing frequency. Radio astronomers everywhere would have sensitive equipment tuned to that band. If they wanted to make contact with other radio astronomers, it would be logical to transmit a signal on or near that frequency. But, perhaps those far superior extraterrestrials, if they exist, tried that thousands of years ago and gave up. Nonetheless, the hydrogen frequency appears to be a good place to start a search.

In fact, the entire microwave band from 1,000 megahertz to 10,000 megahertz is a good place to begin searching. In that frequency range, cosmic background noise is low and Earth's atmosphere is most transparent. But, there are 9,000 million frequencies to examine in that band! We have narrowed the problem from 100 billion to 9 billion, still a sizeable problem.

Where do we look? There are at least 100 billion stars in our galaxy. How do we choose? Young stars may be a poor choice because it takes millions, perhaps billions of years for advanced life forms to evolve. Old stars from the origin of the universe would also be a poor choice because in the beginning the universe consisted of only hydrogen and helium. There were no heavy elements such as oxygen, nitrogen, iron, silicon, and all the others from which a planet could form. Best choice seems to be sun-like stars at the middle or toward the end of their life cycle. Even so, we cannot be sure which stars have planets. Even with careful observations, no planets around other stars have ever been confirmed.

What kind of signal do we look for? In Chapter 6, we noted that electromagnetic waves are emitted by everything at a temperature above absolute zero. The universe is filled with such waves from natural sources, but they are random and without pattern, like the "sizzling" noise you hear as you tune the FM radio between stations. If an extraterrestrial intelligent being produces electromagnetic waves, they will contain a pattern, a recognizable message. Like searching for a needle in a haystack, you will know when you find it.

Two types of signals would stand out against the cosmic background noise: radio, television, and radar type transmissions which unintentionally spill out into the universe, and transmissions specially designed for interstellar contacts. Signals would be recognizable initially from their signal strength, then by their modulation, the rhythmic shifting of frequency or changing strength. Decoding and finding the message in the signal may be time consuming, but there would be no doubt that it was from an intelligent source.

The first SETI project, Ozma, was by Frank Drake in 1960 using an 85-foot dish antenna at Green Bank, West Virginia. Drake spent 400 hours looking at two nearby sun-type stars in a 400,000 hertz band around 1,420 megahertz. Receiving equipment was not automated, so it was a tedious task to scan the possible frequencies.

Since then there have been over 30 attempts, some for only a few hours and some continuing year after year. The former Soviet Union and a half dozen other countries have made serious efforts.

The most comprehensive SETI survey was begun in 1986 by Paul Horowitz of Harvard University. He had built a computerized receiver-signal processor which listens to 128,000 frequencies simultaneously and tested it at Arecibo in 1982, probing 250 stars in only 75 hours. The equipment was so successful that Peterson and Chen of Stanford University designed a signal analyzer that can handle 8.4 million channels at the same time. The search continues using the new analyzer and an 84-foot dish antenna in Massachusetts. The antenna is set at a specified height above the horizon and as the Earth rotates it scans a half-degree wide band around the sky. Next day the antenna is shifted upward slightly and allowed to scan another overlapping half-degree band in the next 24 hours. In 7 months the northern hemisphere sky is scanned from horizon to pole.

Even with automated equipment and computerized analyzers the SETI investigators require years to scan the entire sky at all the desired frequencies. In the first 18 months the equipment processed 15,000 billion numbers. It automatically discarded all but about a million per month which needed further examination. These fell into four categories: noise, earthly radio interference, equipment malfunction, and others. Only three signals looked promising, but on searching that area of the sky many times over, the signals were not seen again.

Horowitz's search is sponsored by The Planetary Society and funded with private donations, including $100,000 from Stephen Spielberg. NASA also started a SETI program in October 1992, which would use a 100-foot antenna to scan the entire sky at a number of wavelengths and to study individual sun-like stars. The project has often been ridiculed by the media and by opponents as a silly waste of money to search for ET. Congressional funding for planning and engineering SETI equipment had been intermittent for many years and was finally terminated in September 1993 after less than a year of operation. Several million dollars in private funding has managed to save that part of the program that studies individual stars. So the search goes on.

What Do You Say to an Extraterrestrial?

During the last century people had no doubt that other celestial bodies were inhabited, that the Moon and Mars were populated by intelligent beings. Radio was unknown as a means of communications, but several scholars proposed other means of letting the extraterrestrials know we are here. One suggested planting a forest of pine trees in Siberia in the shape of a right triangle large enough to be visible from the Moon. Another suggested draping a black cloth over a large white surface of Earth and moving it back and forth. From Mars it would seem to blink. Yet another said to dig a 20 mile wide trench in some geometric shape across a desert, fill it with kerosene, and ignite it.

Now radio is the obvious means of communication. But what should we say? They would not understand any Earth language, may not have ten fingers on which to base a decimal counting system, and probably have a completely different biology. But, an advanced civilization would have one language in common with us. Mathematics and the sciences of chemistry, physics, and astronomy are universal.

A message intended to initiate communications with another world would be designed specifically so it could be easily decoded. A simple message might consist of the first ten *prime numbers*, 1, 2, 3, 5, 7, 11, 13, 17, 19, 23, transmitted in sequence using on-off pulses: one pulse, pause, two pulses, pause, three pulses, pause, etc., to the end, long pause, begin again. Such a sequence would not occur randomly in a natural radio noise source, and would be immediately recognized by an extraterrestrial mathematician. It could serve as a beacon, in effect saying, "We are here."

A much more complex message was transmitted from the radio telescope antenna at Arecibo, Puerto Rico, in November 1974. The antenna is hung over a bowl-shaped valley in the hill country, far from man-made noise sources. Shown in Figure 12.6, it has an area of 20 acres, more than all other radio telescopes put together. The message consisted of a series of on-off pulses, 1,679 of them, displayed in Figure 13.7, which, when arranged in the proper pattern, produce a picture. A smart extraterrestrial would know how to arrange them by noticing that 1,679 is the product of two

```
0000001010101000000000000101000001010000000 1
0010001000100010011011001010101010101010010
0100000000000000000000000000000000000011000
0000000000000000011010000000000000000011010
0000000000000000101010000000000000000001111
1000000000000000000000000000000001100001 1100
0110000110001000000000000110010000110100011
0000110000110101111011111101111101111000000 00
0000000000000000001000000000000000001000000
0000000000000000000100000000000000001111
1100000000000001111100000000000000000000 1
1000001100001110010000100000001000000000
0110100001100011100110101111101111101111011
1110000000000000000000000000010000001100000 0
0001000000000000110000000000000010000011000 0
0000001111110000011000000111110000000001100
0000000000100000000100000000100000100000011
0000000100000001100001100000001000000001100
0100001100000000000001001100000000000011
0001000011000000000110000110000010000000100
0000100000000100000100000001100000001000100
0000001100000000010001000000000100000001000 00
1000000010000000100000001000000000011000 00
0000110000000011000000000100011101011000000 0
0000100000000000000000000010000011111000000
0000001000010111010010110110000000100111000 0
1111111011100001110000011011100000000001010 00
0011101100100000010100000111110010000001010
0000110000001000011011000000000000000000
0000000000000011100000100000000000001110101
0001010101010100111000000001010101000000000
0000000101000000000000001111100000000000000
0111111110000000000001110000000111000000000
1100000000000110000000110100000010110000000
1100110000000110011000010001010000010100100
0010001001000100100010000000100010100010000
0000000010000100001000000000010000000000100
0000000000001001010000000000011110011111010 0
1111000
```

Figure 13.7 The Arecibo message.

prime numbers, 23 times 73. If they are arranged in 23 rows of 73 pulses each, they do not produce anything of obvious intelligence (Figure 13.8), but ordering them into 73 rows of 23 pulses each creates the pictogram as shown in Figure 13.9. (We have used the symbol * for ones and blanks for

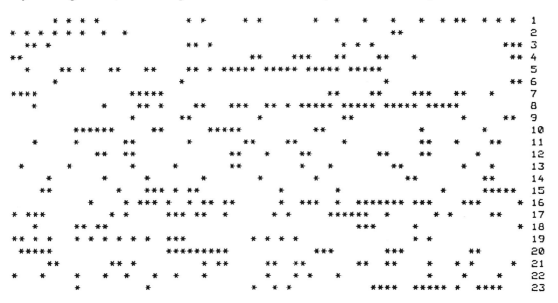

Figure 13.8 The Arecibo message incorrectly decoded.

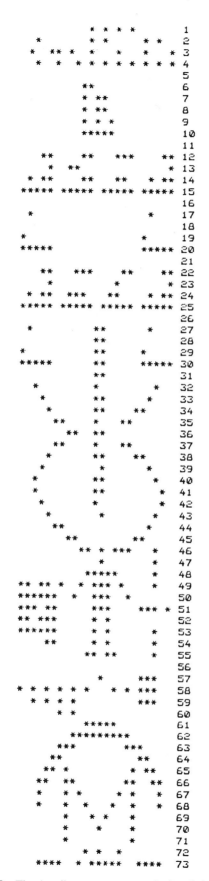

Figure 13.9 The Arecibo message correctly decoded.

zeros so the pattern shows up more clearly.) Notice the human figure at lines 46 to 55. The top of the pictogram, lines 1 to 4 reading from right to left, starts with the binary numbers from one to ten, followed by the atomic numbers for the elements of life and formulas for the organic molecules of life, all using binary numbers. At the bottom is the shape and size of the Arecibo radio telescope. In between is a diagram of the Solar System and the shape of a DNA helix.

This message was transmitted toward Ml3, a great globular cluster of about half a million stars in the constellation Hercules some 24,000 light years away. If beings are intelligent enough to pick up the message, they may be intelligent enough to decode it. If they send an answer back, travel time will be another 24,000 years, to be received by our descendents in the year 49,974! Who can tell where the human species will be at that time.

The Arecibo transmission was not so much an attempt to contact another civilization as it was an exploration of the possibilities in interstellar communication, getting scientists to think about what is involved and how to go about it.

Pioneer 10 and Pioneer 11, leaving the Solar System, carry metal plates with an engraved message from Earth (Figure 13.10). Each plate shows a human couple in front of an outline of the spacecraft, drawn to scale to show the size of humans. To the left is a diagram locating the Sun with respect to 14 pulsars and at the bottom is a sketch of the Solar System showing the path of Pioneer.

The Voyager spacecraft carry recordings of pictures, voices, and a diagram message from Earth to any creature intelligent enough to intercept it and translate it. The recording is something like a videodisc which plays music (Bach,

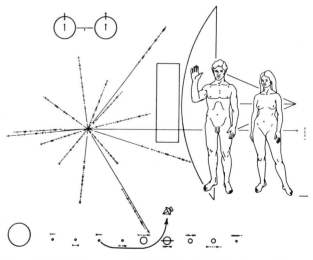

Figure 13.10 Plaque carried out of the Solar System on Pioneer 10 and 11 spacecraft. Short vertical lines and dashes are binary numbers. An extraterrestrial smart enough to catch the spacecraft should be smart enough to understand the numbers. *Courtesy of NASA.*

Beethoven, Chuck Berry, primitive drums), sounds of Earth (baby crying, singing, waves), pictures of Earth, and for international acceptance of the project, the names of leaders of all countries of the world. Instructions for playing the record are on the cover (Figure 13.11) along with diagrams similar to those on Pioneer. If extraterrestrial space travelers catch Voyager and figure out how to make it run, their first response back to Earth will probably be, "Send more Bach!"

The final question: if we do get a message from a highly advanced extraterrestrial civilization, should we respond? If they are like us, but further advanced with the capability for interstellar travel, maybe it would be unwise to let them know of our presence. Human beings have spread over the entire Earth, the stronger oppressing the weaker as they migrated. Would the extraterrestrials have the same goal? They may already know we are here, but do not want to have anything to do with us. Or, if their population is growing exponentially, if they seek greener pastures, if their motivation is simply to conquer, then perhaps we should keep to ourselves. More likely, if their society has survived technological advancement without self-destruction, then they have probably learned to be peaceful and benevolent. We would learn a great deal from such a society.

DISCUSSION QUESTIONS

1. What effect would the discovery of intelligent life elsewhere in the universe have on human society?

Figure 13.11 Record attached to outside of Voyager spacecraft. It is protected by a cover carrying instructions on how to play it along with the pulsar map from Pioneer 10 and 11. In addition, a small sample of pure uranium-238 is attached; an extraterrestrial will be able to determine approximately when the spacecraft was launched using radiometric dating techniques. *Courtesy of NASA.*

2. If intelligent beings were found on some other planet, in what ways would they be like humans? In what way would they be different?

3. If we find anything living on Mars, should we terraform the planet, even if it means death to the native organisms? What if the native life is only bacteria-like?

4. Make up a message to send to extraterrestrials different than the Pioneer, Voyager, and Arecibo messages. Be sure it is decipherable by nonhumans.

5. Watch a TV program with the sound turned down as an extraterrestrial might see it and try to interpret it as an extraterrestrial might.

6. Why is it true that when we look out into space we look back in time?

ADDITIONAL READING

Bioastronomy News. "Bioastronomy: The Search for Extraterrestrial Life." A quarterly newsletter published by The Planetary Society.

Feinberg, G., and R. Shapiro. *Life Beyond Earth.* Morrow and Co., 1980.

Harwit, Martin. "Listening Out." *Air and Space, Smithsonian.* June-July 1994. Editorial on SETI.

Heidmann, Jean. "Pulsar-aided SETI Within a Hundred Light Years." *Acta Astronautica*, February 1992. Technical paper suggests using pulsar frequencies to determine search frequencies.

Horowitz, Norman. "The Search for Life on Mars." *Scientific American*, November 1977. Viking experiments and results.

Huang, Su-Shu, "Life Outside the Solar System." *Scientific American*, April 1960. Fundamentals.

NASA. *SETI, Search for Extraterrestrial Intelligence.* NASA NP-114. U.S. Government Printing Office, 1990. Easy to read pamphlet.

Planetary Society. *The Planetary Report, Life in the Universe.* November/December 1987. Bimonthly magazine.

Ponnamperuma, Cyril, and A. G. W. Cameron. *Interstellar Communication: Scientific Perspectives.* Houghton Mifflin Co., 1974. Complete, technical, readable.

Poynter, Margaret, and Michael J. Klein. *Cosmic Quest: Searching for Intelligent Life Among the Stars.* Athenium, 1984. Simple, easy reading.

Sagan, Carl, and Frank Drake. "The Search for Extraterrestrial Intelligence." *Scientific American*, May 1975. Readable summary of SETI to 1975.

Sagan, Carl, et al. *Murmurs of Earth, The Voyager Interstellar Record.* Random House, 1978. Details of images and music on Voyager.

Scientific American. "Life in the Universe," October 1994. A special issue.

NOTES

Glossary

ABM Antiballistic missile, as in ABM weapon or ABM Treaty.

acceleration A change in velocity; a change in either the speed or direction of a moving object.

aerobraking A maneuver in which a spacecraft accomplishes an orbital change by dropping into the upper atmosphere of a planet to reduce its energy instead of firing retrorockets.

airlock An airtight chamber used to move between spaces of different pressure, from the orbiter to the cargo bay, for example.

alpha particle The nucleus of a helium atom consisting of two protons and two neutrons.

amino acid One of the fundamental organic molecules which includes groups containing two atoms of hydrogen bonded to one atom of nitrogen; the structural unit of all proteins.

aphelion The point which is farthest from the Sun on an elliptical orbit around the Sun.

apogee The point which is farthest from Earth on an elliptical orbit around Earth.

apogee kick Firing rockets at apogee to decrease the eccentricity of the orbit, to make it more nearly circular.

apsides The two extreme ends of the major axis of an elliptical orbit; see also **line of apsides**.

asteroid A piece of rock, metal, or frozen gases in space, larger than a meteoroid but smaller than a planet.

atom The smallest particle of an element, composed of electrons, protons, and neutrons.

atrophy Waste away because of disuse.

attitude The orientation or position of a vehicle.

ballistic Referring to the path followed by a vehicle which is affected only by the forces of gravity and air friction.

bends See **decompression sickness**.

booster rocket A rocket used to provide large initial thrust for launch, usually dropped off after it has served its purpose.

burn Rocket engine firing.

cardiovascular deconditioning Weakening of the heart and blood vessels.

centripetal force The force required to keep an object moving in a curved path; it is directed toward the center of the curve.

chromosphere The lower atmosphere of the Sun.

circadian rhythm A natural periodicity in human activity or bodily function, such as the 24 hour sleep-wake cycle.

cislunar space Referring to the space between Earth and the Moon.

clevis A fitting with a U-shaped end.

combustion The chemical process in which a fuel oxidizes, that is, combines with oxygen, usually accompanied by heat and fire.

comet A ball of dust and frozen gas travelling in an orbit of high eccentricity arround the Sun; as it approaches the Sun, solar heat and the solar wind cause particles to evaporate, break off, and trail into space producing a visible tail.

convection The process by which heat is transferred in a fluid by the bulk motion of warmer fluid of lesser density rising in cooler surroundings of greater density.

core Referring to the Sun, the central region in which nuclear fusion is taking place.

corona The outermost part of the solar atmosphere which extends out into the Solar System past the Earth.

coronagraph A telescope designed for viewing the Sun's corona.

cosmic ray Nucleus of an atom travelling at very high speed through space. Protons coming from the Sun are called **solar cosmic rays** while heavier nuclei coming from outside the Solar System are called **galactic cosmic rays**.

cryogenic Involving very low temperatures.

decompression sickness Illness produced by a too rapid decrease in air pressure surrounding a body, causing bubbles of nitrogen in the blood to expand painfully.

dehydrate Remove the water from a substance, particularly for the preservation of food.

deorbit Fire a rocket engine to reduce the energy of a vehicle so it leaves orbit and returns to Earth.

directed energy weapon An antimissile weapon which uses an intense beam of electromagnetic energy, such

253

as a laser beam, or of small particles, such as hydrogen atoms, to burn through the shell of a missile or damage its electronics.

diurnal   Referring to a daily occurrence.

DNA   The organic molecule which is part of all living cells, containing the genetic "instructions" for the cell to reproduce.

docking   Joining together two orbiting vehicles.

drag   Frictional force which decelerates a vehicle, particularly atmospheric friction.

dysbarism   Decompression sickness. The bends.

eccentricity   A number which tells the elongation of an ellipse.

ecliptic   The plane of the Earth's orbit extended to the celestial sphere. It is also the annual path of the Sun through the stars. The orbits of all the planets except Pluto lie within a few degrees of the ecliptic.

electromagnetic spectrum   The range of wavelengths of electromagnetic waves including, in order of decreasing wavelength: radio, microwaves, infrared, light, ultraviolet, x-rays, and gamma rays.

electromagnetic waves   The means of transmitting energy at the speed of light, produced by the oscillation or acceleration of electrically charged particles.

electron   A small particle of matter carrying a negative electrical charge, one of the three constituent particles of an atom.

electrophoresis   A process by which different materials suspended in a fluid are separated from one another by applying an electrical field.

element   A basic substance consisting of only one kind of atom.

ELINT   Electronic intelligence gathering.

ellipse   A closed oval-shaped curve.

end effector   Robotics term for a grasping device.

ephemeris   A set of numbers which specify the location of a celestial body or satellite in space.

ergonomics   The study of human factors in space operations.

escape velocity   The velocity an object must have to escape the gravitational attraction of a planet (or other celestial body) and neither fall back nor enter an orbit around it.

ET   The Space Shuttle's external tank.

EVA   Extravehicular activity, working outside a spacecraft.

external tank   Referring to the Space Shuttle, the propellant tank which carries liquid hydrogen and liquid oxygen for the orbiter main engines.

filaments   Referring to the Sun, dark streaks on the Sun which are actually **prominences**, masses of hot gas suspended in the atmosphere, which appear dark against the bright disk of the Sun.

flare   A sudden, explosive emission of electromagnetic radiation in the solar chromosphere, sometimes accompanied by the ejection of high speed protons, called solar cosmic rays.

force   A push or a pull which accelerates an object.

fossil fuels   Coal, oil, and natural gas, the fuels formed from organisms which lived millions of years ago and became buried in sediments.

fuel   A substance which burns in the presence of oxygen.

fuel cell   A device which produces electricity by combining hydrogen and oxygen into water.

fusion   A nuclear process by which smaller atomic nuclei fuse together to form larger nuclei, such as the fusion of hydrogen into helium.

g   Symbol for the acceleration of gravity at the surface of the Earth.

galaxy   A basic structure of the universe consisting of billions of stars, dust, and gas, held together by their mutual gravitational attraction, usually in a spherical, elliptical, or spiral shape.

gamma rays   Electromagnetic waves with wavelength shorter than x-rays.

geostationary orbit   See **geosynchronous orbit**.

geosynchronous orbit   An orbit approximately 22,300 miles above the equator with a period of one day, the same as the rotational period of the Earth; a satellite in geosynchronous orbit appears to remain stationary in the sky as viewed from the Earth.

grain   The segment of solid propellant enclosed in a rocket engine.

grapple   A fixture attached to a satellite or other payload so the Space Shuttle remote manipulator arm can grasp it.

gravitation   The natural force of attraction between any two objects.

Hohmann transfer   Transfer orbit which requires the least amount of energy.

hydrogen-alpha (H-alpha)   A particular wavelength of red light produced by glowing hydrogen gas; solar flares are readily seen by their H-alpha light.

hyperbola   An open curve, not closed as an ellipse or circle.

hypergolic   Referring to the spontaneous ignition of a fuel and oxidizer when they come into contact.

hypoxia   Illness produced by lack of oxygen.

ICBM   Intercontinental ballistic missile.

inclination   Referring to an orbit, the angle between the orbital plane and the plane of the Earth's equator measured at the point where a spacecraft crosses from the southern hemisphere to the northern hemisphere.

inertial upper stage (IUS)   A rocket booster attached to a satellite carried in the cargo bay of the Space Shuttle orbiter to boost the satellite to a higher orbit. See also **payload assist module**.

infrared   Electromagnetic waves with wavelengths in the range 1 to 1,000 micrometers, longer than light but shorter than radio waves.

infrastructure   The basic facilities, installations, and equipment needed for an organization or a system to operate.

inhibitor   A coating over the areas of a solid propellant grain which are not supposed to burn.

injector   A mechanical device which forces or sprays fuel and oxidizer into the rocket engine combustion chamber.

intelligence   Referring to the military, gathering information about a potential adversary.

ion   An atom or combination of atoms which carries an electrical charge due to the addition or removal of electrons.

ionize   To produce an ion by adding or removing electrons from a neutral atom or combination of atoms.

ionosphere   Region of a planet's atmosphere in which the atoms are ionized by solar x-ray and ultraviolet radiation.

irradiate   Expose to radiation, particularly to destroy the bacteria in food to preserve it.

IUS   See **inertial upper stage**.

kinetic energy   The energy of a moving object.

kinetic energy weapon   An antimissile weapon which fires a projectile at the missile to destroy it by collision.

Lagrangian points   See **libration points**.

laser   A device which produces an intense beam of electromagnetic waves of a single wavelength, most commonly light waves of a single color.

LEM   See **LM**.

LEO   Low Earth orbit.

libration points   A location in space where the gravitational attraction of two or more celestial bodies add to zero, that is, a neutral point in the gravitational field; also called "Lagrangian point," after the mathematician who first investigated the phenomenon.

light   Electromagnetic waves to which the human eye is sensitive, with wavelengths of approximately 0.5 micrometer.

light year   A measure of interstellar distance; the distance light travels in a year, about 6 trillion miles.

lignin   A substance produced by plants which helps give them their structural strength so they stand upright.

limb   Referring to the Sun, the edge of the disk.

line of apsides   Major axis of an elliptical orbit.

LM   Lunar module, the part of the Apollo spacecraft that landed on the Moon.

LOX   Liquid oxygen.

Mach 1   The speed of sound. Similarly, **Mach 2** is twice the speed of sound, and so on.

magnetosphere   Region around a planet in which the planet's magnetic field is sufficiently strong to prevent or impede the solar wind particles from entering.

manned maneuvering unit   Device equipped with thrusters which an astronaut in a space suit can attach to and fly freely, untethered away from the orbiter.

mass   Most simply, the amount of matter contained in an object. Mass varies with velocity, increasing significantly as it approaches the speed of light. More precisely, a measure of a body's resistance to acceleration.

mass driver   An electrically powered device which accelerates projectiles to very high velocities.

MECO   Main engine cutoff, referring to the Space Shuttle.

metabolism   The chemical processes which take place in an organism to maintain life.

meteor   Bright streak of light in the night sky produced by a meteoroid falling toward Earth and burning up from the heat of friction with the atmosphere.

meteorite   A meteoroid which has survived the fall through the atmosphere and lands on the surface of the Earth.

meteoroid   A piece of rock, metal, or frozen gas in space, varying in size from microscopic to asteroid size.

meter   Metric unit of length equal to about a yard, 39.37 inches to be precise.

micrometeoroid   A microscopically small meteoroid.

micrometer   One-millionth of a meter, also called a micron.

microwaves   Electromagnetic waves with wavelengths in the range 1 millimeter to 30 centimeters (1/25 inch to 1 foot).

missile   An object that is fired or thrown at a target; a projectile.

MMU   See **manned maneuvering unit**.

module   A self-contained, standardized component of a spacecraft, space station, or habitat, designed for ease of assembly with other components.

muscular deconditioning   Weakening of the muscle tissue due to lack of use while weightless.

NASA   National Aeronautics and Space Administration.

neutron   A small particle of matter with no electrical charge, one of the three constituent particles of an atom.

NORAD   North American Aerospace Defense Command.

nozzle   The exhaust duct of a rocket combustion chamber through which the combustion gases are accelerated to higher velocity.

OMS   See **orbital maneuvering system**.

OMV   See **orbital maneuvering vehicle**.

orbit   Closed path followed by a satellite.

orbital injection   Firing rocket engines to transfer a spacecraft from a ballistic trajectory into a closed orbit.

orbital maneuvering system   A pair of rocket engines located in pods next to the tail of the Space Shuttle orbiter, used for orbital insertion, changing orbit, and deorbiting.

orbital maneuvering vehicle   A remotely operated spacecraft which delivers satellites from a space station to higher orbits and returns them for repair and maintenance.

orbital mechanics   The science and mathematics of orbits.

orbiter   The Space Shuttle's rocket-powered aircraft and spacecraft.

organic   Referring to a substance containing carbon as the basic atom of its molecular structure.

O-ring   A rubberlike seal between segments of a solid rocket engine.

oxidation   The chemical process by which a substance is combined with oxygen. May also refer to other chemical processes which involve the loss of electrons, but do not necessarily involve oxygen, such as the reaction between hydrogen and fluorine.

oxidizer   A substance which provides oxygen or other substance to support the combustion of a fuel. An oxidizer does not necessarily involve oxygen. See **oxidation**.

paleontologist   A scientist who specializes in the study of fossils and ancient life forms.

PAM   See **payload assist module**.

payload assist module (PAM)   A solid rocket booster attached to a payload in the Shuttle orbiter's cargo bay for the purpose of boosting the payload to a higher orbit. See also **inertial upper stage**.

perihelion   The point which is closest to the Sun on an elliptical orbit around the Sun.

perigee   The point which is closest to Earth on an elliptical orbit around Earth.

perigee kick   Firing rockets at perigee to increase the eccentricity of the orbit.

period   Referring to an orbit, the time required for a spacecraft to make one complete orbit as viewed from space.

photochemical process   A chemical reaction that is caused by electromagnetic radiation, particularly light and ultraviolet.

photosphere   The visible surface of the Sun.

photosynthesis   The process by which plants convert carbon dioxide from the air, light from the Sun, and water from the Earth into carbohydrates with oxygen as a by-product.

photovoltaic cell   A semiconductor device which converts light directly into electricity; called a solar cell when used in sunlight.

phytoplankton   Tiny algae plants which float in the ocean; the basic organism of the oceanic food chain.

pitch   Up and down rotational motion of the nose and tail of an aircraft or spacecraft. See also **roll** and **yaw**.

pixel   Contraction of "picture element," one bit of an image.

plage   On the Sun, a bright region of hot gas in the chromosphere.

plasma   A gas composed of a mixture of neutral and ionized atoms.

polarimeter   An instrument for measuring the polarization of light waves.

prime number   A number which is evenly divisible only by itself and one.

progressive burn   The burn of a solid propellant grain in which the thrust increases with time.

prominence   Referring to the Sun, a cloud of hot gas in the solar atmosphere appearing bright when seen on the edge of the disk against the dark background of space, and appearing as a dark filament when seen against the bright surface of the Sun.

propellants   The fuels and oxidizers burned in rocket engines to produce thrust.

proton   A small particle of matter carrying a positive electrical charge, one of the three constituent particles of an atom.

proton event   Referring to the Sun, an outburst of high energy protons during a flare which reach the vicinity of the Earth.

rad   A unit of absorption of radiation in a living organism; 600 rads in a day would be a lethal dose for most humans.

radar   A device which transmits a microwave pulse, receives the returned pulse which is reflected from an object, and calculates the distance and direction to the object.

radiation   The transfer of energy from place to place by means of electromagnetic waves. Also, a somewhat ambiguous term applied to high energy waves and particles coming from the Sun, interstellar space, or radioactive substances.

radiation sickness   Illness caused by exposure to large doses of radiation, in severe cases involving nausea, diarrhea, vomiting, dehydration, destruction of blood cells, and perhaps death.

ramjet   A reaction engine in which only the air entering the front intake at high speed prevents exhaust gases from leaving in that direction.

RCS   The Space Shuttle orbiter's reaction control system, 44 thrusters which control its orientation in space.

reaction engine   An engine that produces thrust by the directed expulsion of mass, usually hot gases.

reconnaissance   Gathering images and other data about an area of interest, usually applied to military intelligence.

reentry   The return of a spacecraft into the Earth's atmosphere.

refraction   Bending of electromagnetic waves from a straight line path as they pass from one medium into another.

regolith   Broken and powdered rock on the surface of a planet. On the Moon, regolith is produced by constant bombardment of the surface by meteoroids, large and small.

regression   In orbital mechanics, the rotation of the major axis of an elliptical orbit due to variations of gravity in different parts of the parent body.

regressive burn   The burn of a solid propellent grain during which the thrust decreases with time.

rehydrate   Add water to a dehydrated food to prepare it for eating.

remote sensing   Learning something about an object at a distance from the electromagnetic waves emitted and reflected from it.

rendezvous   Come together at a specified time and place.

resolution Referring to remote sensing, the size of the smallest object that can be distinguished from its surroundings.

retrofire To fire a rocket in the direction of motion in order to reduce the energy of the orbit.

retrorocket A rocket that fires in the direction of motion.

rocket A reaction engine that carries both fuel and oxidizer so it can operate in the absence of air.

roll Rotation of an aircraft or spacecraft around the axis through the nose and tail. See also **pitch** and **yaw**.

satellite An object in orbit around a larger object. In space technology, a man-made object in orbit around Earth, usually one which has some useful purpose. Also, moons are satellites of planets; planets are satellites of the Sun.

scramjet Supersonic ram jet.

SDI Strategic Defense Initiative, the U.S. program for development of a defense against ballistic missiles.

sensor A device which responds to electromagnetic waves impinging on it, usually by producing or modifying an electrical current.

SETI Search for extraterrestrial intelligence.

signature Referring to remote sensing, the characteristic intensities of various wavelengths of electromagnetic radiation coming from an object which uniquely identify that object.

sinter Compress and fuse together by heat.

solar cell See **photovoltaic cell**.

solar wind A steady flow of particles, mostly protons and electrons, from the Sun into interplanetary space, an extension of the solar atmosphere.

solid rocket booster A booster rocket using solid propellants.

space adaptation syndrome Illness similar to seasickness, experienced by many astronauts during the first day or two of weightlessness.

spacecraft Any vehicle designed to operate in space. An unmanned spacecraft in a closed orbit around Earth is usually called a satellite.

Space Shuttle A four part, mostly reuseable vehicle consisting of an orbiter-spacecraft-aircraft, two solid rocket boosters, and an external fuel tank.

space station A habitat-workshop-laboratory permanently in orbit, with crews and supplies brought from Earth as needed.

specific impulse A number which indicates the effectiveness of a fuel-oxidizer combination; the time it takes to burn one pound of fuel while it is producing one pound of thrust.

spectral bands A range of wavelengths which a particular sensor responds to or which has some particular meaning for interpreting the image.

spectrum See **electromagnetic spectrum**.

SRB The Space Shuttle's solid rocket booster.

sunspot A "cool," dark region in the photosphere of the Sun where magnetic fields retard the flow of heat from the interior to the surface.

Sun-synchronous An orbit around Earth which crosses the equator at precisely the same local time on each pass.

tang The part of a fitting that fits into a clevis.

teleoperate Operate equipment or instruments by remote control using a television monitor and radio link.

terminator On a planet, the line which divides daylight from darkness.

terraform Modify the environment of a planet to make it more Earth-like and habitable by humans.

tetrahedron A solid object with four equilateral triangles for sides.

thermostabilize Cook food at temperatures which preserve it by destroying the bacteria.

thrust Force produced by a rocket engine.

thruster A small rocket engine or gas jet used to control the attitude and, to a lesser extent, the orbital speed of a spacecraft.

trajectory Path followed by a vehicle.

transfer orbit Segment of an orbit which a spacecraft follows to move from one orbit to another.

troposphere The lowest six miles, approximately, of the Earth's atmosphere, the region in which all weather occurs.

turbojet A reaction engine in which air is taken in at the front and compressed, fuel is mixed and burned, and the exhaust gases are ejected out the back.

ultraviolet Electromagnetic waves with wavelengths shorter than visible light but longer than x-rays.

valence An integer number which tells how one element combines chemically with another; refers to the number of outermost electrons in an atom.

Van Allen belts Regions in the Earth's magnetosphere where solar electrons and protons become trapped by the magnetic field.

velocity The motion of an object described by its speed and direction.

vernier engine A small rocket or thruster used to make fine adjustments in speed or attitude.

vestibular apparatus Organ in the inner ear which gives a sense of balance and motion.

wavelength The distance between two adjacent crests or troughs of a wave.

weight The force of gravity acting on a mass.

weightlessness Condition of free fall or zero g in which objects in a spacecraft are weightless.

x-rays Electromagnetic waves with wavelength in the range from one-hundredth to one-millionth of a micrometer.

yaw Left and right rotational motion of the nose and tail of an aircraft or spacecraft. See also **pitch** and **roll**.

zero g A misleading term implying the absence of gravity, but meaning weightlessness.

Index